NEW WORLD ORDER

NEW WORLD ORDER

Dixe Wills

ICON BOOKS

Published in the UK in 2007 by
Icon Books Ltd, The Old Dairy,
Brook Road, Thriplow,
Cambridge SG8 7RG
email: info@iconbooks.co.uk
www.iconbooks.co.uk

Sold in the UK, Europe, South Africa and Asia
by Faber & Faber Ltd, 3 Queen Square,
London WC1N 3AU
or their agents

Distributed in the UK, Europe, South Africa and Asia
by TBS Ltd, TBS Distribution Centre, Colchester Road,
Frating Green, Colchester CO7 7DW

This edition published in Australia in 2006
by Allen & Unwin Pty Ltd,
PO Box 8500, 83 Alexander Street,
Crows Nest, NSW 2065

Distributed in Canada by
Penguin Books Canada,
90 Eglinton Avenue East, Suite 700,
Toronto, Ontario M4P 2YE

ISBN-10: 1-84046-810-6
ISBN-13: 978-1840468-10-6

Typesetting and design by Simmons Pugh Ltd

Printed and bound in the UK by Creative Print and Design

CONTENTS

Introduction:
Welcome to the New World Order 7

New World Order League Table 11

Maps of the World 14

Key 20

COUNTRIES 21

NON-COUNTRIES 221

Tables and charts:
New World Ordered 235

Acknowledgements 238

WELCOME TO THE NEW WORLD ORDER

Countries, eh? There probably isn't one of us who hasn't at some point put their head in their hands and cried: 'Is the nation state an intrinsic good or a necessary evil and why, oh why can't I tell the difference?' Tragically, it's heartfelt appeals like this that too often detonate whole cluster bombs of angst-ridden enquiries in the questioner's brain, viz. can the term 'failed state' be applied to countries other than France?; if I wrap myself in the flag, am I more likely to be buried at sea?; is Antarctica the one on the top or the one on the bottom?, etc., etc.

In a nutshell, these queries are merely the outward manifestation of the latent desire within us all to know, once and for all: *Which are the Good Countries and which are the Rotten Ones?*

But who can you trust to tell you? Certainly not the folk at the tourist board of [*insert your choice of nation here*] who are only too willing to extol the virtues of their particular country's sunny resorts and shiny happy people while neglecting to mention that those same locals gleefully massacre each other at the slightest provocation, and that the beaches have only just been cleared of live munitions (give or take the odd anti-personnel mine the child soldiers have disguised as a stepping stone).

Certainly not anyone who doesn't work for a tourist board either, because all these other people lie to you as well. They're friendly enough – they laugh at your jokes and will occasionally return your calls – but they never really look you in the eye, do they?

That's where *New World Order* comes in, with its trust-worthy ink and dependable paper wrapped in a good solid cover of honest wood pulp/petro-chemical conglomerate. Every word is guaranteed to sift not only the wheat from the chaff but also, crucially, the wheatear from the chiffchaff, in order to bring you the *real* New World Order with a definitive league table of all countries everywhere.[1]

Unsure of whether the Swedes are more morally consistent than the Swiss? Need to know what the time is in Gabon right now?[2] Can't work out just what it is about Laos? Look no further than the following pages. And in this case 'no further' really means *no further*. Do not undermine the truth contained herein with so-called 'second opinions' from self-styled 'reference books' or, worse still, acquaint yourself with *soi-disant* 'facts' first hand. You will only confuse yourself and cause unnecessary suffering to others who had hitherto looked up to you.

Above all, remember this: just like you, all nations are a mistake from which they are trying to recover. Have some pity.

[1] A word of caution – there are also many countries that aren't really countries at all. Normally, you could overlook these, but in the permanently evolving political climate in which we find ourselves, today's Christmas Islander is tomorrow's marauding Visigoth. For this reason there is a section devoted to the inhabited quasi-nations that lurk and simper on this, our broiling globe.

[2] It's 4.15 pm.

TECHNICAL STUFF

The New World Order League Table works by the simple expedient of awarding every nation a mark out of twenty for each of five hand-picked categories. These scores are added up by professional mathematicians who arrive at a total out of a hundred for each country. Each one is then allotted a place in the world league table according to the score it has obtained. This process allows you, the reader, to discover – probably for the first time ever – why Bulgaria is more deserving of its place in the family of nations than Tuvalu, but not quite so deserving as Kyrgyzstan.

The League Table is guaranteed infallibile up to, but not including, the moment Iceland launches a tactical nuclear strike against Denmark sometime in 2011.

COUNTRIES OF THE WORLD (in alphabetical order):

Afghanistan
Albania
Algeria
Andorra
Angola
Antigua & Barbuda
Argentina
Armenia
Australia
Austria
Azerbaijan
Bahamas
Bahrain
Bangladesh
Barbados
Belarus
Belgium
Belize
Benin
Bhutan
Bolivia
Bosnia & Herzegovina
Botswana
Brazil
Brunei
Bulgaria
Burkina Faso
Burma
Burundi
Cambodia
Cameroon
Canada
Cape Verde Islands
Central African Republic
Chad
Chile
China
Colombia
Comoros
Congo
Congo, Democratic Republic of
Costa Rica
Cote D'Ivoire
Croatia
Cuba
Cyprus
Czech Republic
Denmark
Djibouti
Dominica
Dominican Republic
East Timor
Eastern Gabon
Ecuador
Egypt
El Salvador
England
Equatorial Guinea
Eritrea
Estonia
Ethiopia
Fiji
Finland
France
Gabon
Gambia
Georgia
Germany
Ghana
Greece
Grenada
Guatemala
Guinea
Guinea-Bissau
Guyana
Haiti
Honduras
Hungary
Iceland
India
Indonesia
Iran
Iraq
Ireland
Israel
Italy
Jamaica
Japan
Jordan
Kazakhstan
Kenya
Kiribati
Korea (North)
Korea (South)
Kuwait
Kyrgyzstan
Laos
Latvia
Lebanon
Lesotho
Liberia
Libya
Liechtenstein
Lithuania
Luxembourg
Macedonia
Madagascar
Malawi
Malaysia
Maldives
Mali
Malta
Marshall Islands
Mauritania
Mauritius
Mexico
Micronesia
Moldova
Monaco
Mongolia
Montenegro
Morocco
Mozambique
Namibia
Nauru
Nepal
Netherlands
New Zealand/Aotearoa
Nicaragua
Niger
Nigeria
Northern Ireland
Norway
Oman
Pakistan
Palau
Panama
Papua New Guinea
Paraguay
Peru
Philippines
Poland
Portugal
Qatar
Romania
Russia
Rwanda
St Kitts & Nevis
St Lucia
St Vincent & the Grenadines
Samoa
San Marino
Sao Tome & Principe
Saudi Arabia
Scotland
Senegal
Serbia
Seychelles
Sierra Leone
Singapore
Slovakia
Slovenia
Solomon Islands
Somalia
South Africa
Spain
Sri Lanka
Sudan
Suriname
Swaziland
Sweden
Switzerland
Syria
Taiwan
Tajikistan
Tanzania
Thailand
Togo
Tonga
Trinidad & Tobago
Tunisia
Turkey
Turkmenistan
Tuvalu
Uganda
Ukraine
United Arab Emirates
United Kingdom
United States of America
Uruguay
Uzbekistan
Vanuatu
Vatican City
Venezuela
Vietnam
Wales
Yemen
Zambia
Zimbabwe

NEW WORLD ORDER LEAGUE TABLE

Country	Score	Ranking	Country	Score	Ranking
Netherlands	85	1	Bhutan	69	29
Guatemala	81	2	Bulgaria	69	29
Brazil	80	3	Australia	68	31
Belgium	79	4	Iceland	68	31
Chile	78	5	Peru	68	31
Kiribati	76	6	San Marino	68	31
Slovenia	76	6	St Lucia	68	31
Czech Republic	75	8	Sweden	68	31
Syria	75	8	Afghanistan	67	37
Micronesia	74	10	Cape Verde Islands	67	37
Jordan	73	11	Iraq	67	37
Latvia	73	11	Jamaica	67	37
Northern Ireland	73	11	Kazakhstan	67	37
Greece	72	14	Nepal	67	37
Mali	72	14	Vatican City	67	37
Congo	71	16	Bahamas	66	44
East Timor	71	16	Botswana	66	44
El Salvador	71	16	Cambodia	66	44
Guyana	71	16	Comoros	66	44
Morocco	71	16	Libya	66	44
Vanuatu	71	16	Tuvalu	66	44
Barbados	70	22	China	65	50
Dominica	70	22	Ireland	65	50
Japan	70	22	Macedonia	65	50
Kyrgyzstan	70	22	Tanzania	65	50
Nicaragua	70	22	Algeria	64	54
Slovakia	70	22	Andorra	64	54
Switzerland	70	22	Croatia	64	54

Country	Score	Ranking
Niger	64	54
Norway	64	54
Somalia	64	54
Trinidad & Tobago	64	54
Bolivia	63	61
Costa Rica	63	61
Gabon	63	61
Guinea	63	61
Hungary	63	61
Moldova	63	61
Russia	63	61
Equatorial Guinea	62	68
Georgia	62	68
Cameroon	61	70
Egypt	61	70
Germany	61	70
Lesotho	61	70
Papua New Guinea	61	70
Wales	61	70
France	60	76
India	60	76
Indonesia	60	76
Liberia	60	76
Namibia	60	76
Poland	60	76
Senegal	60	76
Singapore	60	76
Zambia	60	76
Belize	59	85
Djibouti	59	85
Ecuador	59	85
Italy	59	85
Malawi	59	85
Scotland	59	85
Chad	58	91
Denmark	58	91
Luxembourg	58	91
New Zealand	58	91
Palau	58	91
Taiwan	58	91
Armenia	57	97
England	57	97
Montenegro	57	97
Suriname	57	97
Azerbaijan	56	101
Burundi	56	101
Finland	56	101
Ghana	56	101
Guinea-Bissau	56	101
Lebanon	56	101
Lithuania	56	101
Pakistan	56	101
Samoa	56	101
Belarus	55	110
Malta	55	110
Mozambique	55	110
Rwanda	55	110
Burkina Faso	54	114
Ethiopia	54	114
Kenya	54	114
Madagascar	54	114
South Africa	54	114
Argentina	53	119
Cuba	53	119
Eastern Gabon	53	119
Korea (North)	53	119
Malaysia	53	119
Venezuela	53	119
Gambia	52	125
Oman	52	125
Uganda	52	125
USA	52	125

Country	Score	Ranking
Yemen	52	125
Benin	51	130
Cyprus	51	130
Laos	51	130
São Tomé & Prín.	51	130
St Kitts & Nevis	51	130
Tajikistan	51	130
Bosnia & Herz.	50	136
Estonia	50	136
Iran	50	136
Kuwait	50	136
Haiti	49	140
Liechtenstein	49	140
Mongolia	49	140
Serbia	49	140
Spain	49	140
Thailand	49	140
Dem. Rep. of Congo	48	146
Fiji	48	146
Seychelles	48	146
United Kingdom	48	146
Bangladesh	47	150
Colombia	47	150
Grenada	47	150
Marshall Islands	47	150
Panama	47	150
Romania	47	150
Vietnam	47	150
Austria	46	157
Eritrea	46	157
Mauritius	46	157
Sri Lanka	46	157
Tonga	46	157
Turkey	46	157
Bahrain	45	163
Korea (South)	45	163
Portugal	45	163
St Vincent & Gren.	45	163
Mexico	44	167
Nauru	44	167
Solomon Islands	44	167
Uruguay	44	167
Albania	42	171
Canada	42	171
Dominican Republic	42	171
Angola	41	174
Mauritania	41	174
Honduras	40	176
Ukraine	40	176
Philippines	39	178
C. African Republic	38	179
Cote D'Ivoire	38	179
Brunei	36	181
Sierra Leone	36	181
Togo	36	181
Israel	35	184
Sudan	35	184
Uzbekistan	33	186
Antigua & Barbuda	32	187
Zimbabwe	32	187
Tunisia	31	189
Burma	30	190
Monaco	29	191
Qatar	29	191
Maldives	28	193
Turkmenistan	28	193
Saudi Arabia	27	195
Swaziland	27	195
U. Arab Emirates	27	195
Paraguay	25	198
Nigeria	21	199

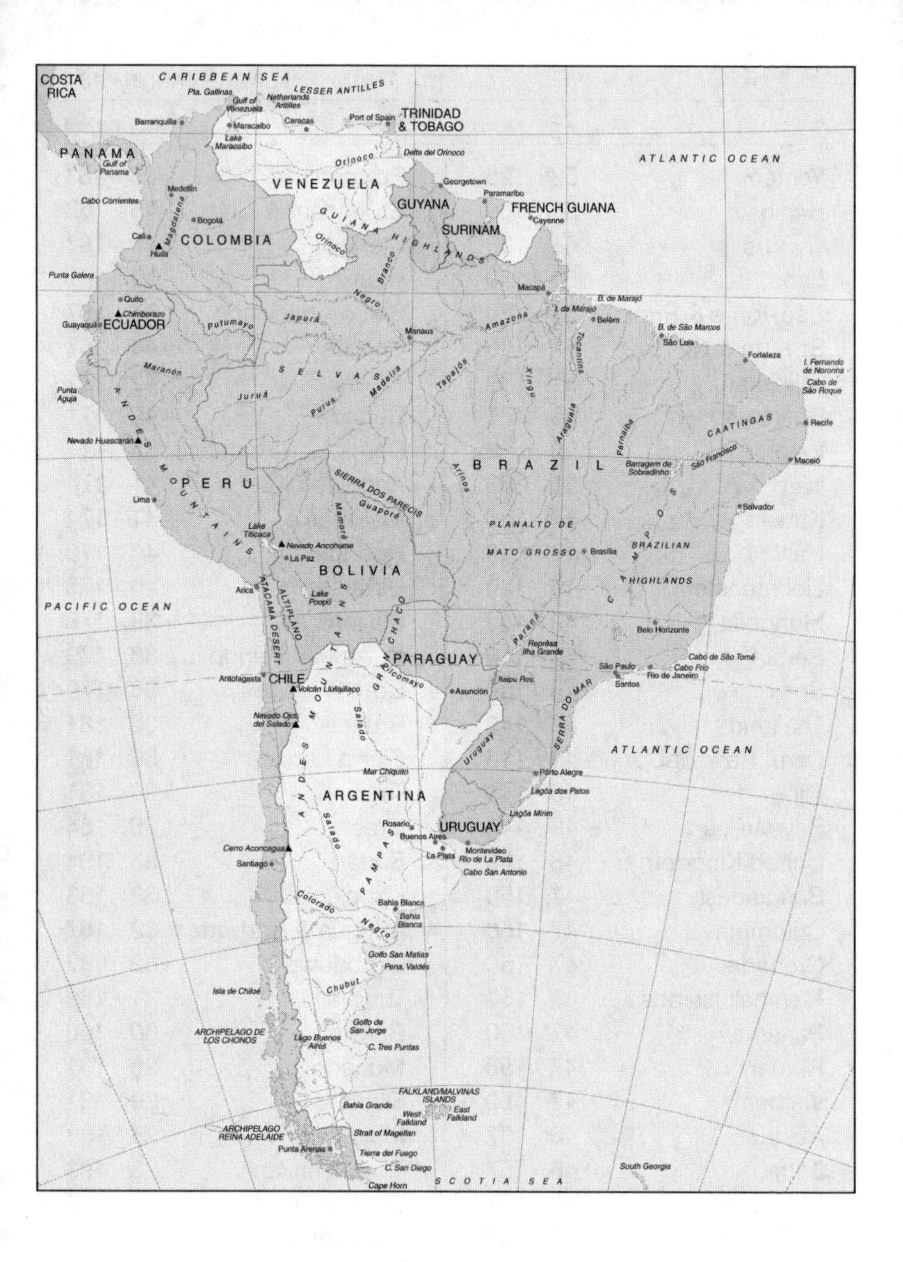

South America

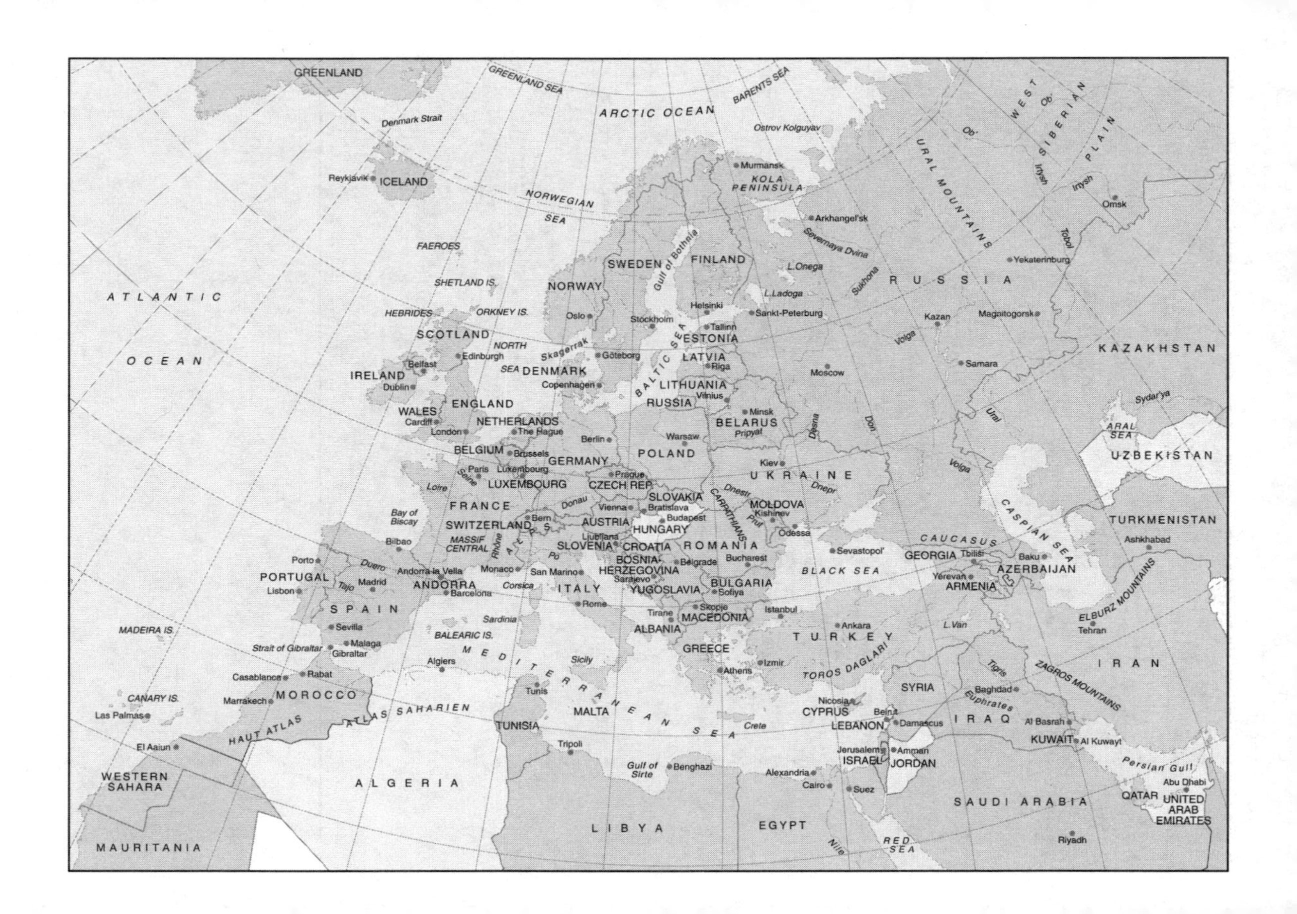

GREENLAND
GREENLAND SEA
ARCTIC OCEAN
BARENTS SEA
Denmark Strait
Ostrov Kolguyev
Reykjavik
ICELAND
NORWEGIAN SEA
Murmansk
KOLA PENINSULA
Arkhangel'sk
Severnaya Dvina
URAL MOUNTAINS
Ob'
WEST SIBERIAN PLAIN
Irtysh
Omsk
Tobol
FAEROES
SHETLAND IS.
ATLANTIC OCEAN
HEBRIDES
ORKNEY IS.
NORWAY
SWEDEN
Gulf of Bothnia
FINLAND
L.Onega
L.Ladoga
Sukhona
RUSSIA
Yekaterinburg
Oslo
Stockholm
Helsinki
Sankt-Peterburg
Kazan
Magnitogorsk
Volga
SCOTLAND
NORTH SEA
Skagerrak
Tallinn
ESTONIA
Edinburgh
Göteborg
LATVIA
Riga
Samara
KAZAKHSTAN
Belfast
IRELAND
Dublin
DENMARK
Copenhagen
BALTIC SEA
LITHUANIA
Vilnius
RUSSIA
Moscow
Sydar'ya
ENGLAND
WALES
Cardiff
London
NETHERLANDS
The Hague
Minsk
BELARUS
Pripyat
Desna
Don
Ural
ARAL SEA
UZBEKISTAN
Berlin
Warsaw
POLAND
BELGIUM
Brussels
Paris
Seine
Luxembourg
GERMANY
Prague
Kiev
UKRAINE
Loire
LUXEMBOURG
CZECH REP.
Dnestr
Dnepr
FRANCE
Donau
SLOVAKIA
Vienna
Bratislava
CARPATHIANS
MOLDOVA
Kishinev
Prut
CASPIAN SEA
TURKMENISTAN
Bay of Biscay
SWITZERLAND
Bern
AUSTRIA
Budapest
HUNGARY
Odessa
MASSIF CENTRAL
Rhône
ALPS
Ljubljana
SLOVENIA
CROATIA
ROMANIA
Sevastopol'
CAUCASUS
GEORGIA
Tbilisi
Baku
Ashkhabad
Bilbao
Porto
Duero
Andorra la Vella
Monaco
San Marino
Po
BOSNIA-HERZEGOVINA
Belgrade
Bucharest
BLACK SEA
AZERBAIJAN
PORTUGAL
Tajo
Madrid
ANDORRA
Corsica
ITALY
Sarajevo
YUGOSLAVIA
BULGARIA
Sofiya
Yerevan
ARMENIA
Lisbon
Barcelona
Rome
Skopje
SPAIN
Tirane
MACEDONIA
Istanbul
ELBURZ MOUNTAINS
Sardinia
ALBANIA
Ankara
L.Van
Tehran
MADEIRA IS.
Sevilla
BALEARIC IS.
TURKEY
Malaga
Strait of Gibraltar
Gibraltar
MEDITERRANEAN SEA
GREECE
Algiers
Sicily
Izmir
Tigris
ZAGROS MOUNTAINS
IRAN
Casablanca
Rabat
Athens
TOROS DAGLARI
SYRIA
Baghdad
CANARY IS.
Marrakech
MOROCCO
Tunis
Nicosia
Euphrates
Las Palmas
ATLAS SAHARIEN
HAUT ATLAS
TUNISIA
MALTA
CYPRUS
Beirut
Crete
LEBANON
Damascus
IRAQ
Al Basrah
El Aaiun
Tripoli
Jerusalem
Amman
KUWAIT
Al Kuwayt
ISRAEL
JORDAN
Gulf of Sirte
Benghazi
Alexandria
Persian Gulf
WESTERN SAHARA
ALGERIA
Cairo
Suez
SAUDI ARABIA
QATAR
Abu Dhabi
UNITED ARAB EMIRATES
LIBYA
EGYPT
Nile
RED SEA
Riyadh
MAURITANIA

Europe

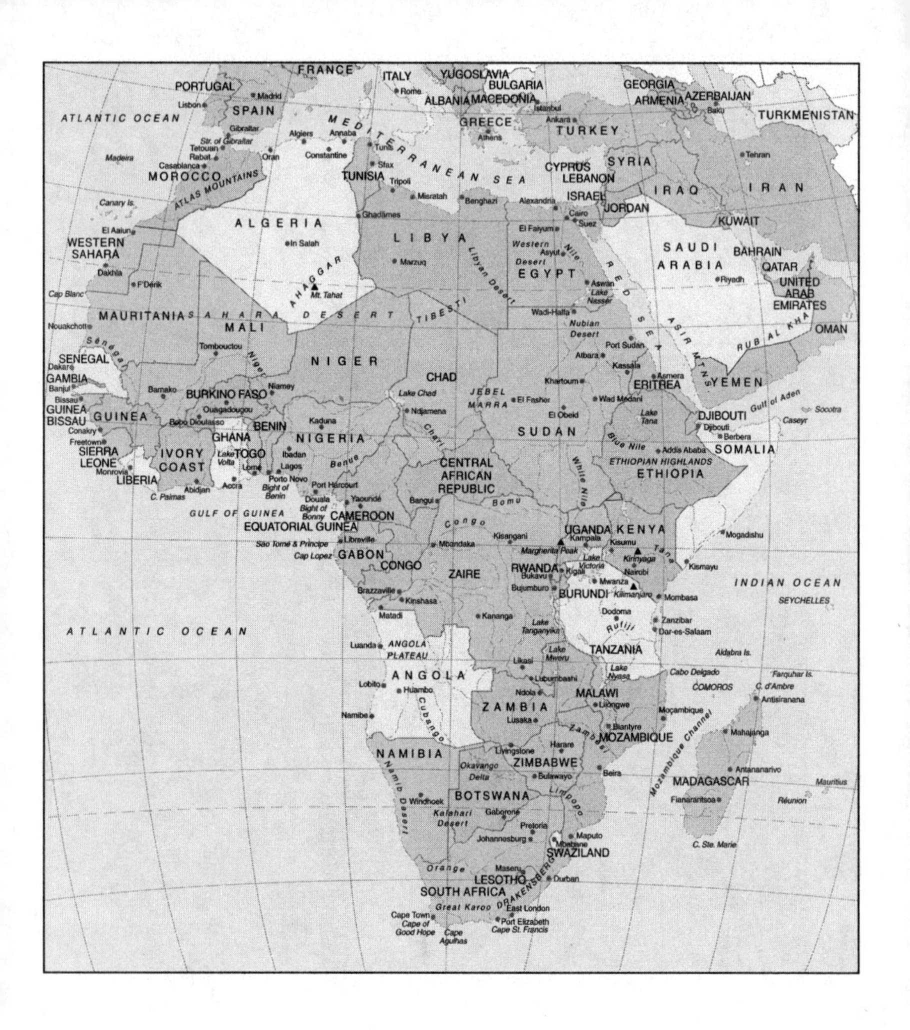
FRANCE
ITALY
YUGOSLAVIA
BULGARIA
PORTUGAL
Madrid
Rome
ALBANIA
MACEDONIA
GEORGIA
ARMENIA
AZERBAIJAN
Lisbon
SPAIN
Istanbul
Baku
TURKMENISTAN
ATLANTIC OCEAN
Gibraltar
Str. of Gibraltar
Tetouan
Rabat
Casablanca
Algiers
Annaba
MEDITERRANEAN SEA
GREECE
Athens
Ankara
TURKEY
Madeira
Oran
Constantine
Tunis
Sfax
Tehran
MOROCCO
ATLAS MOUNTAINS
TUNISIA
Tripoli
CYPRUS
LEBANON
SYRIA
Misratah
Benghazi
ISRAEL
IRAQ
IRAN
Canary Is.
Ghadāmes
Alexandria
Cairo
Suez
JORDAN
El Aaiun
ALGERIA
LIBYA
El Faiyum
KUWAIT
WESTERN SAHARA
In Salah
Libyan Desert
Western Desert
Asyut
Nile
SAUDI ARABIA
BAHRAIN
QATAR
Dakhla
Marzuq
EGYPT
RED SEA
Riyadh
UNITED ARAB EMIRATES
F'Derik
AHAGGAR
Mt. Tahat
Aswan
Lake Nasser
Cap Blanc
MAURITANIA
SAHARA DESERT
TIBESTI
Wadi-Halfa
Nubian Desert
ASIR MTNS.
RUB AL KHALI
OMAN
Nouakchott
MALI
Sénégal
Tombouctou
Port Sudan
SENEGAL
Atbara
Dakar
NIGER
Kassala
GAMBIA
Niger
CHAD
Asmera
Banjul
Bamako
Niamey
Khartoum
ERITREA
YEMEN
Bissau
BURKINO FASO
Lake Chad
JEBEL MARRA
El Fasher
Wad Medani
Gulf of Aden
Socotra
GUINEA BISSAU
Ouagadougou
El Obeid
Lake Tana
DJIBOUTI
Caseyr
GUINEA
Bobo Dioulasso
BENIN
Kaduna
Ndjamena
Djibouti
Conakry
NIGERIA
SUDAN
Berbera
Freetown
GHANA
Chari
Blue Nile
Addis Ababa
SOMALIA
SIERRA LEONE
IVORY COAST
Lake Volta
TOGO
Ibadan
Benue
CENTRAL AFRICAN REPUBLIC
ETHIOPIAN HIGHLANDS
ETHIOPIA
Monrovia
Lome
Lagos
White Nile
LIBERIA
Abidjan
Accra
Porto Novo
Bight of Benin
Port Harcourt
C. Palmas
Douala
Yaoundé
Bangui
Bomu
GULF OF GUINEA
Bight of Bonny
CAMEROON
EQUATORIAL GUINEA
Congo
UGANDA
KENYA
Libreville
Kisangani
Kampala
Kisumu
Mogadishu
Sao Tomé & Principe
Mbandaka
Margherita Peak
Kirinyaga
Cap Lopez
GABON
Lake Victoria
Tana
CONGO
ZAIRE
RWANDA
Kigali
Nairobi
Kismayu
Bukavu
Mwanza
Bujumburo
BURUNDI
Kilimanjaro
INDIAN OCEAN
Brazzaville
Kinshasa
Mombasa
SEYCHELLES
Matadi
Kananga
Dodoma
Rufiji
Zanzibar
ATLANTIC OCEAN
Lake Tanganyika
Dar-es-Salaam
Luanda
ANGOLA PLATEAU
Lake Mweru
TANZANIA
Aldabra Is.
Likasi
ANGOLA
Lubumbashi
Lake Nyasa
Cabo Delgado
Farquhar Is.
Lobito
Huambo
COMOROS
C. d'Ambre
Ndola
MALAWI
Antsiranana
Cubango
Lilongwe
Namibe
ZAMBIA
Moçambique
Lusaka
Blantyre
Mahajanga
Zambezi
MOZAMBIQUE
Mozambique Channel
Livingstone
Harare
NAMIBIA
ZIMBABWE
Antananarivo
Okavango Delta
Beira
Bulawayo
MADAGASCAR
Mauritius
Namib Desert
Windhoek
BOTSWANA
Limpopo
Fianarantsoa
Réunion
Kalahari Desert
Gaborone
Pretoria
Maputo
Johannesburg
Mbabane
C. Ste. Marie
SWAZILAND
Orange
Maseru
LESOTHO
DRAKENSBERG
Durban
SOUTH AFRICA
Great Karoo
East London
Cape Town
Cape of Good Hope
Port Elizabeth
Cape St. Francis
Cape Agulhas

Africa

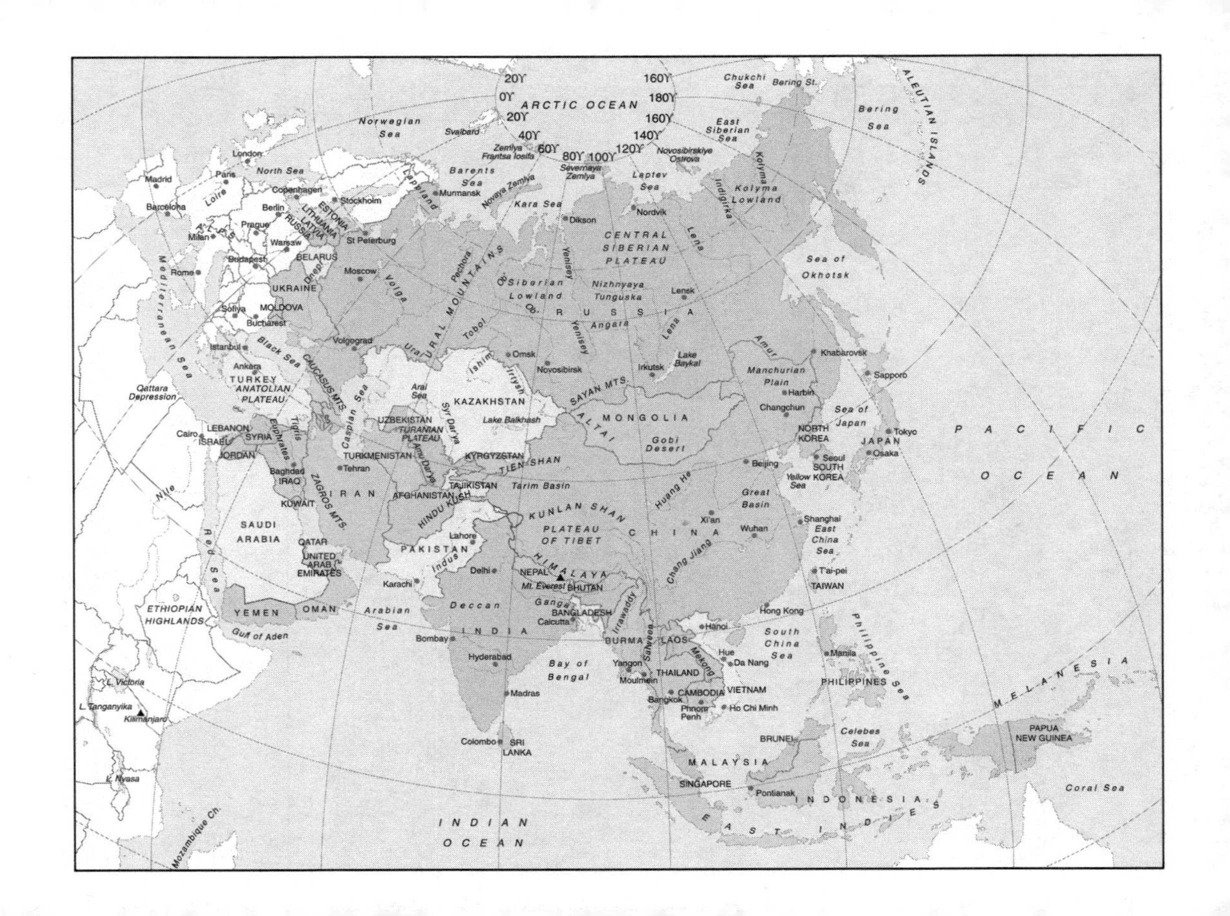

ALEUTIAN ISLANDS
Bering Sea
Sea of Okhotsk
P A C I F I C
O C E A N
M E L A N E S I A
Coral Sea
PAPUA NEW GUINEA
ARCTIC OCEAN
Chukchi Sea
Bering St.
East Siberian Sea
Kolyma
Kolyma Lowland
Indigirka
Novosibirskiye Ostrova
Laptev Sea
Lena
Nordvik
CENTRAL SIBERIAN PLATEAU
Nizhnyaya Tunguska
Lensk
Lake Baykal
Irkutsk
Angara
R U S S I A
Khabarovsk
Sapporo
Sea of Japan
Tokyo
JAPAN
Osaka
Amur
Manchurian Plain
Harbin
Changchun
NORTH KOREA
Seoul
SOUTH KOREA
Yellow Sea
Beijing
Shanghai
East China Sea
Tai-pei
TAIWAN
Philippine Sea
Manila
PHILIPPINES
Celebes Sea
I N D O N E S I A
E A S T I N D I E S
Great Basin
Wuhan
Xi'an
C H I N A
Chang Jiang
Huang He
M O N G O L I A
Gobi Desert
ALTAI
SAYAN MTS.
Severnaya Zemlya
Dikson
Yenisey
Novosibirsk
Ob
Siberian Lowland
Kara Sea
Novaya Zemlya
Zemlya Frantsa Iosifa
Barents Sea
Svalbard
Murmansk
Pechora
URAL MOUNTAINS
Irtysh
Omsk
Ishim
Tobol
KAZAKHSTAN
Lake Balkhash
TIEN SHAN
KYRGYZSTAN
Tarim Basin
KUNLAN SHAN
PLATEAU OF TIBET
HIMALAYA
NEPAL
Mt. Everest
BHUTAN
BANGLADESH
Calcutta
Ganga
BURMA
Irrawaddy
Salween
Yangon
Moulmein
LAOS
THAILAND
Bangkok
Mekong
CAMBODIA
VIETNAM
Phnom Penh
Ho Chi Minh
Hanoi
Hue
Da Nang
Hong Kong
South China Sea
MALAYSIA
BRUNEI
Pontianak
SINGAPORE
Bay of Bengal
Madras
SRI LANKA
Colombo
INDIAN OCEAN
I N D I A
Deccan
Hyderabad
Delhi
Bombay
Lahore
PAKISTAN
Indus
Karachi
Arabian Sea
HINDU KUSH
AFGHANISTAN
TAJIKISTAN
Syr Dar'ya
Amu Dar'ya
UZBEKISTAN
TURANIAN PLATEAU
TURKMENISTAN
Aral Sea
Ural
Caspian Sea
I R A N
Tehran
ZAGROS MTS.
Lappland
Norwegian Sea
Volga
Moscow
St Peterburg
Stockholm
ESTONIA
LATVIA
LITHUANIA
RUSSIA
BELARUS
Volgograd
CAUCASUS MTS.
Dnepr
UKRAINE
MOLDOVA
Black Sea
North Sea
Copenhagen
Berlin
Warsaw
Prague
Budapest
Bucharest
Sofiya
London
Paris
Loire
ALPS
Milan
Rome
Madrid
Barcelona
Mediterranean Sea
Istanbul
Ankara
TURKEY
ANATOLIAN PLATEAU
Tigris
Euphrates
SYRIA
LEBANON
ISRAEL
JORDAN
Baghdad
IRAQ
KUWAIT
QATAR
UNITED ARAB EMIRATES
OMAN
SAUDI ARABIA
YEMEN
Gulf of Aden
Red Sea
Cairo
Nile
Qattara Depression
ETHIOPIAN HIGHLANDS
L. Victoria
Kilimanjaro
L. Tanganyika
L. Nyasa
Mozambique Ch.

Asia

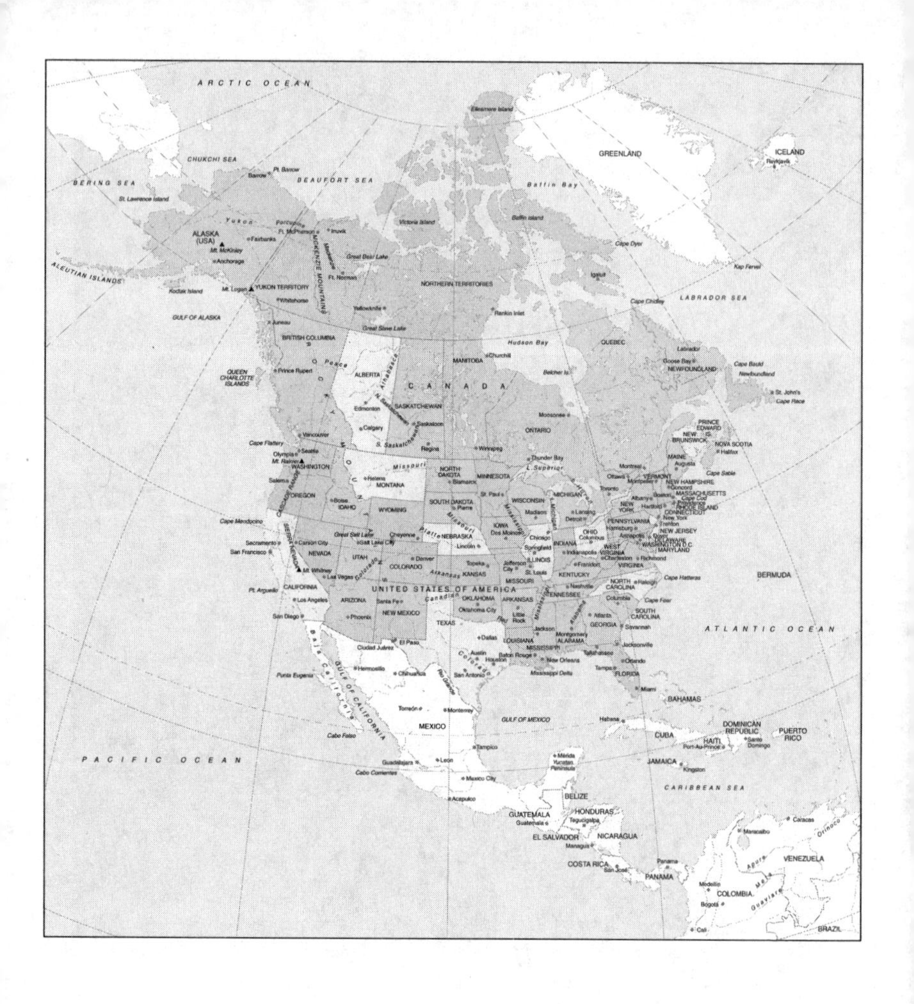

North and Central America

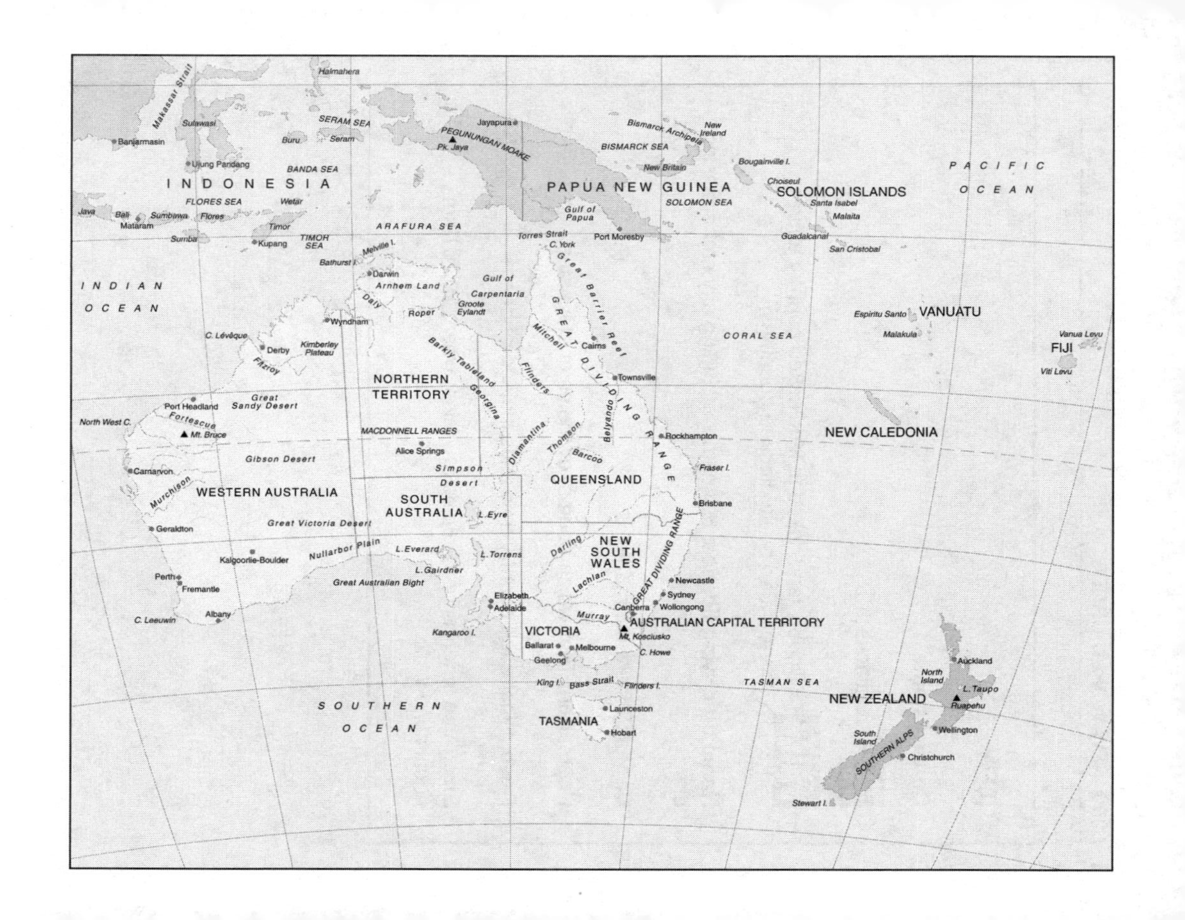

Australasia

KEY

Country name

 = Landlocked

 = Doubly landlocked (a landlocked country entirely surrounded by landlocked countries)

Name on driving licence: Full name, including any embarrassing bits

Capital: Where the money is

Population: That which booms, dwindles or stagnates

Dosh: What each country now uses instead of beads

Size: Area in terms of multiples of Wales, where 20,761 km^2 equals one Wale

COUNTRIES

ALL THE WORLD'S NATIONS

Afghanistan

ASIA

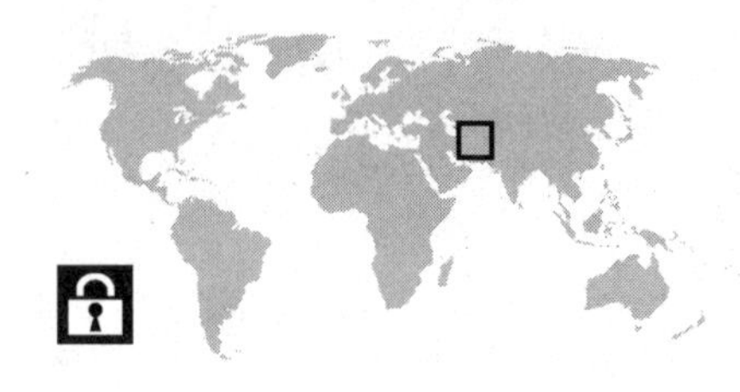

colour key: 1 = turquoise 2 = blue 3 = green 4 = red 5 = yellow 6 = orange 7 = pink 8 = purple 9 = brown

Name on driving licence: Islamic Republic of Afghanistan
Capital: Kabul
Population: 28.5 million
Dosh: Afghani = 100 puls
Size: 647,500 km^2 (31 Wales)

Complete history: Afghanistan began in ancient times and has been beginning over and over again ever since. For a country with not much in it once you take away the mountains, it has still been alluring enough for Aryans, Persians, Greeks, Macedonians, Central Asians, Britons and Russians to want it for their own. The nation's modern history began in 1747 when local tribes stopped fighting each other for a bit to take on whoever were that week's invaders.

Top spot: The Hindu Kush – not only was it once ruled by the university-challengetastic Genghis Khan but it also has a name that, when sounded, has the same texture as a really good falafel, and apparently that's not a coincidence. **[15]**

Customs to treasure: Afghan farmers are flower-loving sentimentalists at heart and have been known to pull up whole fields of crops and replace them with poppies so that they always have something pretty on hand to give their wives and girlfriends. **[12]**

Flag fact: Between 1901 and 2002, Afghanistan heroically went through nineteen different national flags, a rate of a new one every five years or so. Even with their pitifully low life expectancy, the average Afghan can expect to see off eight different flags before poverty, disease, landmines or car bombs have them pushing up the poppies. **[19]**

Motto: 'Welcome. Ah, I see you've brought your army with you.' **[13]**

Opening lines of national anthem that are, if nothing else, informative:
'This land is Afghanistan/It is the pride of every Afghan.' **[8]**

SCORE: 67 **WORLD RANKING:** 37

EUROPE

Albania

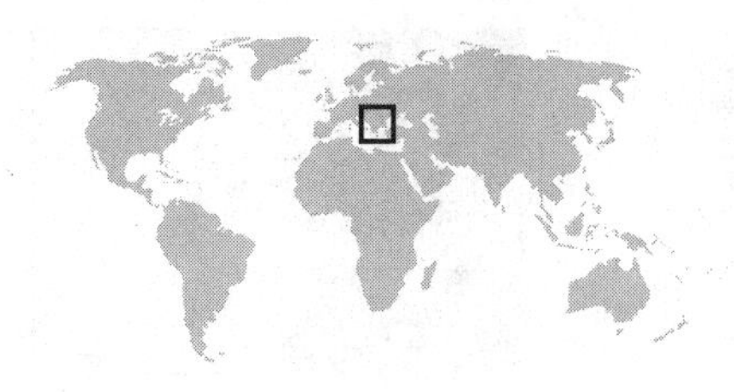

Name on driving licence: Republic of Albania
Capital: Tirana
Population: 3.5 million
Dosh: Lek = 100 qindarka
Size: 28,750 km^2 (1.3 Wales)

Complete history: Once part of something called Illyria – as featured in Shakespeare's *Twelfth Night* – Albania got unaccountably bored and in 167 BC voted to join the Roman Empire. When things went pear-shaped *chez* Caesar, the Albanians jumped ship for the Byzantine Empire before succumbing to the inevitable ravaging by Goths, Bulgarians, Slavs and, most embarrassingly, Normans. Recovered in time to sign up for the Serbian and Ottoman Empires. Declared itself independent in 1912, presumably to attract the attention of an empire. Duly subsumed into the Third Reich in 1943. Currently self-governing, though no reasonable offer refused.

Crowning achievement: June 1982, the single 'Albania, Albania' by Alexei Sayle (masquerading as the Albania World Cup Squad) is released in the UK to almost total indifference. **[2]**

Top spot: Berat, the so-called 'City of a Thousand Windows' (though admittedly there are few cities around nowadays with fewer than a thousand windows, so perhaps this is not such a great boast). **[3]**

Customs to treasure: Cheekily, Albanians nod their heads to mean 'no' and shake them to mean 'yes'. They also have 27 words for moustache, which, lets face it, does not bode well. **[5]**

Religious affiliations: Decades of Communist rule have left a widespread disbelief in God. This void has been filled to some extent by a belief in Norman Wisdom. **[12]**

National anthem entirely free of hubris: 'For the Lord Himself has said/That the nations shall vanish from the earth/But Albania will live on.' **[20]**

SCORE: 42 WORLD RANKING: 171

Algeria

AFRICA

colour key: 1 = turquoise 2 = blue 3 = green 4 = red 5 = yellow 6 = orange 7 = pink 8 = purple 9 = brown

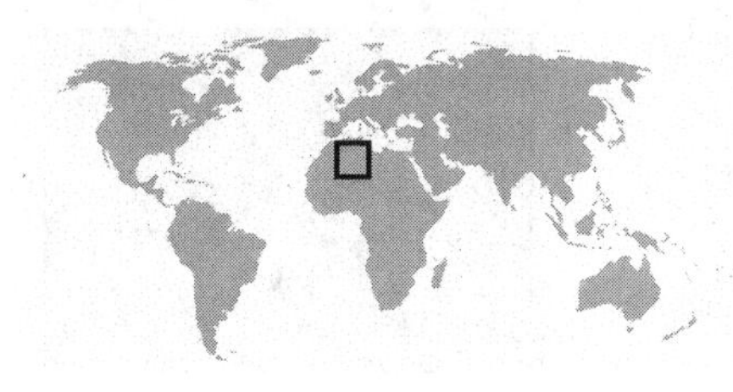

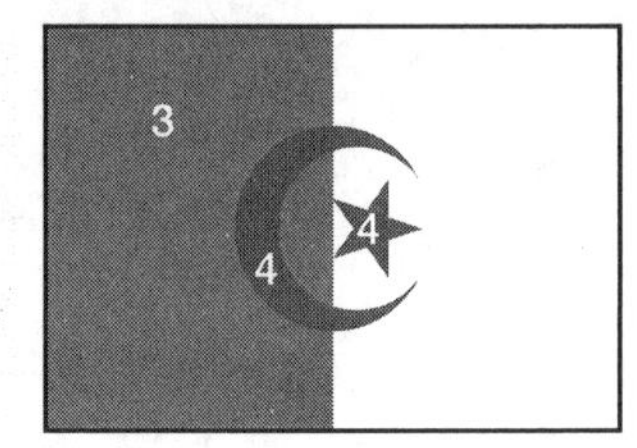

Name on driving licence: The People's Democratic Republic of Algeria
Capital: Algiers
Population: 32 million
Dosh: Algerian dinar = 100 centimes
Size: 2,380,000 km^2 (115 Wales)

Complete history: Overrun at various times by Phoenicians, Carthaginians, Romans and Vandals (which explains why so few pumps work in Algeria). Arabs were next in the queue, breezing in during the long hot nights that characterised the 7th century and setting about intermarrying with the local Berbers. The French dropped by in 1830, claiming they just needed to borrow *une tasse de sucre*. When they were still there in 1962, the locals became suspicious and asked them to leave.

Top spot: The Atlas Mountains, so called because you can find them on one. **[17]**

Gift to the world: Albert 'Cammy' Camus – goalkeeper, bon viveur and sometime existentialist philosopher. Best known for writing French A-Level syllabus-friendly novels and for joking that the most absurd way to die would be in a car crash. **[15]**

Historical low point: Albert Camus' death in a car crash (1960). **[4]**

Made in Algeria: Raï, a form of subversive protest pop, now largely appropriated by the Egyptians in a kindly bid to take the heat off white people who feel guilty about stealing the music of black people. **[10]**

Opening lines of national anthem: 'We swear by the lightning that destroys/By the streams of generous blood being shed.' The remainder of the anthem is devoted to all the other things they'd like to do with the French. **[18]**

SCORE: 64 **WORLD RANKING: 54**

Andorra

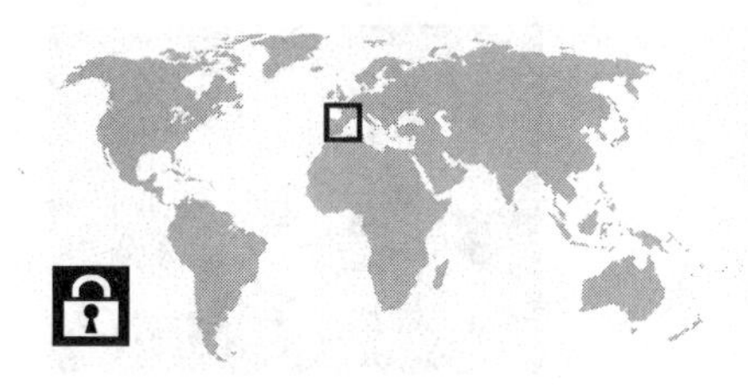

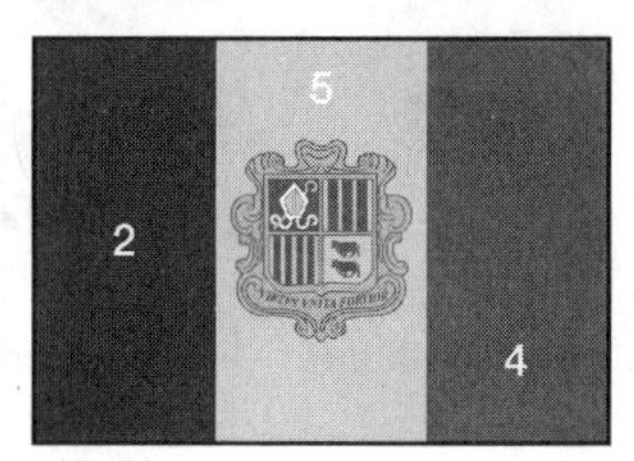

Name on driving licence: Andorra
Capital: Andorra La Vella
Population: 68,000
Dosh: Euro = 100 cents
Size: 470 km² (0.02 Wales)

Complete history: Andorra is far and away the oddest form of nation state in Europe, and arguably the world. From the Middle Ages, the 'two princes' that made the cowpat-sized country into a principality were not proper princes who caught the eyes of maidens in high towers and went forth on a variety of deluded crusades against largely imaginary evils, but the Catholic bishop of ho-hum Spanish town La Seul d'Urgell, yoked in unholy alliance with the French head of state (first kings, then puffed-up emperor types, then a series of mad, lacklustre or excitingly corrupt presidents). It was only in 1993 that the Andorrans recognised the folly of this arrangement and abruptly turned themselves into a 'parliamentary co-princedom' which, the French had assured them, was what all the European countries were doing thenadays.

Top spot: The natural spa at Escaldes-Engordany is apparently Europe's largest. Biggest isn't always best, of course. **[5]**

Crowning achievement: Every year, Andorra welcomes 8 million tourists, or 351 tourists for each ethnic Andorran (who make up just a third of the population), which is why they all have such big houses. **[14]**

Gifts to the world: Cheap skiing, Catalan, tobacco. **[8]**

Motto: 'Someday, our prince will come. With any luck, we'll be out.' **[18]**

Opening lines of national anthem: 'The great Charlemagne, my Father, liberated me from the Saracens/And from heaven he gave me life from Meritxell, the great Mother.' Apparently this is true of all Andorrans. **[19]**

SCORE: 64 **WORLD RANKING:** 54

Angola

AFRICA

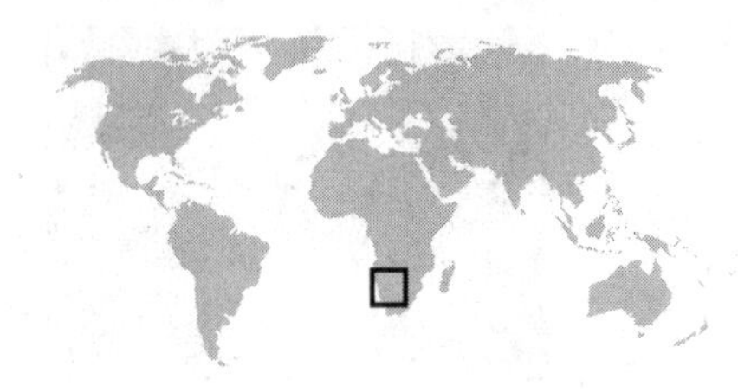

Name on driving licence: Republic of Angola
Capital: Luanda
Population: 11 million
Dosh: Readjusted kwanza = 100 lwei
Size: 1,247,000 km^2 (60 Wales)

Complete history: While further up country three wise men were wrapping up the gold, frankincense and myrrh, and looking forward to hearing some first-rate cattle a-lowing, Bantu-speaking types were busy heading south to settle Angola. It wasn't until the late 16th century, however, that the Portuguese introduced themselves. According to the European custom of the time, this meant that they now owned the country, which was useful since it meant they could ship the local populace to Brazil to replace the ungrateful *indios* who insisted on dying off in large numbers. By 1975, the colonisers couldn't be bothered with Angola any more, having rather carelessly lost Brazil in the meantime. Their parting gift was one of those jolly civil wars the Cold War super powers were so happy to foment. Peace broke out in the country in 2002, though by then there wasn't much country left.

Top spot: Like all proper countries (see Russia and Azerbaijan), Angola comes in two ready-to-assemble parts. Due to a lack of strong ropes, independence-seeking Cabinda can now be found squeezed between the two Congos. [15]

Gift to the world: Babies. The country has the highest fertility rate on the planet. [16]

Sad fact: Life expectancy for those babies: 37.25 years for girls, 36 years for boys. [2]

Pub fact: There are more landmines in Angola than people. [3]

Opening lines of national anthem: 'O Fatherland, we shall never forget/The heroes of the Fourth of February.' Heaven forbid. [5]

SCORE: 41 WORLD RANKING: 174

colour key: 1 = turquoise 2 = blue 3 = green 4 = red 5 = yellow 6 = orange 7 = pink 8 = purple 9 = brown

AMERICAS Antigua & Barbuda

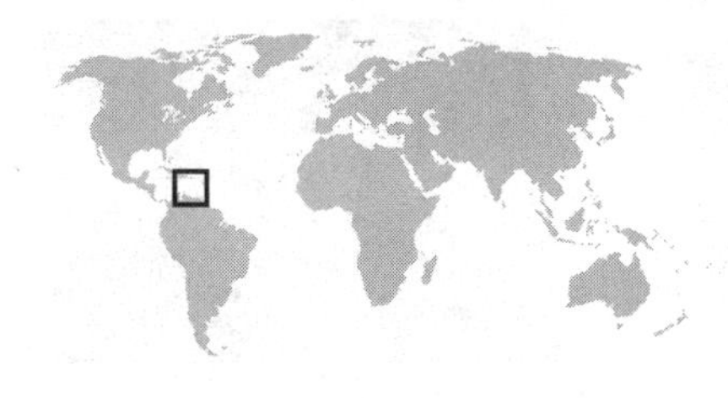

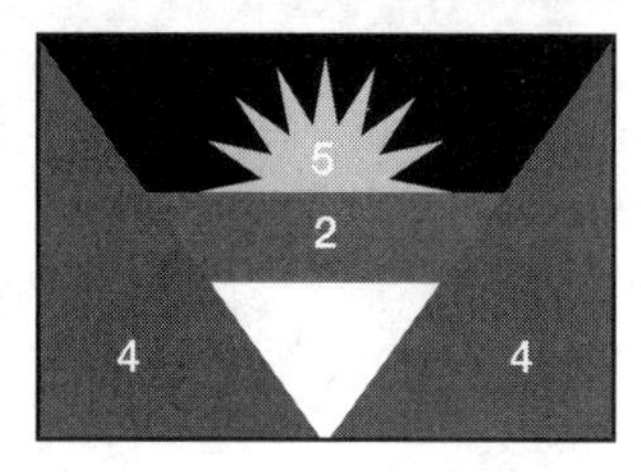

Name on driving licence: Antigua and Barbuda
Capital: St John's
Population: 77,000
Dosh: East Caribbean dollar = 100 cents
Size: 440 km² (0.02 Wales)

Complete history: Despite its name, Antigua and Barbuda is actually three islands, the pleasingly circular Redondo failing to get name-checked on the trifling pretext that no one lives there. Mind you, only 1,400 people live on Barbuda, which is a game reserve rather than a proper island, so it can count itself lucky to get a billing. It was the stone-wielding Ciboney who got to the islands first, around 2400 BC, but as we all know it's who's holding the parcel when the music stops that counts, and the Arawaks, Caribs and English have all had to pass it on. Independence finally arrived in 1981, but in a world where determinism is king, can any of us truly claim to be independent?

Historical high: 1936 – Robert Graves publishes *Antigua, Penny, Puce* – a novel about a cheap stamp. **[7]**

Top spot: There are allegedly 365 beaches on Antigua, which seems a suspiciously neat figure. No doubt every leap year it is proclaimed that there are 366. **[3]**

Pub fact: Antigua has no rivers or forests but makes up for this with the world's highest marriage rate. **[18]**

Sad fact: Future column owner Horatio Nelson built the Antigua docks but described the island as a 'vile place' and a 'dreadful hole' and chose to sleep on his ship moored in the harbour, thus blowing his chances of making a bit of spare cash doing voice-overs for the Antigua tourist board. **[1]**

National anthem bemoaning a lack of seating: 'Fair Antigua and Barbuda!/We thy sons and daughters stand.' **[3]**

SCORE: 32 **WORLD RANKING: 187**

Argentina

AMERICAS

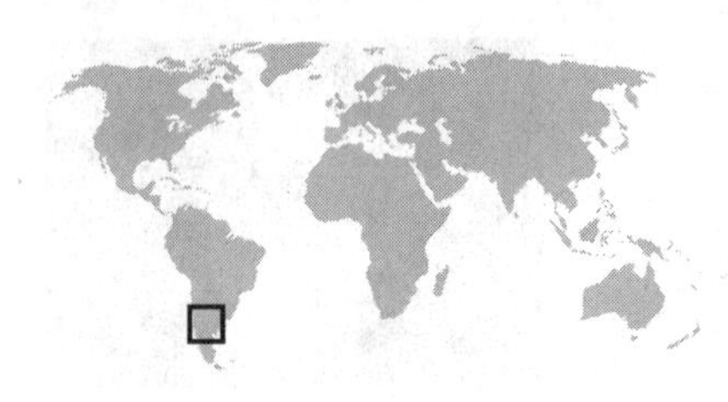

Name on driving licence: The Argentine Republic
Capital: Buenos Aires
Population: 39 million
Dosh: Peso = 10,000 australs
Size: 2,767,000 km^2 (134 Wales)

Complete history: The 'Italy of South America', Argentina is a land of gauchos roaming free o'er the pampas (just like Italy). First settled by Amerindians, things were messed up considerably by the arrival of Spaniards in 1516, and then by Italians (and more Spaniards) from 1870 onwards, even though the latter wave of invaders tried awfully hard not to slaughter the natives the second time around. Never ones to do things by halves, it took Argentina less than a century to go from being an extremely prosperous nation to one in which the rubber band round a wad of trillion peso notes was more valuable than the money within it.

Crowning achievement: In Patagonia, Argentina boasts the largest Welsh-speaking community outside Wales. **[15]**

Top spot: In Patagonia, Argentina boasts the largest Welsh-speaking community outside Wales to have much use for the word 'pengwin'. **[13]**

Sad fact: The term *desaparecido*, to denote someone the authorities have 'disappeared', was invented here in the 1970s during the 'dirty war'.

Historical low: Winning the 1978 World Cup. The hosts were in a bit of a fix and needed to win their last group game against Peru by four goals to progress to the final. Peru, who had only conceded six goals in their previous five games, somehow contrived to lose 6-0, so nothing suspicious there. **[1]**

Motto: 'We like the Falklands but you hang onto them for the time being.' **[14]**

Opening lines of national anthem: 'Mortals! Hear the sacred cry/Freedom! Freedom! Freedom!' **[10]**

SCORE: 53 **WORLD RANKING:** 119

ASIA

Armenia

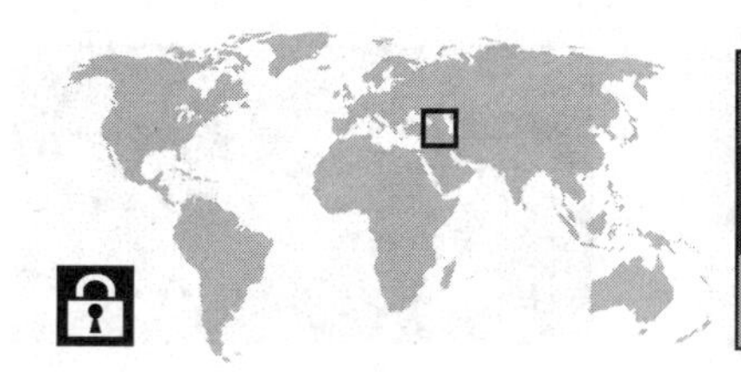

4
2
6

Name on driving licence: Republic of Armenia
Capital: Yerevan
Population: 2.9 million
Dosh: Dram = 100 couma
Size: 29,800 km² (1.4 Wales)

Complete history: Armenia was already an entity and busy at the cutting edge of iron and bronze technology over 2,500 years ago, little knowing that the big money was going to be in gold. Alexander the Great soon muscled in, followed inevitably by the Romans. After a quick stab at independence, Armenia caught the eye of regional bullies the Mongols. The Ottomans and Persians managed to rip the country in twain, the latter accidentally losing their half to Russia. The Turks held onto theirs by the less than subtle ruse of massacring hundreds of thousands of Armenians and then pretending ever after that they'd done no such thing. By 1918, the nation was in a bit of a mess, having somehow lost a new bit to Iran. What was left became part of the Transcaucasian Republic of the Soviet Union, only becoming Armenia again in 1936 and leaving the Soviet Union in 1991. Currently scrapping intermittently with neighbours Azerbaijan.

Crowning achievement: Yerevan (founded around 800 BC) is one of oldest continuously inhabited cities in the world, though don't mention it to the Damascenes or they'll laugh you to scorn. **[15]**

Top spot: The hill resort of Dilizhan is often likened by Armenians to heaven, despite the fact that few of them have been to both. **[16]**

Sad fact: Earthquakes in 1984 and 1988 killed over 80,000 people. **[0]**

Pub fact: Armenia was the first state within the Roman Empire to make Christianity its official religion. **[11]**

National anthem to cheer the heart: 'Death is everywhere the same/Man is born just once to die.' **[15]**

SCORE: 57 WORLD RANKING: 97

Australia

OCEANIA

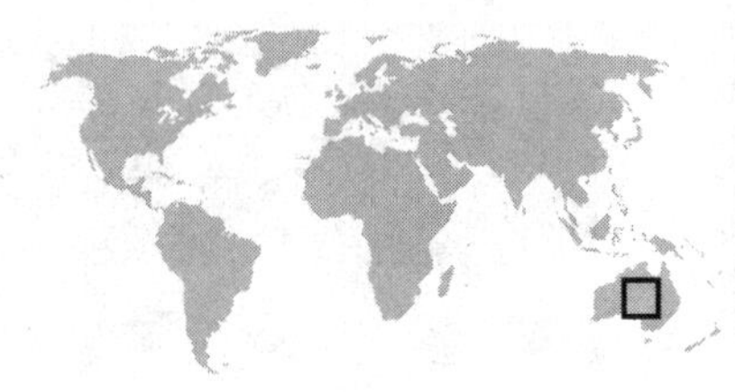

Name on driving licence: Commonwealth of Australia
Capital: Canberra
Population: 19.9 million
Dosh: Australian dollar = 100 cents
Size: 7,687,000 km^2 (373 Wales)

Complete history: Aboriginal people were quite happy living their aboriginal lives from the time they arrived from South-East Asia (about 50,000 years ago) until the arrival of Captain Cook (1770). After that, it all went a bit haywire – not surprising really if, like Britain, your idea of perfect colonists are civil-service types, prison guards and 12-year-olds who have pickpocketed handkerchiefs from unwary ladies in the backstreets of Spitalfields. The main point of Australia today is its use as a vital early stronghold from which to win the game of Risk.

Crowning achievement: Holders of the record for the most comprehensive victory in an international football match, having somehow contrived to win 31-0 against American Samoa in a World Cup Qualifier on 11 April 2001. Endearingly, the Socceroos still failed to qualify for the finals. **[20]**

Where to avoid: Australia is a member of a small, slightly elitist, band of countries (including Canada and Switzerland), all of which have chosen underdogs for capitals. Underdogs, of course, are a good thing. Usually. **[3]**

Pub fact: Australia's capital used to be Melbourne. It was demoted in 1908 in order to convert the city into a set for soap operas. **[7]**

Popular misconception: 'Australia is a nation of Pom-haters.' It is not. They love the English. **[20]**

The Rod Laver Award for most creative use of the word 'girt' in a national anthem: 'We've golden soil and wealth for toil/Our home is girt by sea.' **[18]**

SCORE: 68 **WORLD RANKING:** 31

Austria

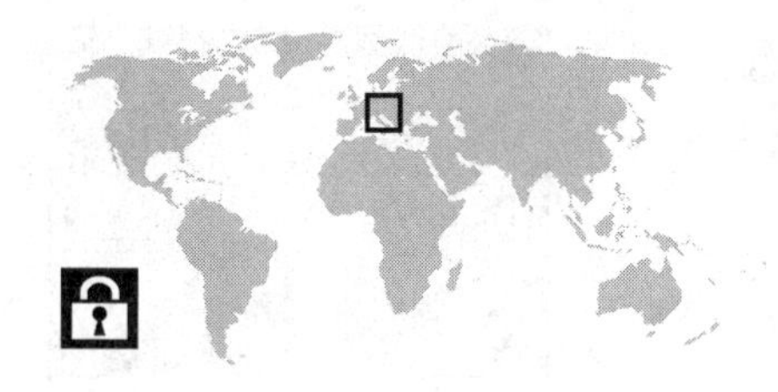

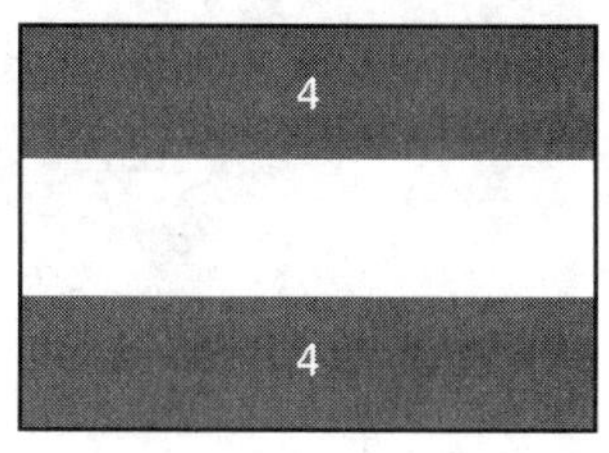

Name on driving licence: Republic of Austria
Capital: Vienna
Population: 8.1 million
Dosh: Euro = 100 cents
Size: 83,900 km² (4 Wales)

Complete history: Just made it into the Roman Empire, before being overrun by consecutive waves of Asians, Germanics, Slavs, King Charlemagne and Magyars. German rule, under the baton of the loveable King Otto I, was followed by eight centuries of the Holy Roman Empire. Brief spell of Austro-Hungarian sabre-rattling until the alliance joined the wrong side in World War I. Coerced into making similar mistake for follow-up. Undeterred, Austrians still vote in vast numbers for unhinged right-wing demagogues.

Crowning achievement: That moment in *The Third Man* when Orson Welles' face looms out of the shadows. **[14]**

Religious affiliations: The cult of Amadeus has been inculcated into every aspect of Austrian life. It's a wonder they don't just change the name of the country to Mozart and have done with it. **[3]**

Historical low: 1907/08 – the Viennese Academy of Fine Arts rejects the applications of would be student Adolf Hitler on the grounds of his 'unfitness for painting'. **[0]**

Made in Austria: Shrödinger's Cat, as brought to life (or death) by Vienna-born quantum mechanic Erwin Shrödinger. Simply leave a cat in a box with a canister of poison gas, a Geiger counter with a trigger mechanism, and a radioactive atomic nucleus that has a 50 per cent chance of decaying. The cat has two possible states – alive or dead – when you open the box again. A game still popular at children's parties in the more louche districts of Salzburg. **[15]**

National anthem in praise of hammers: 'Land of hammers, with a rich future/You are the home of great sons.' **[14]**

SCORE: 46 **WORLD RANKING: 157**

Azerbaijan

ASIA

colour key: 1 = turquoise 2 = blue 3 = green 4 = red 5 = yellow 6 = orange 7 = pink 8 = purple 9 = brown

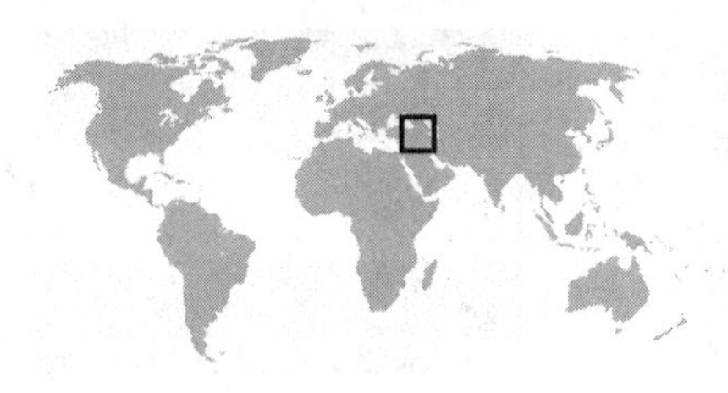

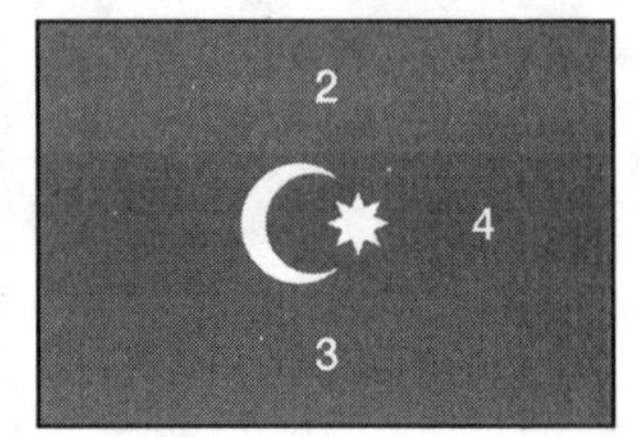

Name on driving licence: Republic of Azerbaijan
Capital: Baku
Population: 7.8 million
Dosh: Azerbaijani manat = 100 gopik
Size: 86,600 km^2 (4.2 Wales)

Complete history: Yet another sorry tale of invasions sweeping over the land like bitter waves on the seashore of regret: Arabs, Persians, Turkics, Mongols, Persians again and finally Russians all came, saw and conquered. From here on, the nation's history takes on much the same hue as that of their best mates, the Armenians. First there's some business with the Transcaucasian Republic, in which Azerbaijan joins hands briefly with Armenia and Georgia under the life-giving sun of the Soviet Union. Then nothing much happens until 1991 when the life-giving sun inexplicably goes out. Since then, some separatist unrest aside, it's all been about fighting a losing battle with Armenia over the territory of Nagorno-Karabakh. As a result, a fragment of Azerbaijan has become separated and is now the other side of Armenia, which is a bit careless.

Crowning achievement: The 4,000 Neolithic petroglyphs of Qobustan. These depict Stone Age folk hunting, running about like mad things and throwing some shapes on the dance floor. Fab. **[16]**

Top spot: The Ateshgah Fire Temple. Home to flames that leap out of the ground on a whim and which are not at all the product of the natural gas vent on which the temple is built. **[17]**

Pub fact: In the days when it could be bothered, Azerbaijan produced over half the world's oil. **[4]**

Religious affiliations: Zoroastrianism originated in Azerbaijan, a full 2,500 years before the birth of Zorro. **[13]**

National anthem at ease with itself: 'Azerbaijan! Azerbaijan!/O Great Land, your children are heroes.' **[6]**

SCORE: 56 WORLD RANKING: 101

Bahamas

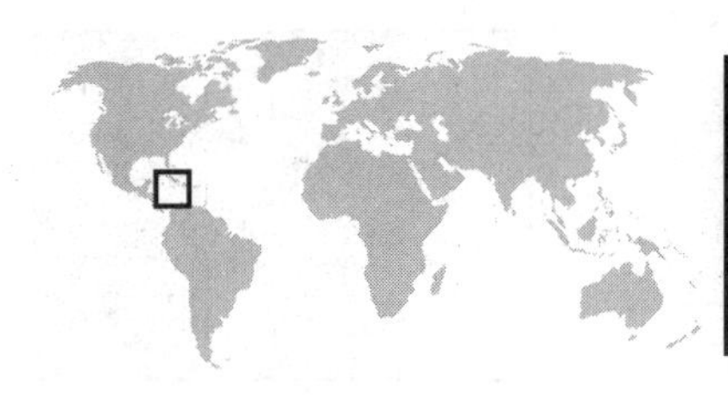

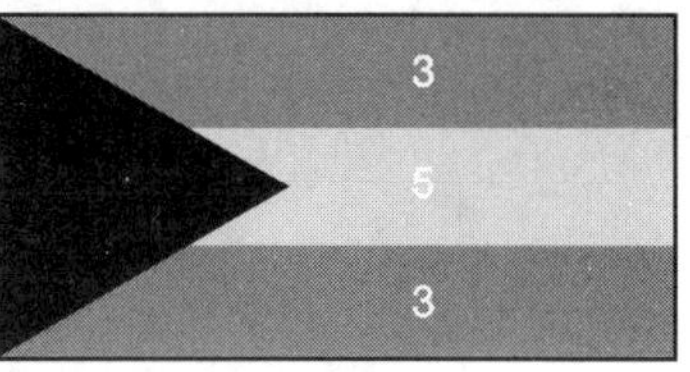

Name on driving licence: Commonwealth of the Bahamas
Capital: Nassau
Population: 297,000
Dosh: Bahamian dollar = 100 cents
Size: 5400 km^2 (0.7 Wales)

Complete history: Christopher Columbus is reputed to have first laid his dread foot on the sands of 'West India' at San Salvador, one of 700 Bahamian islands. Perhaps the Lucayans he met were only too happy to have the opportunity to see a bit of the world (within two decades, the entire population of the Bahamas had been shipped off to Hispaniola as slaves). Then again, perhaps not. The islands were resettled in 1650 by English types known excitingly as Eleutherian Adventurers. Usual trouble and strife thereafter. Independence in 1973.

Crowning achievement: Talking Heads' seminal scratch-us-and-we're-all-Africans-underneath album *Remain in Light* was recorded at Compass Point studios, Nassau. Other Compass Point artistes include such legends as James Brown, Roxy Music and (ahem) Flock of Seagulls. **[18]**

Top spot: Possessor of the world's largest open-air aquarium (approximately fourteen acres, if you were thinking of replicating it in your back garden). Next to it is a huge casino where, presumably, you can bet on which of the goldfish dies next. **[7]**

Made in the Bahamas: Sidney Poitier, star of *To Sir, With Love*, an eerily prescient title given his knighthood in 1974. **[15]**

Customs to treasure: The Goombay Summer Festival. Not to be confused with the annual Junkanoo celebration, which is a different thing altogether, or the Goombay Dance Band, whose existence is best consigned to the part of the brain marked 'Do Not Disturb'. **[12]**

Opening lines of national anthem: 'Lift up your head to the rising sun, Bahamaland/March on to glory, your bright banners waving high.' *Bahamaland*? **[14]**

SCORE: 66 **WORLD RANKING:** 44

Bahrain

MIDDLE EAST

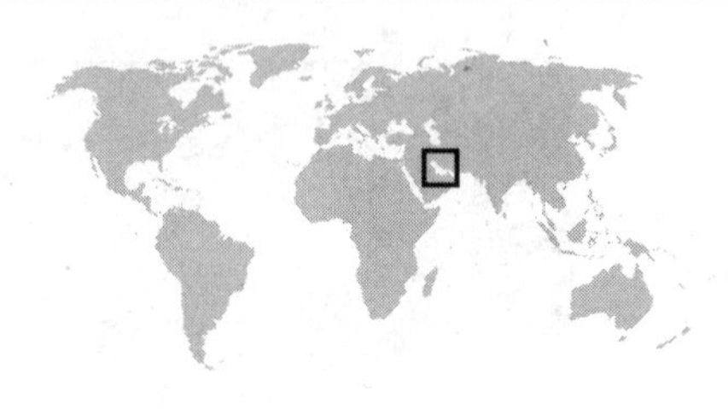

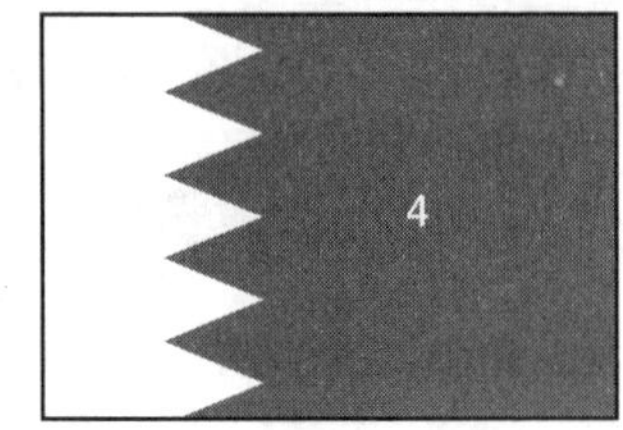

Name on driving licence: Kingdom of Bahrain
Capital: Manama
Population: 678,000
Dosh: Bahraini dinar = 1000 fils
Size: 620 km^2 (0.03 Wales)

Complete history: Bahrain was once part of a civilisation called Dilmun, a name better suited to an office worker in an unfunny middlebrow tabloid newspaper cartoon strip in which our hero is unable to understand why his colleagues chaff him for referring to the young women in the typing pool as 'dolly birds'. (And how come his company has a typing pool anyway? Is it because the cartoonist himself hasn't seen the inside of an office for thirty years?) Conquered by the usual suspects: Arabs (7th century), Portugal (1521), Persians (1603), Saudi Arabia (1782). The British, ever helpful in a time of crisis, assisted the locals in warding off full annexation by the Saudis. Remained under Britannia's protective wing until 1971.

Historical low: Bahrain now hosts a race in the Formula 1 championship, each round of which apparently uses enough petrol to have doused and immolated 500,000 witches in the Middle Ages. **[3]**

Top spot: Old is overrated. 'New' is the new 'new'. Manama is so new, bits of it are still in the box. **[14]**

Pub fact: Bahrain means 'two seas', a reference to the sub-sea springs that shoot freshwater into the saltwater above. The name is also useful as an expression of disgust at the onset of a heavy shower (see also 'Romania', a country obsessed with oarsmanship). **[13]**

Sad fact: Destinations served by Concorde (RIP): Britain, France, USA, err, Bahrain. **[10]**

Opening lines of national anthem: 'Our Bahrain/Country of security.' We'll wake you up when it's over. **[5]**

SCORE: 45 **WORLD RANKING: 163**

ASIA

Bangladesh

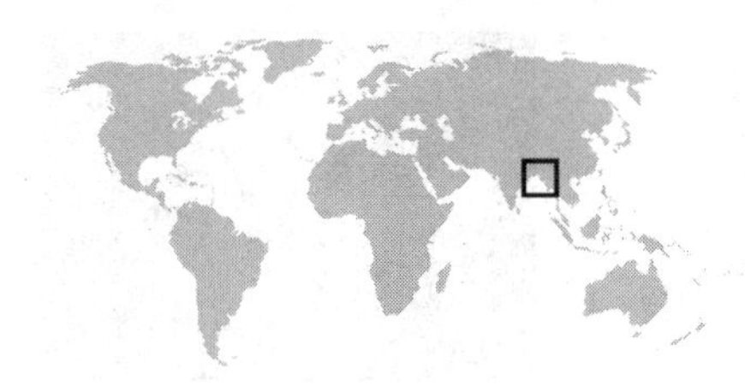

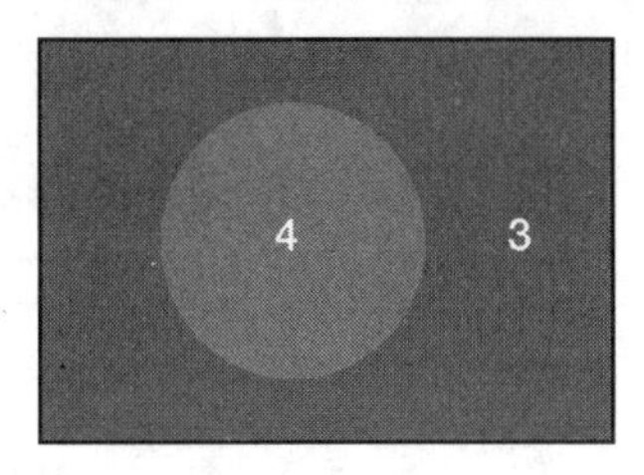

Name on driving licence: The People's Republic of Bangladesh
Capital: Dhaka
Population: 141 million
Dosh: Taka = 100 paise
Size: 144,000 km^2 (7 Wales)

Complete history: In the days when 'Bag handles' (anag) was merely the eastern end of Bengal, it was ruled by Buddhists. Later it was ruled by Muslims. After the Battle of Plessey in 1757, it was effectively ruled by a bunch of London traders (aka the East India Company) which, on reflection, must have been mortifying. Redeemed itself by becoming (part of) British India, then East Pakistan, then Free Bengal, and finally Bangladesh or, to give it its full name, Concert for Bangladesh.

Top spot: The Royal Bengal tiger lurks in the swampy mangrove forests of the Sundarbans region. The tigers apparently earned the sobriquet 'royal' because of their predilection for making embarrassing remarks about the locals at state functions. **[16]**

Popular misconception: 'Dhaka serves as the capital of Bangladesh and Senegal.' It does not. Senegal has its own capital which it calls Dakar. In today's global economy few countries are likely to succeed if their capital city is in another continent, something the Senegalese cottoned on to very early before too much damage had been done. **[12]**

Sad fact: In 1878, Raj District Officer John Beames wrote of the country-in-waiting: 'I have never seen so lovely a place to look at, nor one so loathsome to live in.' So, not entirely won over. **[10]**

Pub fact: According to Transparency International, Bangladesh shares with Chad the distinction of being the world's most corrupt country. **[2]**

Opening lines of national anthem: 'My Bengal of gold, I love you/Forever your skies, your air set my heart in tune.' **[7]**

SCORE: **47** WORLD RANKING: **150**

Barbados AMERICAS

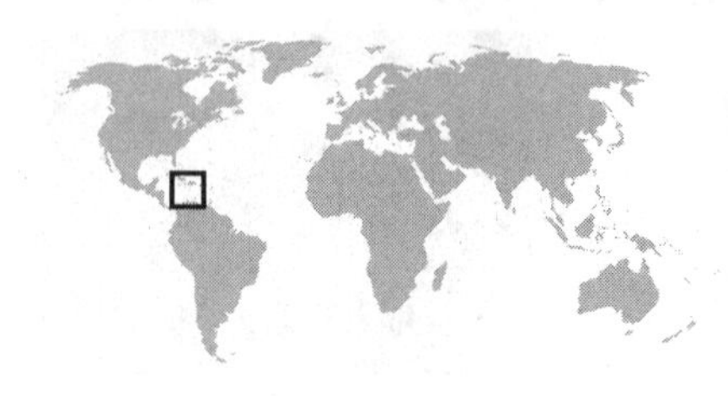

Name on driving licence: Barbados
Capital: Bridgetown
Population: 277,000
Dosh: Barbadian dollar = 100 cents
Size: 430 km² (0.02 Wales)

Complete history: The first settlers, Amerindians from Venezuela, arrived in 1623 BC and were nothing if not stout fellows and fellowesses. They came in dugout canoes – much the same as those sung about so stirringly by Rolf Harris in his 1965 hit 'War Canoe', though less fitted for war – braving the heart-stoppingly swift Caribbean currents that have since done for many an ocean-going vessel. The British, attempting a sort of chronological symmetry, settled in AD 1627, only to discover through carbon dating the canoes of the original inhabitants that they were four years too late. As can be imagined, many of the colonisers never recovered from this blow and the British vowed to leave the islands as soon as one of the home countries won the World Cup.

Crowning achievement: Barbados gained independence in 1966 in return for a promise that their cricketers would 'go easy' on the England team that year. The West Indies, breezing to a 3-0 lead, duly let their hosts win the final test, Bridgetown boy Gary Sobers getting himself out first ball, just to make sure. **[18]**

Sad fact: The population density is a how-many-students-can-you-get-in-a-phone-box 644 people per km². This compares with 324 in India, and 245 in the 'overpopulated' UK. **[9]**

Top spot: Bridgetown possesses a miniature version of Trafalgar Square, though without the pigeons. **[14]**

Pub fact: Barbados boasts the world's third-oldest parliamentary democracy (est. 1639). **[15]**

National anthem sounding suspiciously as though it has been culled from a fairy tale: 'In plenty and in time of need/When this fair land was young.' **[14]**

SCORE: 70 **WORLD RANKING:** 22

colour key: 1 = turquoise 2 = blue 3 = green 4 = red 5 = yellow 6 = orange 7 = pink 8 = purple 9 = brown

Belarus

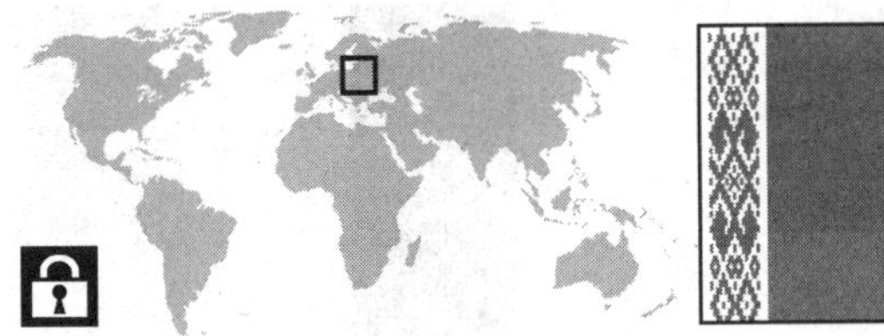

Name on driving licence: Republic of Belarus
Capital: Minsk
Population: 10.3 million
Dosh: Belarusian rouble = 100 kopeks
Size: 207,600 km² (10 Wales)

Complete history: The Slavic people toiled away so that by the 9th century the area had become the mighty Kievan Rus. So mighty, in fact, that they only lost *part* of the country to the invincible Mongol hordes. They then cosied up to the Lithuanians and Poles for a while until the Ruskies took over. Come the Revolution, Belarus became an independent state for about five minutes until the Russians noticed and promptly invaded again, renaming the now Communist country Byelorusia (White Russia). The Nazis only stayed 1941–4, but somehow still managed to kill off a quarter of the population. Independent (and Belarus) again in 1991. Dodgy president.

Crowning achievement: On 26 April 2004 – a date revered by Belarusians – national heroine Alesya Goulevich swung a world record 99 hula hoops for three full revolutions. If you're in Belarus on 26 April (Alesya Day), be sure to join in the celebrations by taking a hula hoop and shaking your hips. (Do make sure though that Alesya still holds the record or your actions could see you facing charges of 'mocking the Belarusian nation' and a possible twenty-year sentence.) **[13]**

Top spot: The bison roam wild in Belovezhskaya Pushcha. **[16]**

Gift to the world: Female Olympic athletes who look suspiciously like they should be competing in the events for men. **[5]**

Motto: 'No, that's Belarusian with *one s*.' **[12]**

National anthem penned by a speech writer: 'We maintain generous friendship and gain our powers/Within the industrious free family.' **[9]**

SCORE: 42 **WORLD RANKING: 171**

Belgium

EUROPE

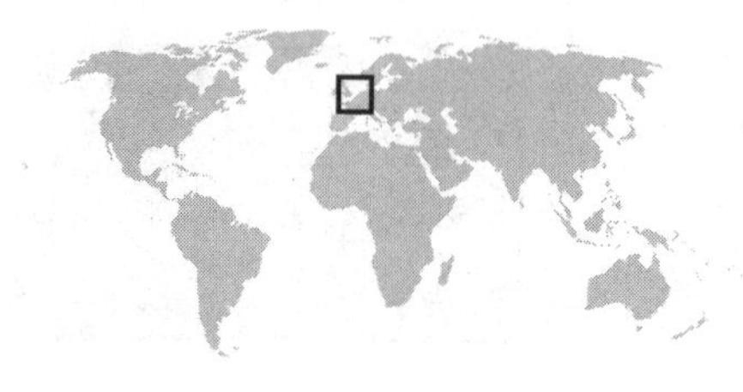

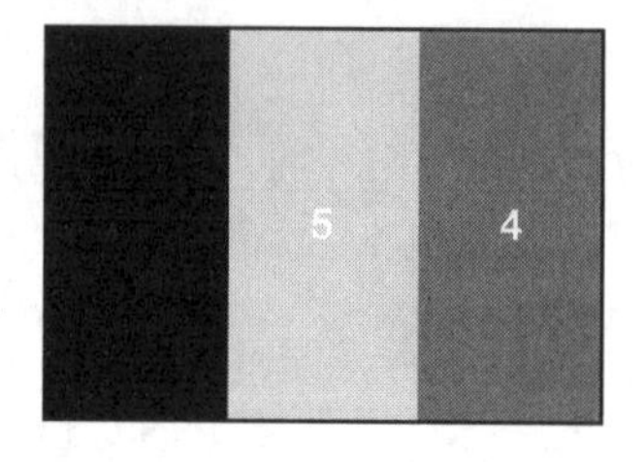

colour key: 1 = turquoise 2 = blue 3 = green 4 = red 5 = yellow 6 = orange 7 = pink 8 = purple 9 = brown

Name on driving licence: Kingdom of Belgium
Capital: Brussels
Population: 10.3 million
Dosh: Euro = 100 cents
Size: 30,500 km^2 (1.5 Wales)

Complete history: Still in little pieces in the Middle Ages, Belgium was united by the dukes of Burgundy and enjoyed a pleasant spell of prosperity before the 'Cockpit of Europe' succumbed, with heart-sinking inevitability, to a succession of Austrian, Spanish and French ne'er-do-wells. After the Napoleonic Wars, Belgium spent fifteen years in union with the Netherlands as the self-abasing Low Countries. Proceeded to take over bits of Africa and make them more hellish than even the English and French had managed, which was really quite an achievement. Nazis 1940–4. Cockpit of Europe again from 1957.

Crowning achievement: Successfully converting the name of its capital into a *Daily Mail* term of abuse (see also *asylum seeker* and *single mother*). **[20]**

Made in Belgium: Walloons. **[18]**

Popular misconception: 'Belgium is famous for not producing famous people.' This is a myth. Here is a list just off the top of your head: Hergé, Jean-Claude Van Damme, René Magritte, Plastic Bertrand, and the King of Belgium. Furthermore, the *chansonnier* Jacques Brel, rabidly claimed by the French, was very satisfyingly born in the outskirts of Brussels. Odder still, so was Audrey Hepburn. **[15]**

Pub fact: One of the oldest surviving oil paintings in the world, Jan Van Eyck's 'De Aanbidding van het Lams God' ('Something Something of the Lamb of God'), hangs in St Baaf's Cathedral, Ghent. **[10]**

National anthem promoting the idea of donating body parts to one's country: 'Noble Belgium – forever a dear land/We give you our hearts and our arms.' **[16]**

SCORE: 79 **WORLD RANKING:** 4

Belize

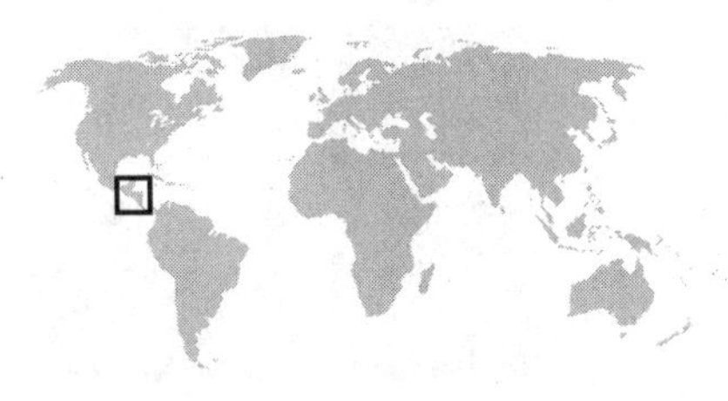

Name on driving licence: Belize
Capital: Belmopan
Population: 273,000
Dosh: Belizean dollar = 100 cents
Size: 23,000 km² (1.1 Wales)

Complete history: In 1783, after slapping a few colonies along the coast of what was then Guatemala, the perfidious English signed an accord in which the Spanish government gave them logging rights in return for the building of a road between Belize City and Guatemala City. The English cleverly interpreted this to mean that they could annex the land – which they renamed British Honduras – in return for doing nothing at all. They only started pulling out in 1973, when the country became Belize, and grudgingly granted independence in 1981. The patriotic appeals for the return of Belize made by successive Guatemalan presidents have been found to be in direct proportion to the magnitude of the corruption scandals they face.

Top spot: Though 300 km long, the Belize Barrier Reef is still pipped by that one off Australia. **[15]**

Popular misconception: 'The coastal town of Blair Atholl is the same Blair Atholl that play in the Perthshire Amateur League and that won the North Perthshire Cup in 2005/06.' It is not. The Belizean Blair Atholl *does* have a football team, but not one nearly good enough to capture the North Perthshire Cup, retrospectively or otherwise. **[12]**

Pub fact: Belize is so breezy that the tallest structure in the country is the Mayan-built Canaa Pyramid, which clocks in at a measly 43 metres. **[8]**

Actual motto: '*Sub umbra floreo*' ('I flourish in the shade'). **[12]**

Opening lines of national anthem: 'O land of the free by the Carib Sea/Our manhood we pledge to thy liberty!' **[12]**

SCORE: **59** WORLD RANKING: **85**

Benin

AFRICA

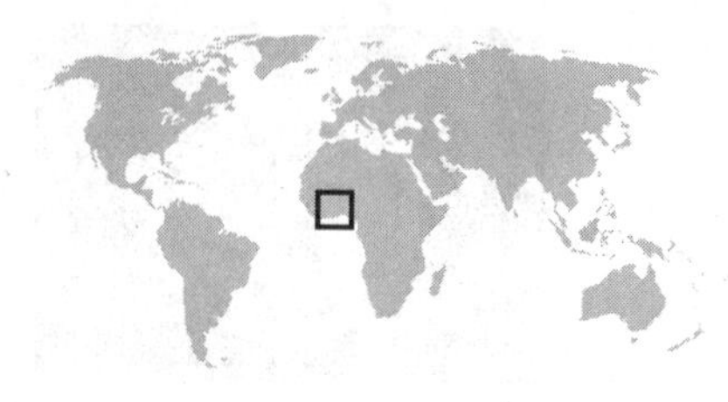

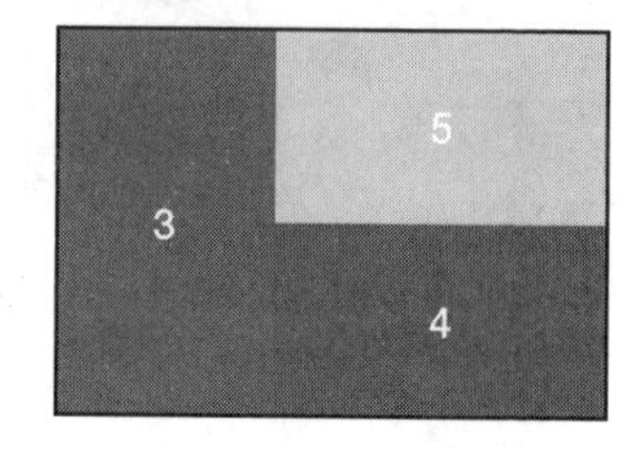

Name on driving licence: Republic of Benin
Capital: Porto-Novo
Population: 7.3 million
Dosh: CFA franc = 100 centimes
Size: 112,600 km^2 (5.4 Wales)

Complete history: Benin has spent most of its life as Dahomey, or at least as a sliver of the once vast West African kingdom. Regrettably, several Dahomean kings were more than happy to sell their subjects to European slave traders. This is why democratically elected heads of state are sometimes A Good Thing. When slavery at last came to an end, the French muscled in, eventually taking over the country and adding it to their numerous acquisitions in the region. Since they were even more reluctant to relinquish their empire than the British, independence was only granted in 1960, the name change coming fifteen years later. Currently the only African country shaped like a '99' in which someone has bitten off the top third of the flake.

Top spot: Abomey, nowadays a weird pokey town of fetish temples on the long road north from the coast, used to be the capital of the kingdom of Dahomey. **[14]**

Crowning achievement: Ganvié (population 15,000) is the largest lake village in Africa. The inhabitants live in houses on stilts and whizz about in pirogues, a boat that only Africans can look cool in, or indeed manoeuvre. **[17]**

Sad fact: One of the exhibits in a museum in Abomey is a throne made out of human skulls. **[6]**

Religious affiliations: Ronald Reagan had voodoo economics, the Beninois just have voodoo. **[2]**

Opening lines of national anthem: 'Formerly, at her call, our ancestors/Knew how to engage in mighty battles.' **[12]**

SCORE: 51 **WORLD RANKING:** 130

colour key: 1 = turquoise 2 = blue 3 = green 4 = red 5 = yellow 6 = orange 7 = pink 8 = purple 9 = brown

Bhutan

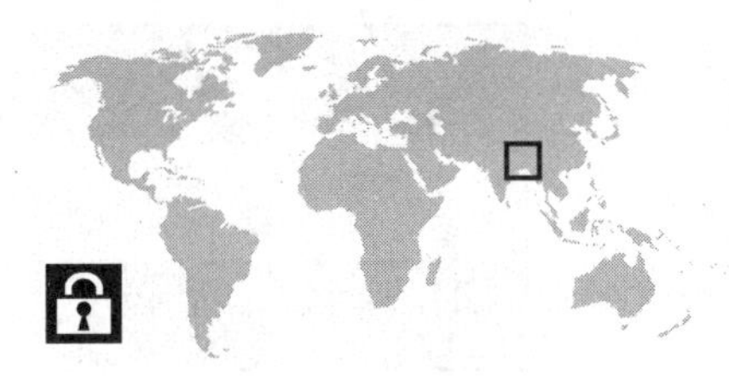

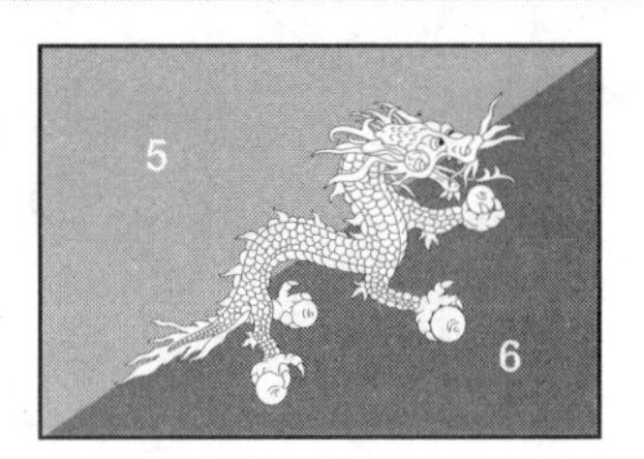

Name on driving licence: Kingdom of Bhutan
Capital: Thimpu
Population: 2.2 million
Dosh: Ngultrum = 100 chetrum
Size: 47,000 km² (2.3 Wales)

Complete history: If the Bhutanese do ever get a yen for world domination, their first round opponents will be on the tough side: India to the south, and China to the north. This explains Bhutan's fondness for Buddhism, a religion more concerned with personal enlightenment than naked military expansionism. For all that, the Bhutanese do enjoy a bit of physical activity and spent hundreds of years divided into tiny fiefdoms all fighting one another. It wasn't until the early 1600s that the Tibetan lama Shabdrung Ngawang Namgyal declared that Bhutan should become a nation (mainly in order that he could rule it). Everyone cooled their heels until Shabdrung died and it was safe for internecine war to break out between rival district governors. Things got so bad that in 1910 the ubiquitous British decided they'd better look after Bhutan's foreign affairs, presumably on the grounds that they had some experience in being foreign and would thus make a good job of it. First ever democratic elections due 2008.

Crowning achievement: The abolition of slavery. Pity it took until 1958. **[5]**

Top spot: The government of Bhutan has created the Sakteng Wildlife Sanctuary specifically to safeguard the habitat of the yeti (or *migoi*), just in case such a creature exists, which is kind. **[18]**

Customs to treasure: Archery is Bhutan's national sport. **[14]**

Pub fact: Bhutan had no television service whatsoever until 1999. **[16]**

Opening lines of national anthem: 'In the Thunder Dragon Kingdom, where cypresses grow/Refuge of the glorious monastic and civil traditions.' **[16]**

SCORE: **69** WORLD RANKING: **29**

Bolivia

AMERICAS

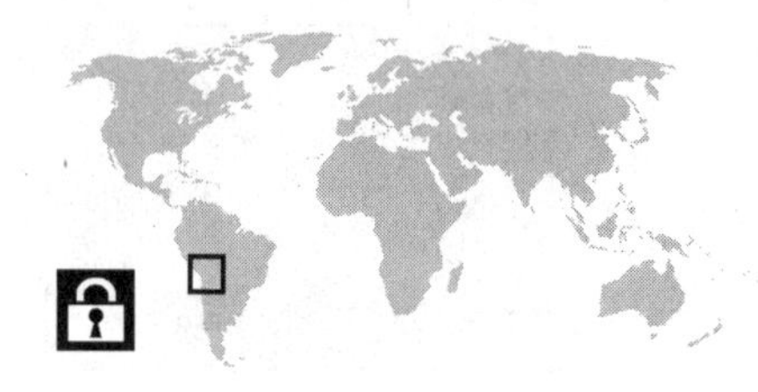

colour key: 1 = turquoise 2 = blue 3 = green 4 = red 5 = yellow 6 = orange 7 = pink 8 = purple 9 = brown

Name on driving licence: Republic of Bolivia
Capital: Sucre/La Paz
Population: 8.7 million
Dosh: Boliviano = 100 centavos
Size: 1,099,000 km^2 (53 Wales)

Complete history: Five things have happened in Bolivia, if you discount the daily *coups d'état*. They are: the arrival of the Incas; the arrival of the conquistadores; the loss of a bit of itself to Paraguay; the loss of its coastline to Chile; and the receipt of a bit of beach given by Peruvian president Alberto Fujimori, so that Bolivians could pretend they had a coastline again.

Top spot: Lake Titicaca, aside from having the world's best name for a lake, is also the Earth's highest navigable one (3,812 metres), making altitude sickness both de rigueur and fun. **[14]**

Pub fact: It's said that the amount of silver sent from Bolivia to Spain by the conquistadores was so vast that the ingots laid end to end could have stretched between the two. This would have inconvenienced shipping in the Atlantic, however, so it was never tried out. **[8]**

Sad fact: The hill of Potosí was so riddled with silver that the Spaniards were able to mine it for over 250 years. During this time, an estimated 2 million slave miners had the temerity to escape the horror of the mines by dying, a course of action the Pentagon has since rightly identified as 'asymmetrical warfare'. **[17]**

Motto: '*¡Oye, gringo! Deja de buitrear en nuestro lago.*' ('Hey, gringo, stop throwing up in our lake.') **[15]**

Opening lines of national anthem: 'Bolivians, a favourable destiny/Has crowned our vows and longings.' **[9]**

SCORE: 63 WORLD RANKING: 61

EUROPE Bosnia & Herzegovina

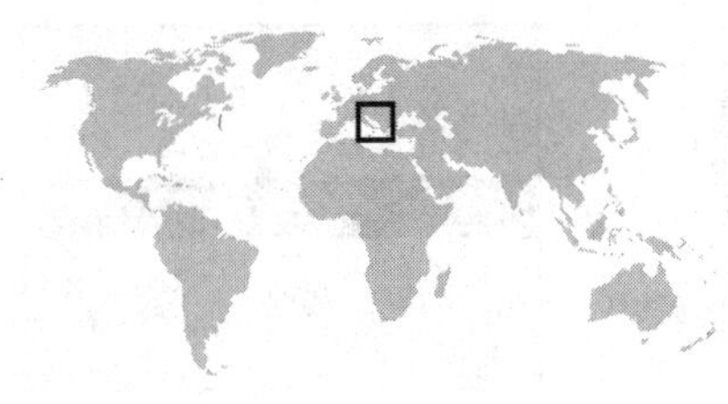

Name on driving licence: Bosnia and Herzegovina
Capital: Sarajevo
Population: 4 million
Dosh: Convertible marka = 100 convertible pfenniga
Size: 51,100 km^2 (2.5 Wales)

Complete history: After the customary settling by the Slavs, and invasions by the Ottoman Turks and Austro-Hungarians, the area was subsumed into Yugoslavia in 1929. When the Nazis arrived, they gave B&H to their friends the Croats. Part of Yugoslavia again by the end of the war under the friendly gaze of General Tito who, whatever else you may say about him, at least prevented the various ethnic communities from massacring each other. B&H declared independence in 1992, unleashing a civil war between local Serbs and a short-lived Croat–Muslim alliance. Peace, if not harmony, established under the Dayton Peace Accord.

Top spot: Saved from landlockness by a mere 20 km of Adriatic coast, or 1 km per 200,000 inhabitants, so the beaches may become crowded in summer. **[12]**

Historical low: Being wholly and utterly responsible for World War I. The Latin Bridge in Sarajevo was the spot chosen by Gavrilo Princip to shoot Archduke Ferdinand. In those days, that was enough to kick off world conflagration. **[3]**

Flag fact: Since 1998, B&H has been saddled with a hopeless and frankly rather unpleasant blue-and-yellow flag clearly based on strips worn by North American Soccer League teams circa 1974. **[3]**

Pub fact: 'Sarajevo roses' are shell holes in the capital that have been filled with red cement. **[15]**

Opening lines of national anthem: The anthem, adopted in 1998, is entirely word free, so that no one might declare themselves offended by whatever words might have been used or the language in which they might have been put. Modern times. **[17]**

SCORE: 50 **WORLD RANKING: 136**

Botswana

AFRICA

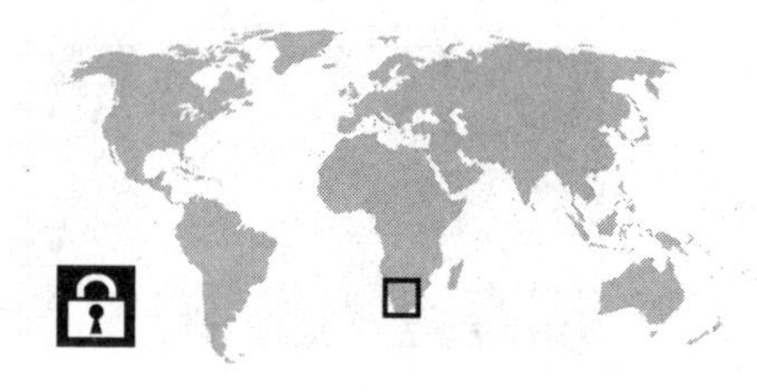

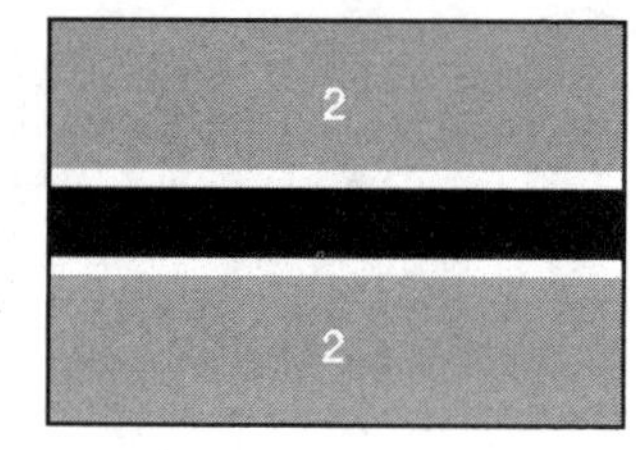

colour key: 1 = turquoise 2 = blue 3 = green 4 = red 5 = yellow 6 = orange 7 = pink 8 = purple 9 = brown

Name on driving licence: Republic of Botswana
Capital: Gaborone
Population: 1.6 million
Dosh: Pula = 100 thebe
Size: 600,400 km^2 (28 Wales)

Complete history: The San people – not to be confused with what posh boys call the staff in the school sanatorium – have been here for as long as anyone can remember. The Tswana, relative newcomers, moved in about a thousand years ago, forcing the San into the Kalahari Desert. The British, inevitably, rocked up somewhat later and pretended to own the place between 1885 and 1966, calling it the Bechuanaland Protectorate.

Popular misconception: 'The Kalahari Desert, which accounts for 85 per cent of Botswana, is a desert in the "Ice Cold in Alex" sense.' *C'est pas vrai*, as the Botswanans would say if only they'd been colonised by the French. It's actually covered by long grass that Laurens van der Post, bless him, described as 'fields of gallant corn'. **[18]**

Pub fact: Botswana is home to the Okovango Delta, the world's largest inland delta. (NB The highest concentration of pub quizzes with specific rounds on 'Deltas – Inland and Coastal' is to be found in East Anglia.) **[14]**

Flag fact: The black-and-white stripe running across the nation's flag represents a zebra *and* racial harmony, which is clever (although separating black and white would seem to send out the wrong message, whereas a uniform dark grey might perhaps have hinted at a more integrated society, but that's a minor quibble). **[16]**

Made in Botswana: The San, aka The Bushmen, still live in the Kalahari despite continuing attempts by the Botswanan government to exterminate them. **[15]**

National anthem guaranteed to please Germaine Greer: 'Awake, awake, O men, awake!/And women close beside them stand.' **[3]**

SCORE: 66 **WORLD RANKING:** 44

Brazil

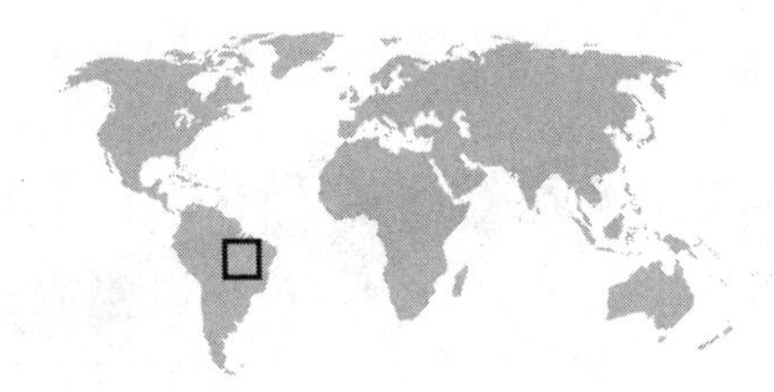

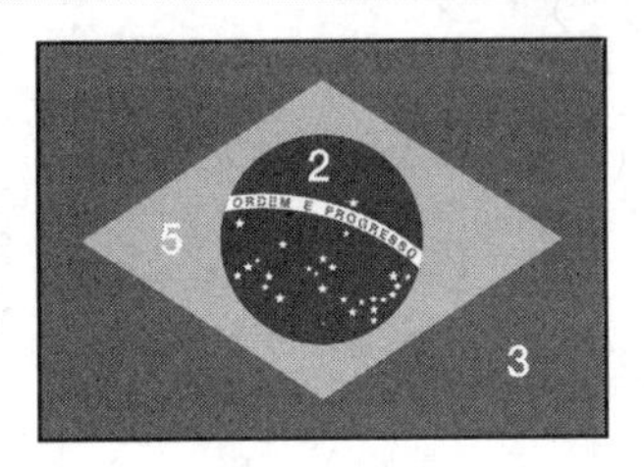

Name on driving licence: Federative Republic of Brazil
Capital: Brasilia
Population: 184 million
Dosh: Real = 100 centavos
Size: 8,512,000 km^2 (410 Wales)

Complete history: People were hanging around in the area now known as Brazil for ages before Portuguese explorer Pedro Alvarez Cabral came along. Portugal and Spain then had a right set to about who owned which parts of South America and eventually split it down the middle, the Portuguese taking everything to the right of the line. Unlike the Spanish, who divvied up their half into manageable chunks like Ecuador and Peru, the Portuguese preferred having one nice big country, probably with an eye to future World Cups, and united the whole of their half in 1808, even moving their royal court from Lisbon to Rio de Janeiro. One could almost make an argument for saying that Portugal thus became part of the Brazilian empire, but not quite.

Crowning achievement: The only country to date to have been named after a nut. **[16]**

Top spot: The capital, Brasilia, was laid out in the form of an aeroplane. Refreshingly bonkers architect Oscar Niemeyer then filled it with buildings made to resemble teacups and such. Too cool for school. **[17]**

Pub fact: The world's fifth-largest country contains the world's largest swamp. Since the Pantanal covers 109,000 km^2 (5.25 Wales), it's likely to hold the record for some time to come, unless it rains an awful lot somewhere that's already a bit mushy, like Burma. **[15]**

Gift to the World: 'The Girl from Ipanema'. **[19]**

Opening lines of national anthem: 'The Ypiranga's placid banks heard/The resounding shouting of a heroic people.' **[13]**

SCORE: 80 **WORLD RANKING:** 3

Brunei

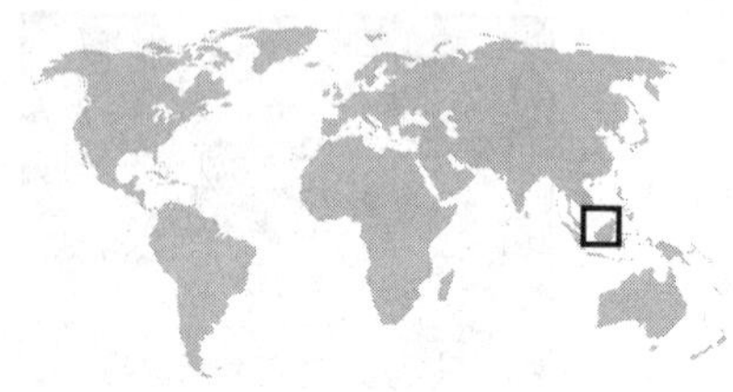

colour key: 1 = turquoise 2 = blue 3 = green 4 = red 5 = yellow 6 = orange 7 = pink 8 = purple 9 = brown

Name on driving licence: Negara Brunei Darussalam
Capital: Banda Seri Begawan
Population: 358,000
Dosh: Bruneian dollar = 100 cents
Size: 5,770 km^2 (0.3 Wales)

Complete history: Like a past Grand National winner boiled down to form the base for an own-brand contact adhesive, Brunei has seen better times. As recently as the 16th century it bossed the entire island of Borneo and the lower Philippines, and its armed forces went on to chalk up a win against a Spanish side full of experienced internationals. It took the British, with their flair for mixing apparent diplomacy with guns full of really big hurty things, to bring the country to its lowest ebb, forcing the sultanate into protectorate status in 1888. Brunei didn't become fully independent again for another century, though by then there wasn't much of it left and no one was quite sure whether it was worth the trouble.

Crowning achievement: The Brunei Revolt (1962), an anti-monarchy rebellion snuffed out by the British (who love a good monarchy) but which helped put the kibosh on the creation of the North Borneo Federation, a weedy name for a country if ever there was. **[15]**

Popular misconception: 'Brunei was named after former national treasure Frank Bruno.' **[8]**

Pub fact: Brunei sports the world's largest palace, which reportedly cost a piffling US$450 million, so presumably some corners were cut. **[5]**

Customs to treasure: A cheeky disregard for democracy (and, indeed, sexual equality). Brunei has been ruled by male sultans from the same family for over 600 years and not one of them has an election win to his name. **[5]**

Opening lines of national anthem: 'God Bless His Majesty/With A Long Life.' Ho-hum. *Yawns.* **[3]**

SCORE: 36 **WORLD RANKING:** 181

Bulgaria

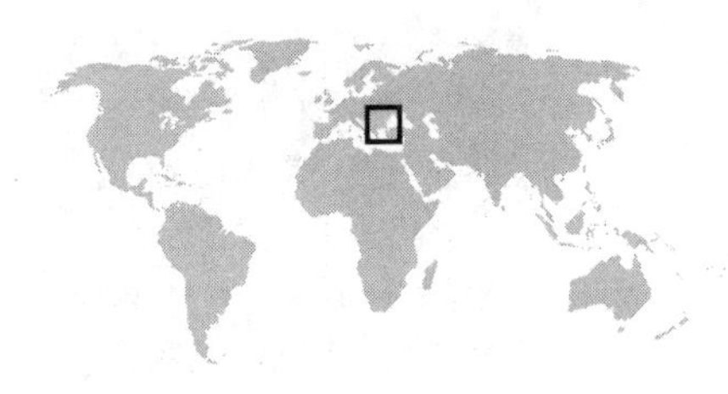

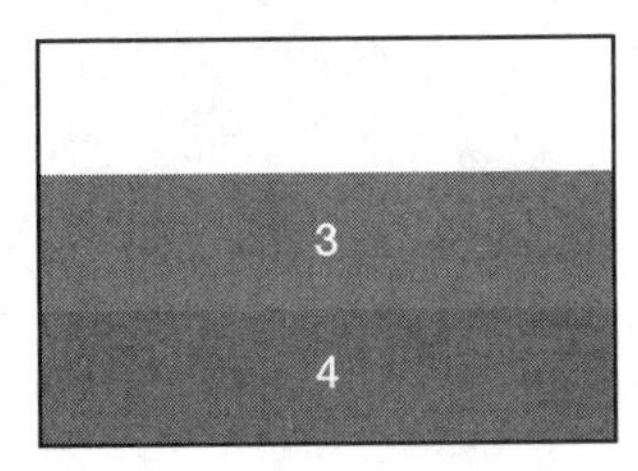

Name on driving licence: Republic of Bulgaria
Capital: Sofia
Population: 7.5 million
Dosh: Lev = 100 stotinki
Size: 110,900 km² (5.4 Wales)

Complete history: Eastern European history is entirely composed of waves of drifters biffing stationary folk about the head until they give in. These drifters (sometimes called 'driftics') then themselves become stationary for a bit, a state called 'empire', until another wave of drifters comes along to biff them about the head. In the case of Bulgaria, the blanks can be filled in if you start with Slavs/Bulgars and insert, consecutively: Byzantines, Ottoman Turks (and sometimes Turkics), Russians and, confusingly, Soviets (this last brought upon the Bulgarians by their insistence on allying themselves with Germany every time there was a war on).

Crowning achievement: Arranging for a former king to become prime minister. This feat was accomplished when King Siméon Saxe-Coburg-Gotha left the country in a huff in 1948, on finding he'd been abolished, only to return 53 years later to triumph at the ballot box. **[15]**

Pub fact: Top womble Uncle Bulgaria was not, in fact, Bulgarian, but some other nationality (see also Tomsk). **[12]**

Gift to the world: Bulgarian folk music boasts a unique array of irregular rhythms (11/8 time being a particularly alarming example) and was a major influence on the young Kate Bush. **[14]**

Popular misconception: 'Bulgaria is the home of bulgar wheat.' Not so, though there are currently no restrictions on the amount of bulgar wheat you can take into the country as long as you can convince customs officials that it's for personal use. **[13]**

National anthem apparently written by Sir John Betjeman on an off day: 'Over Thrace the sun is shining/Pirin looms in purple glow.' **[15]**

SCORE: 69 **WORLD RANKING: 29**

Burkina Faso AFRICA

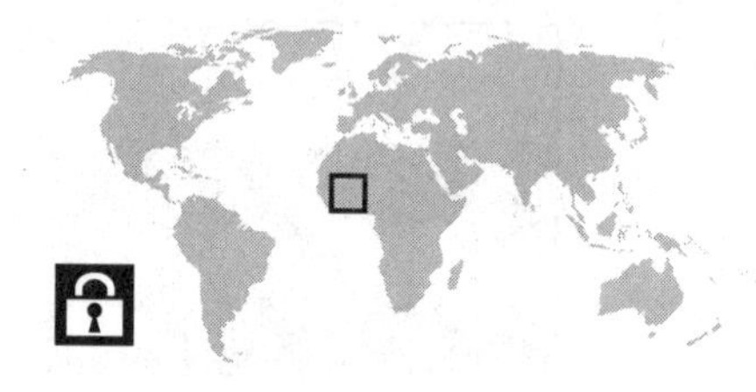

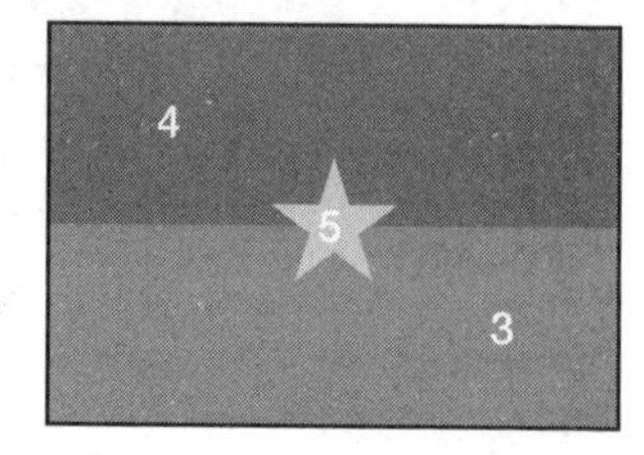

colour key: 1 = turquoise 2 = blue 3 = green 4 = red 5 = yellow 6 = orange 7 = pink 8 = purple 9 = brown

Name on driving licence: The Democratic People's Republic of Burkina Faso
Capital: Ouagadougou
Population: 13 million
Dosh: CFA franc = 100 centimes
Size: 274,200 km² (13 Wales)

Complete history: Burkina Faso is more a land of people than time. The Mossi and the Bobo (both Voltaic peoples, so named because they naturally produce their own electricity) rub along quite happily with another sixty or so other ethnic groups. They've done so for so long that attempts to subjugate them by the Mali and Songhai empires and, much more recently, by the dastardly French, seem but blips in the space-time continuum, which must be some consolation when a cheery legionnaire lobs a grenade through your window.

Crowning achievement: There have been just five *coups d'état* since independence in 1960. This relative political stability has ensured a supply of bottled Guinness to every corner of the land, which is an achievement of sorts. [14]

Top spot: Finding something to do or anywhere to go in Burkina Faso has been the national pastime for as long as anyone can remember. Hopeful rumours that the country was unique in possessing beaches formed from the ground-down bones of beached whales were only dashed when it was discovered that the country was landlocked. [2]

Pub fact: Burkina Faso was previously known as Haute Volte ('High Voltage'). [5]

Popular misconception: 'The capital, Ouagadougou, was named after the catchy novelty hit by singing sensations Black Lace.' [15]

What happens when your national anthem is written by a roundy-uppy sort of history teacher with an axe to grind, rather than by a poet: 'Against the humiliating bondage of a thousand years/Rapacity came from afar to subjugate them for a hundred years.' [18]

SCORE: 54 **WORLD RANKING:** 114

ASIA

Burma (Myanmar)

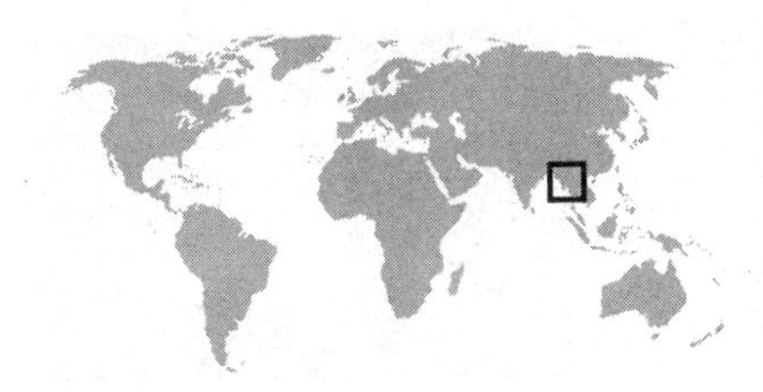

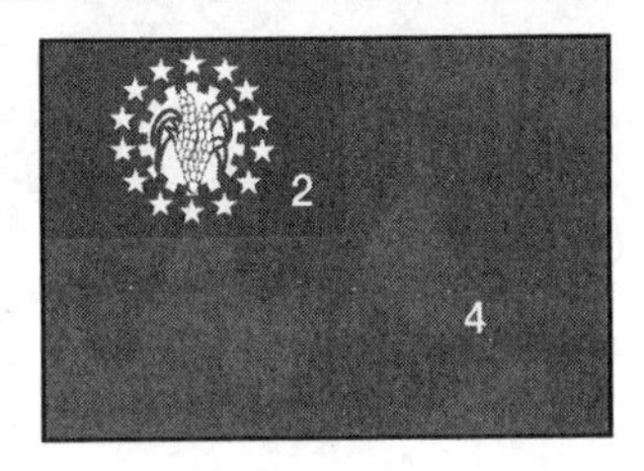

Name on driving licence: Union of Burma/Union of Myanmar
Capital: Pyinmana
Population: 43 million
Dosh: Kyat = 100 pyas
Size: 678,500 km² (33 Wales)

Complete history: Oh dear, oh dear. There's stiff opposition for the title of Official Pariah State nowadays, but Burma has proved equal to the task. Having a military junta that used to call itself SLORC was a good start, and keeping that nice Aung San Suu Kyi locked up in her house more or less sealed it. Previous to all this, the British hung around sweatily. Before them came the Burmans, the Shan and the Mons, sometimes all at once.

Top spot: Mandalay not only made it into *Rebecca* – Daphne Du Maurier's tale of love, revenge and leaky pleasure cruisers – but also Ian Dury's 'Hit Me With Your Rhythm Stick', a feat that remains unequalled by any other city in the world. **[12]**

Made in Burma: Rice. **[8]**

Pub fact: Right-on people who grow unusual facial hair and who claim their clothes are made by unionised dolphins from 100 per cent organic seaweed believe that if you refuse to call the country Myanmar this will in some way cause the downfall of the military junta responsible for dreaming up the name. **[4]**

Popular misconception: 'The capital of Burma is Rangoon.' Uh uh. Just to prove it was still capable of as much perversity as the next mad ruling elite, in 2006 the junta moved the capital to an obscure town called Pyinmana and made it illegal for anyone but government officials to go there. **[0]**

National anthem in which one's country is claimed to be the sun, or at the very least a really powerful torch: 'Fountain of Freedom/Source of Light.' **[6]**

SCORE: 30 **WORLD RANKING: 190**

Burundi

AFRICA

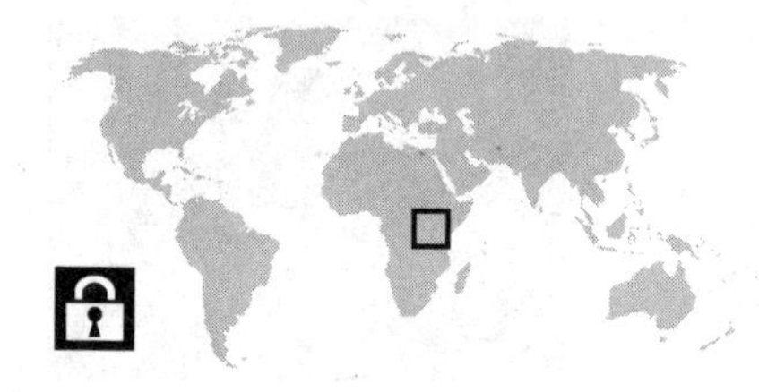

colour key: 1 = turquoise 2 = blue 3 = green 4 = red 5 = yellow 6 = orange 7 = pink 8 = purple 9 = brown

Name on driving licence: Republic of Burundi
Capital: Bujumbura
Population: 6.2 million
Dosh: Burundi franc = 100 centimes
Size: 27,800 km² (1.3 Wales)

Complete history: In the beginning was the Twa, a group of pygmy hunter-gatherers. The Twa would hunt and gather by day, lose things by night, and go out the next day to hunt and gather them again. This went on for thousands of years until the arrival of the Hutu who, being a taller people, banished the Twa and their diurnally gathered possessions into the far wildernesses. Next came the Tutsi who lorded it over the Hutu. In the late 1890s, the Germans, being a blue-eyed and blonde people, elbowed the Tutsi aside. Then, as is well known, our Teutonic cousins lost World War I to Belgium, thus allowing the latter to become top dogs in Urundi (as was). In 1961, the people voted to be ruled instead by Mwami Mwambutsa IV who claimed he had been ruling since 1915 anyway, in the way kings do.

Sad fact: Around 300,000 people (mainly Hutus) have been slaughtered in ethnic conflicts since independence. **[0]**

Troubling scouting fact: The Association des Guides du Burundi became affiliated with the World Association of Girl Guides and Girl Scouts in 1972, yet it took the Association des Scouts du Burundi until 1979 to join their equivalent international body. No wonder things go wrong. **[15]**

Gift to the world: The Master Drummers (think Adam and the Ants minus everything but drums). **[14]**

Pub fact: The line 'Dr Livingstone, I presume?' was first used here in 1873. **[15]**

Opening lines of national anthem: 'Beloved Burundi, gentle country/Take your place in the concert of nations.' **[12]**

SCORE: 56 **WORLD RANKING:** 101

ASIA **Cambodia**

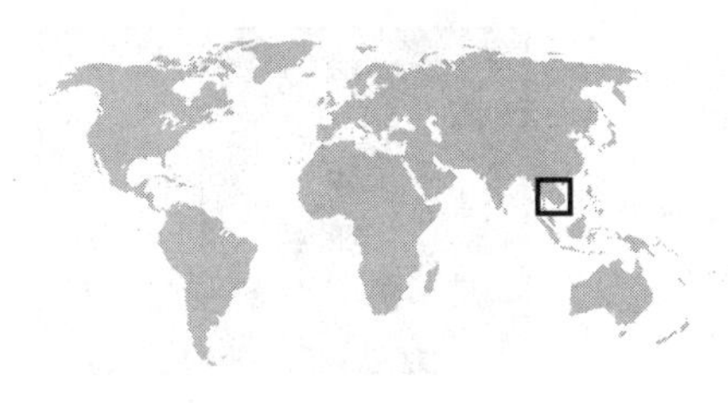

Name on driving licence: Kingdom of Cambodia
Capital: Phnom Penh
Population: 13 million
Dosh: Riel = 100 sen
Size: 181,000 km² (8.7 Wales)

Complete history: For six centuries from 802, the Khmer (not then Rouge but a more Hindu/Buddhist Orange) ruled their empire with a rod of bamboo, this being more readily available and easier to swish around than iron. Sadly, bamboo proved not much use against the harder woods employed by the Thai forces who swept through the country in 1431. Nothing much happened until 1863, when the French came wielding sticks fashioned from the hardest substance known to man, stale baguette. Independence lumbered over the horizon in 1954. Then there was that terrible business with the killing fields.

Crowning achievement: 'Holiday in Cambodia' by punk artistes The Dead Kennedys is the only single to date to mention a brand-name deodorant, chronicle the crimes of Pol Pot, and be released three times (1980/1988/1995). **[15]**

Top spot: Angkor Wat and Angkor Thom together form the world's largest temple complex. Eight hundred years after being built, Angkor at last emulated the Sussex commuter-belt town of Horsham by giving its name to a beer (albeit a lager, but you can't have everything). **[17]**

Flag fact: Cambodia's flag is the only one in the world to feature a building. **[12]**

Popular misconception: 'Dr Haing S. Ngor (Dith Pran in *The Killing Fields*) was the first non-professional actor to win an Oscar.' He was the second. Disappointing really. **[16]**

Opening lines of national anthem: 'Heaven protects our King/And gives him happiness and glory.' This is all well and good if you happen to be king, but very few Cambodians are. **[6]**

SCORE: 66 **WORLD RANKING:** 44

Cameroon

AFRICA

colour key: 1 = turquoise 2 = blue 3 = green 4 = red 5 = yellow 6 = orange 7 = pink 8 = purple 9 = brown

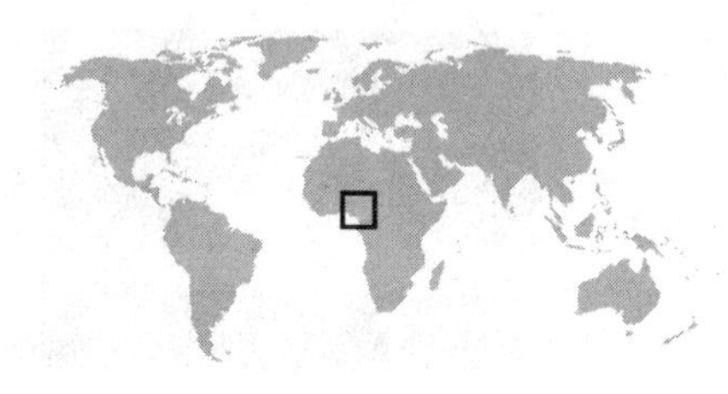

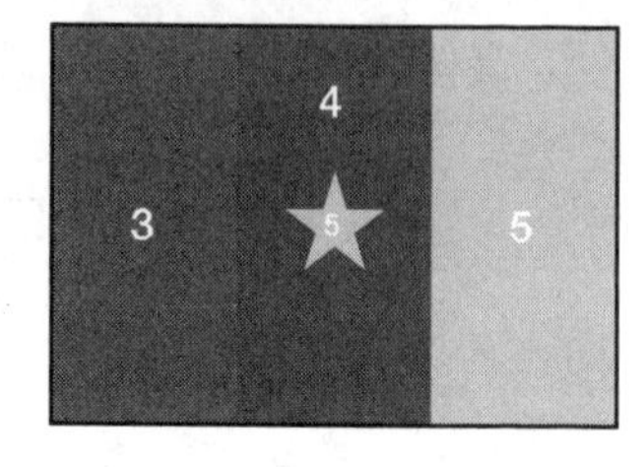

Name on driving licence: Republic of Cameroon
Capital: Yaoundé
Population: 16 million
Dosh: CFA franc = 100 centimes
Size: 475,400 km^2 (23 Wales)

Complete history: All was fine until Portuguese sailors began rocking up and soon it was all slavery and unpleasantness for ages and ages. However, in 1884, the Germans got up really early one morning and threw their towels over the entire country but, in common with so many of their possessions around the world, Germany misplaced Cameroon in 1918. Before the bespectacled Belgians could swipe it, the British and French carved the country up between them, each naming their respective bit after themselves in the time-honoured fashion. In 1961, French Cameroon (by then the Cameroon Republic) joined up with most of British Cameroon, some of which was no longer British Cameroon, having voted to become part of Nigeria, which was no longer British either.

Crowning achievement: Cameroon is variously known as the 'Hinge of Africa' or 'Africa in Miniature' but never the 'Miniature Hinge of Africa', which is just as well, or Africa's door would forever be falling off and incapacitating shipping in the Atlantic. **[15]**

Top spot: Cameroon's coastal belt is one of the world's rainworthiest areas, some parts receiving 10 metres of the stuff per annum, and yet ducks are seldom seen there. It's a mystery. **[16]**

Made in Cameroon: Palm oil, cassava, bauxite. Mix thoroughly before baking. **[11]**

Pub fact: Cameroon is, to date, the only country in the world to have been named after the Portuguese word for prawns (*camarões*). **[12]**

Opening lines of national anthem: 'O Cameroon, thou cradle of our fathers/Proudly rally to defend your liberty.' Why is there always so much fighting in national anthems? Discuss. **[7]**

SCORE: 61 **WORLD RANKING:** 70

Canada

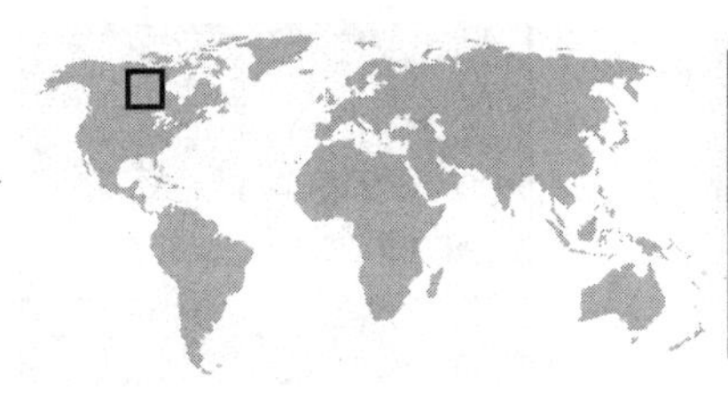

Name on driving licence: Canada
Capital: Ottawa
Population: 32 million
Dosh: Canadian dollar = 100 cents
Size: 9,976,100 km^2 (481 Wales)

Complete history: Most of Canada's recent history involves the English getting the better of the French. Contrary to popular belief, the French – a self-effacing people – love nothing more than an evening in their local *restau* or *estaminet* being regaled with stories by an English holidaymaker of the times their countrymen were outfoxed by General Wolfe and relieved of their vast territories in North America. The Gallic term for such balladeers – '*salauds anglaises*' – can be roughly translated as 'English salads', a high compliment given that the French now openly recognise England's superiority in all matters of the kitchen. Preceding the Europeans by about 40,000 years were the Asians, followed some time later by the Inuits. Note that it shows good breeding to include this in any impromptu lecture on the subject.

Crowning achievement: Becoming the world's second-largest country without anyone really noticing, or indeed caring. **[16]**

Made in Canada: The Mountie. Whisper it, but in some parts it's felt that having a policeman as one's national symbol is rather less than the epitome of cool. **[5]**

Pub fact: 'Esquimaux' was a name made popular by the French and derives from an Algonquian word meaning 'eaters of raw flesh', which is why Eskimos prefer to be called Inuit. **[5]**

Motto: 'No, actually I'm *Canadian*.' **[12]**

Opening lines of national anthem: 'O Canada! Our home and native land!' Since only 2 per cent of Canadians are Amerindian/Inuit, the use of the phrase 'native land' could be ironic, but don't absolutely bank on it. **[4]**

SCORE: 42 **WORLD RANKING:** 171

Cape Verde Islands AFRICA

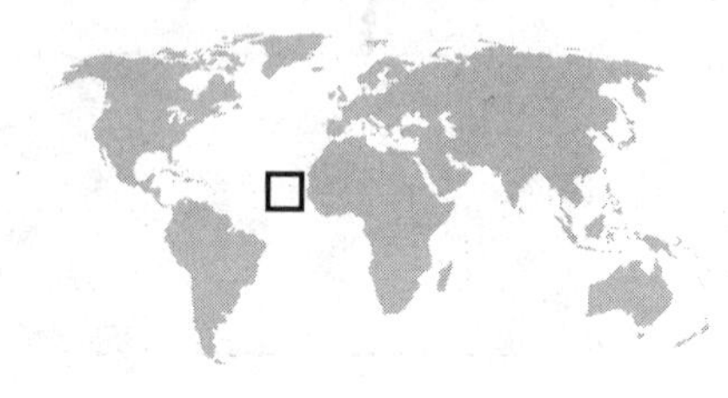

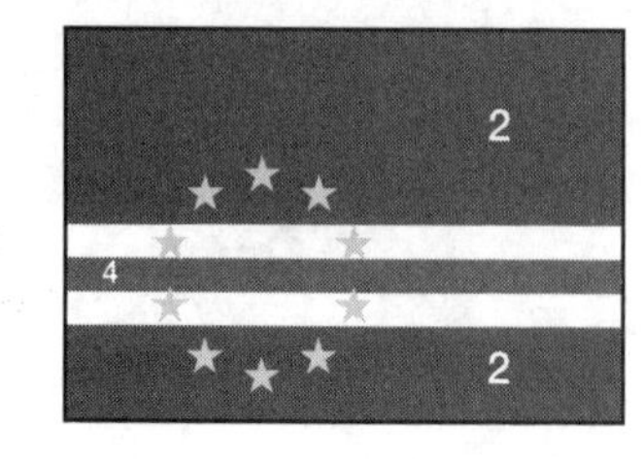

Name on driving licence: Republic of Cape Verde
Capital: Praia
Population: 445,000
Dosh: Cape Verdean escudo = 100 centavos
Size: 4,030 km² (0.2 Wales)

Complete history: Ten major islands, some lesser isles, and a scattering of shrapnel huddled in two groups (windward and leeward, as is the wont of islands), Cape Verde has rarely been a happy place to be. The Portuguese took over in the 15th century and filled the place up with slaves ready for transportation to the Americas. Even independence (1975) was overshadowed by the launch of zeitgeisty UK sitcom *Doctor On The Go*.

Top spot: The peak of the Fogo Volcano is Cape Verde's top spot, if only literally. **[12]**

Crowning achievement: Being surrounded by water yet suffering almost perennial droughts. **[8]**

Pub fact: Cape Verde is Africa's most westerly point. Sadly, this can often lead to unseemly scenes when it transpires that the question master at The Pig and Leper was actually after the most westerly point on the African *mainland* (Dakar, Senegal). To avoid unnecessary violence, do sort out any such ambiguities at the time the question is asked rather than at the announcement of the final scores when, inevitably, you discover your team has been beaten into second place by half a point. **[17]**

Motto: 'Water, water everywhere nor any a drop to drink.' **[10]**

National anthem: 'Sing, brother/Sing, my brother/For Freedom is a hymn/And Man a certainty/With dignity, bury the seed/In the dust of the naked island.' If only more national anthems were like this, strife and anguish would be a thing of the past, and dusty seeds the new black. Ripping. **[20]**

SCORE: 67 **WORLD RANKING:** 37

AFRICA Central African Republic

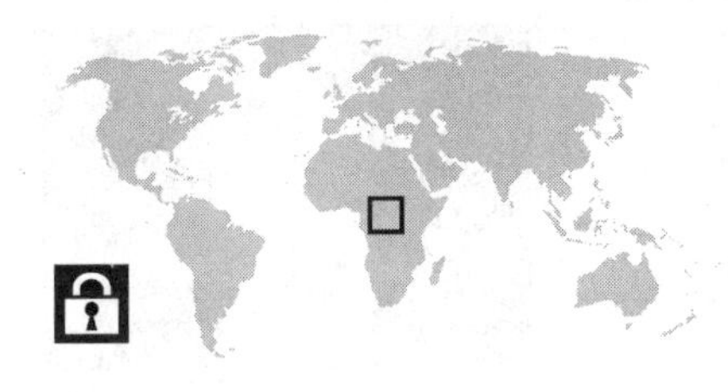

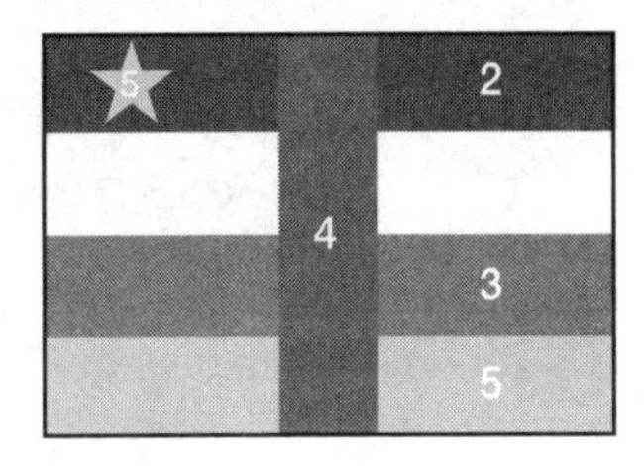

Name on driving licence: Central African Republic
Capital: Bangui
Population: 3.7 million
Dosh: CFA franc = 100 centimes
Size: 623,000 km^2 (30 Wales)

Complete history: Coups are a good deal easier to arrange than elections and less likely to compel the populace to wait in long queues for hours at a time only to be given a ballot slip with an X preprinted against the name of that year's dictator. Since bullets, rather than ballots, are the CAR's preferred method of transferring power, this leaves the inhabitants with more time to do other stuff such as befriending local warthogs and hunting manioc. As for the past, the French slouched about awkwardly here for a while. Independence was declared in 1960 and the indecorous scramble for the honour of misruling the country has been going on ever since.

Crowning achievement: 1976 – the republic becomes the Central African Empire. The empire lasts three years during which its borders are not expanded by a single inch. **[3]**

Sad fact: The coronation of the reportedly cannibalistic 'Emperor' Bokassa (1977) is purported to have cost a quarter of the nation's income for that year. **[2]**

Made in the Central African Republic: Manioc. Warthogs. Gold. Diamonds. In the event of a shortage of gold and diamonds, manioc rings encrusted with warthogs may be purchased at the central market in Bangui. **[7]**

Pub fact: The anniversary of the death of Barthélemy Boganda is a public holiday (29 March: bring your own sadness). **[10]**

National anthem impervious to the charms of the common full-stop: 'Oh! Central Africa, cradle of the Bantu!/Take up again your right to respect, to life!' **[16]**

SCORE: 38 **WORLD RANKING: 179**

Chad

AFRICA

colour key: 1 = turquoise 2 = blue 3 = green 4 = red 5 = yellow 6 = orange 7 = pink 8 = purple 9 = brown

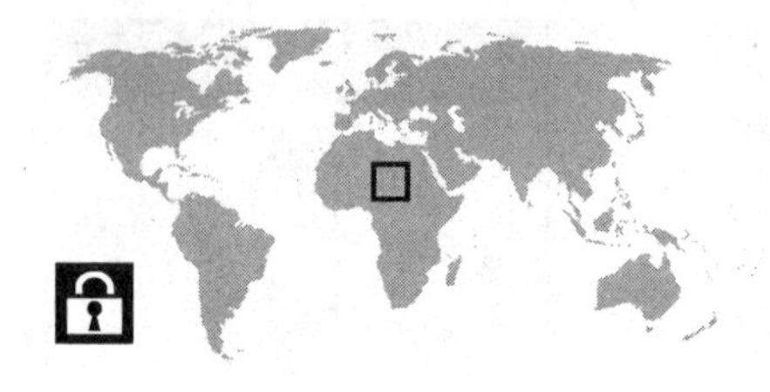

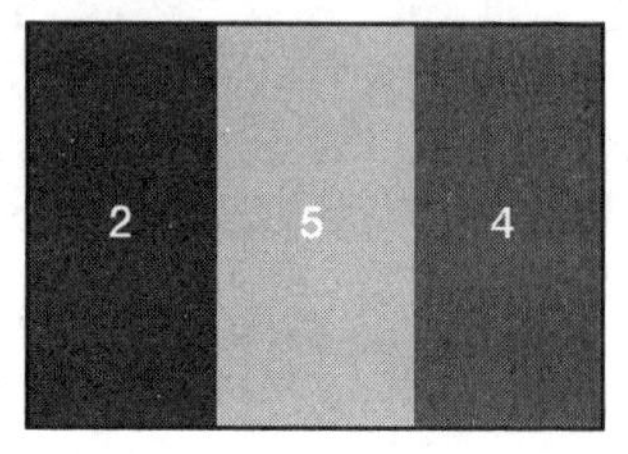

Name on driving licence: Republic of Chad
Capital: Ndjamena
Population: 9.5 million
Dosh: CFA franc = 100 centimes
Size: 1,284,000 km² (62 Wales)

Complete history: A country of two halves rather unfairly burdened with a name that makes it sound like an All American College Boy who has just made the state track-and-field team. The vest half is populated largely by wandering Arabs and Tuaregs; whereas the tracksuit trousers area is taken up by folk who prefer to stand very very still. The country used to be part of the Kanem Empire before being incorporated into the more African-sounding kingdom of Bornu. The Sudanese took over briefly before being biffed by the French in 1900. Since independence in 1960, civil wars have been a popular pastime, while the invasion by Libya in 1980 proved less of a hit.

Crowning achievement: The northern Tibesti Mountains are home to the world's foremost racing camels. **[15]**

Customs to treasure: *Pari-match*, in which a group of women hire a bar on a Sunday and attempt to make some money on the drinks bought by their friends. **[10]**

Made in Chad: Uranium. Cotton. If only someone could invent a cotton bud that cleaned the ear by means of a very localised nuclear explosion, the people of Chad could wave goodbye to poverty. **[13]**

Sad fact: Lake Chad is now only 10 per cent of the size it was in 1970. The remainder is thought to be held in private collections. **[4]**

National anthem that, in today's hysterical climate, might be seen as somewhat suspect: 'Oh, my Country, may God protect you/May your neighbours admire your children.' **[16]**

SCORE: 58 **WORLD RANKING:** 91

Chile

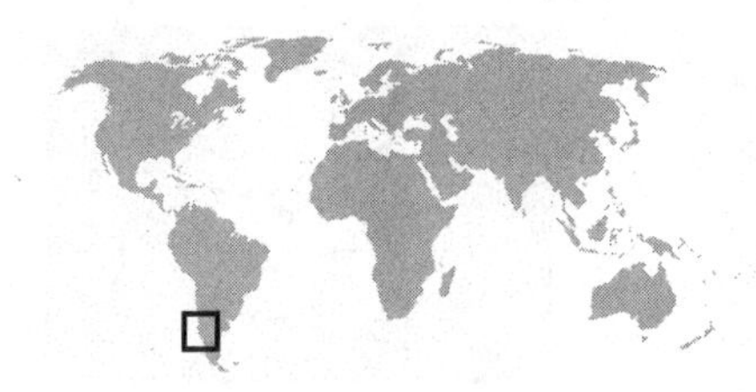

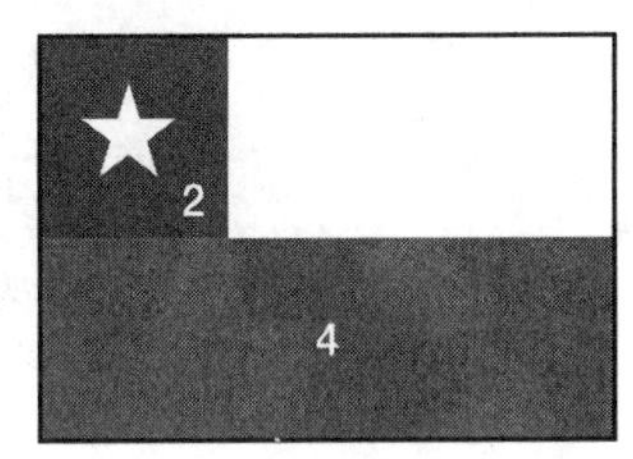

Name on driving licence: Republic of Chile
Capital: Santiago
Population: 16 million
Dosh: Chilean peso = 100 centavos
Size: 756,900 km^2 (36 Wales)

Complete history: Salvador Allende. Everything else is a footnote.[1]

Crowning achievement: Chile is a country of world records, so it's no surprise that in 1973 the nation's voters conspired to bring the planet's first democratically elected Marxist government to power. **[17]**

Top spot: Easter Island, a mere 3,800 km west of the Chilean mainland, is the world's only island with colossal stone heads on it. No coincidence, surely. **[18]**

Pub fact: The Chilean coastline is longer than that of the west coast of the US and Canada combined. Really. **[16]**

Where to avoid: The Atacama Desert, the driest place on Earth, some parts of which have not been rained upon for at least 400 years. **[12]**

The Pablo Neruda Prize for national anthems that rhyme in English but not in their original language: 'Chile, your sky is a pure blue/Pure breezes blow across you too.' ('*Puro, Chile, es tu cielo azulado/Puras brisas te cruzan también*.') **[15]**

[1]Some Amerindians ambled over here about 8,000 years ago. However, few of the pioneers were still around in 1520 when Portuguese explorer Ferdinand Magellan became the first European to catch sight of the coast. Twenty years later, Chile became a Spanish colony, so someone slipped up there. In 1818, independence blew in with a jaunty swagger. One war with Peru and Bolivia later and Chile, the world's spindliest country, became a bit fatter at the top, to everyone's general relief.

SCORE: 78 **WORLD RANKING:** 5

China ASIA

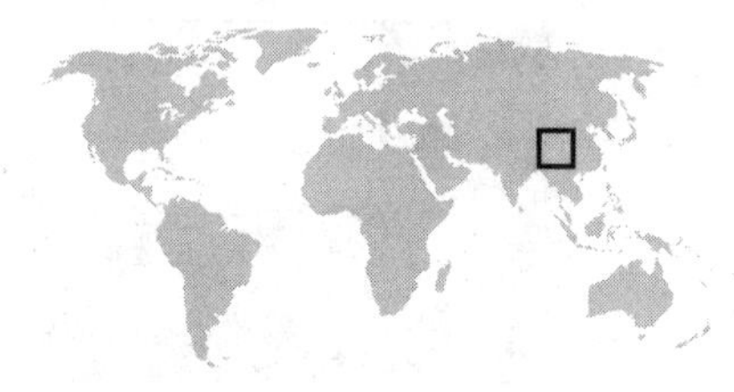

Name on driving licence: The People's Republic of China
Capital: Beijing
Population: 1,314 million
Dosh: Renminbi yuan = 10 jiao = 100 fen
Size: 9,597,000 km² (463 Wales)

Complete history: The Chinese, being the planet's oldest continuous civilisation, have history to burn. Worse still, the Chinese love words, so books like Jung Chang's *Wild Swans* that deal with but a few generations come out the size of haystacks. Thankfully, if you apply Western poetry's mantra – 'Condense, condense, condense' – to the 4,000 years of Chinese history, you can still get it into one sentence (where * denotes a dynasty): Shang*, Zhou*, Qin*, Han* (consecutive); Wei, Shu, Wu kingdoms (concurrent); Tang*; Genghis Khan; Yuan* (Kublai Khan), Ming* (vases), Manchu Qing* (Fu Manchu); The Opium War, The Boxer Rebellion, The Last Emperor, The Kuomintang (Chiang Kai-shek), The Long March, Manchukuo, World War II, Communism, Tibet, The Little Red Book, The Great Leap Forward, The Cultural Revolution, The Gang of Four, Deng Xiaoping, Tiananmen Square, Chris Patten*.

Crowning achievement: The Great Wall of China, famous for being the only man-made object visible from space (erroneously, as it turns out) but still best viewed from Earth. **[16]**

Pub fact: Despite being the most populous country ever, your chances of being born Chinese are still a measly 1 in 5, and even less if you are already alive. **[5]**

Made in China: China. **[16]**

Customs to treasure: Pretending that New Year's Day lands on the day of the second new moon after the winter solstice. Bless. **[13]**

National anthem proposing construction methods that would probably not pass muster with the building inspectors: 'Arise, ye who refuse to be slaves!/With our flesh and blood, let us build our new Great Wall!' **[15]**

SCORE: 65 **WORLD RANKING: 50**

colour key: **1** = turquoise **2** = blue **3** = green **4** = red **5** = yellow **6** = orange **7** = pink **8** = purple **9** = brown

Colombia

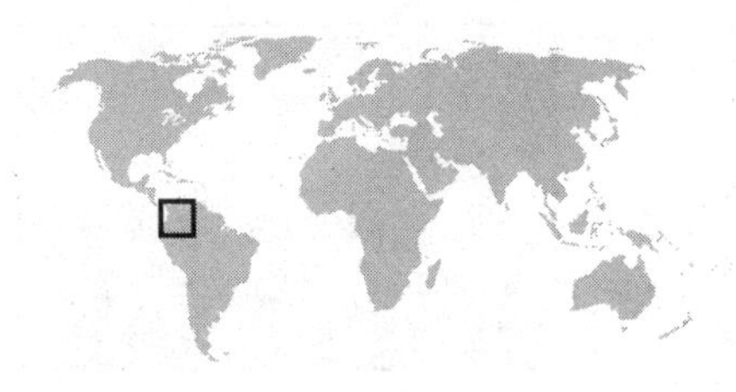

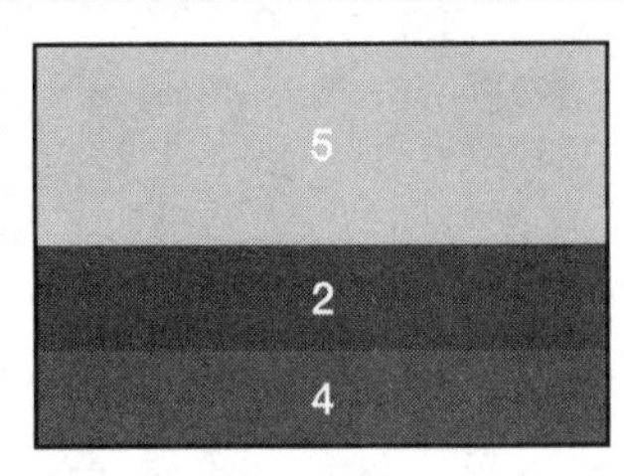

Name on driving licence: Republic of Colombia
Capital: Bogotá
Population: 42 million
Dosh: Colombian peso = 100 centavos
Size: 1,138,900 km² (55 Wales)

Complete history: Yet another sorry tale of folk (the Chibcha in this case) minding their own business for thousands of years before having their sylvan idyll destroyed by men with weedy beards and ridiculous tricorn hats. It fell to an ancestor of the invaders, one Simón Bolívar, to defeat the Spaniards and become president. There's been trouble more or less ever since.

Pub fact: The only South American country to enjoy both an Atlantic and a Pacific coast. Dispiritingly, this is only because Panama sticks out of it like some sort of grotesquely deformed trunk. (Despite Colombians' well-known fondness for elephants, it's better to keep this observation to yourself while travelling in the country.) **[8]**

Crowning achievement: The wettest place on Earth, enjoying over 13 metres of rain per year, is a place called Lloro (lit. 'I cry'). **[17]**

Sad fact: Colombia produces over half the world's cocaine, while being home to the longest-running civil war in the Americas. Not that the two are connected in any way. **[2]**

Gift to the World: Shakira. (Although, of course, if challenged with regard to the merits of the smouldering songstress, you should admit only to an enjoyment of her earlier Spanish-language work which was all about teenage suicide and the awfulness of home life in Bogotá.) **[17]**

Chorus of national anthem: 'Oh unfading glory!/Oh immortal joy!/In furrows of pain/Good is already germinating.' These somewhat cautiously optimistic words are interspersed with no fewer than eleven eight-line stanzas, each one eulogising the struggle for independence, and each one more turgid than the one before. **[3]**

SCORE: 47 WORLD RANKING: 150

Comoros

AFRICA

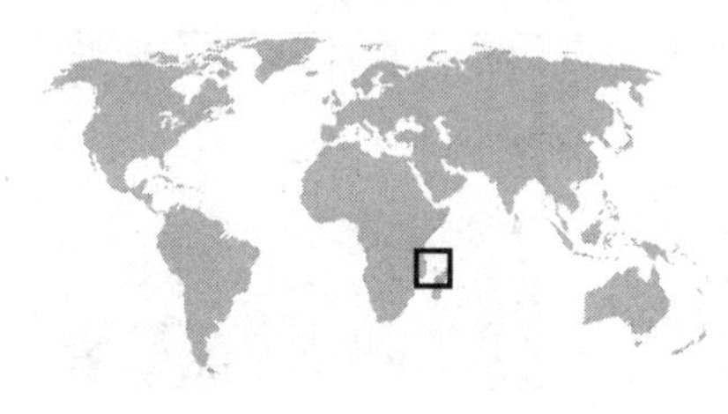

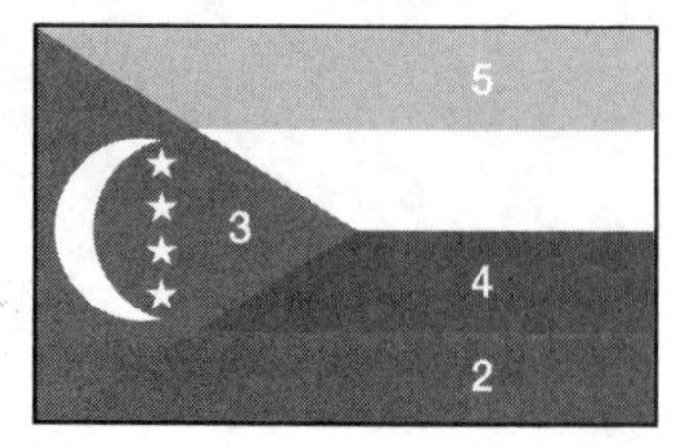

colour key: 1 = turquoise 2 = blue 3 = green 4 = red 5 = yellow 6 = orange 7 = pink 8 = purple 9 = brown

Name on driving licence: Union of the Comoros
Capital: Moroni
Population: 750,000
Dosh: Comoran franc = 100 centimes
Size: 2,150 km^2 (0.1 Wales)

Complete history: Little did the string of volcanoes that erupted eons ago between Mozambique and Madagascar know that some day it would become an entire country with its own seat at the United Nations in between the Colombians and Congolese. The French like a good volcano, of course, and have their own word for it and everything, so the only surprise is that it took them until the 19th century to snaffle these ones. One of the islands, Mayotte, so resembles a baguette that it opted to remain French when the rest of its friends became independent in 1974.

Crowning achievement: Comorans are very insistent that their Mount Karthala is the largest active volcano in the world. Although good for the tourist industry, one can't help feeling that the day might arrive when this is discovered to be something of a two-edged sword. **[13]**

Top spot: The capital, Moroni, aside from being a brave if ultimately mistaken stab at a plural of moron, is rather a peaceful spot in which the minute-by-minute fear of instant molten death is wont to give an added frisson to one's day. **[13]**

Gift to the world: Ylang-ylang. Comoros is the planet's foremost producer of the repetitious little perfume. **[12]**

Pub fact: The Mayottans, not to be outdone by the bragging of the Comorans, claim that they possess the largest lagoon in the world. **[14]**

First three lines of national anthem apparently written to allow for the correct use of a single semi-colon: 'The flag is flying/Announcing complete independence;/The nation rises up.' **[14]**

SCORE: 66 **WORLD RANKING:** 44

Congo

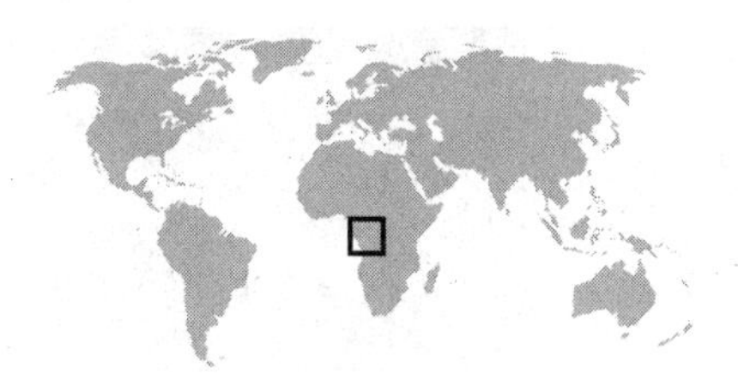

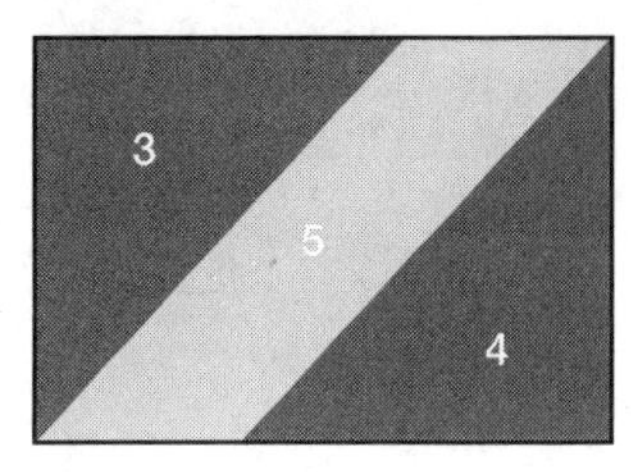

Name on driving licence: Republic of the Congo
Capital: Brazzaville
Population: 3 million
Dosh: CFA franc = 100 centimes
Size: 342,000 km^2 (17 Wales)

Complete history: One of only two African countries to rhyme with 'bongo', Congo is often mistaken for its bigger brother, the clumsily named Democratic Republic of Congo. On the grounds that the inclusion of the word 'democratic' in a nation's name is a sure-fire indication that they've never had an election that hasn't been rigged, the people of the Republic of the Congo (aka Congo-Brazzaville) should be thankful for the relative brevity of their country's moniker. Suffice to say, of course, that they've seen very little democracy either, before, during or after its submersion into French Equatorial Africa.

Customs to treasure: In a successful bid to rile the Americans, Congo became the first communist state in Africa. **[11]**

Top spot: The house built for Charles de Gaulle (the well-known airport who went on to become president of France). It was hastily constructed during World War II, when Brazzaville suddenly found itself the capital of Free France. As a way of thanking the Congolese after the war was over, de Gaulle promised never to live there. **[13]**

Popular misconception: 'The outline of the nation as seen on a map resembles a road-flattened iguana.' A schoolboy error. It resembles a road-flattened iguanodon. **[20]**

Flag fact: It used to have a mattock on it. Good times. **[15]**

National anthem apparently inspired by Bohemian Rhapsody: 'And if we have to die/Does it really matter?' **[12]**

SCORE: 71 **WORLD RANKING: 16**

Congo, Democratic Republic of

colour key: 1 = turquoise 2 = blue 3 = green 4 = red 5 = yellow 6 = orange 7 = pink 8 = purple 9 = brown

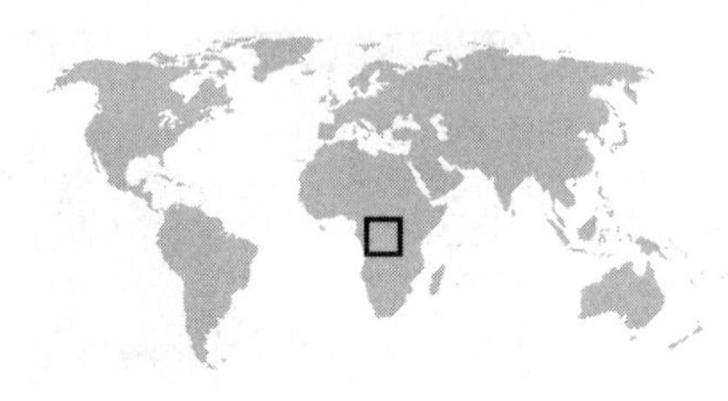

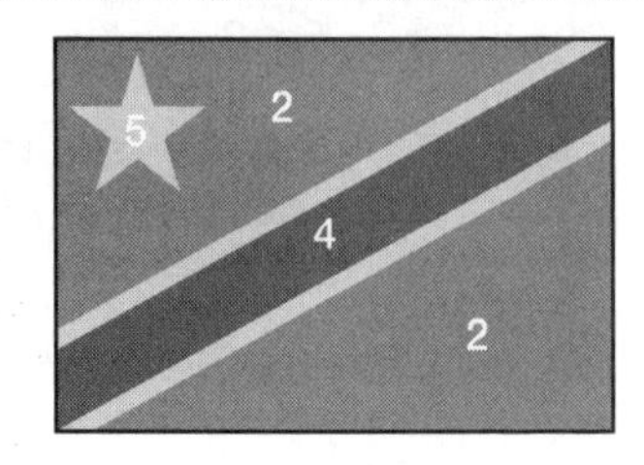

Name on driving licence: Democratic Republic of the Congo
Capital: Kinshasa
Population: 58 million
Dosh: Congolese franc = 100 centimes
Size: 2,345,400 km^2 (113 Wales)

Complete history: The Bantu got here first, the Portuguese came in second, the Belgians took bronze (or would have done, had they found any). The Congolese threw off the Belgian yoke in 1960 principally so that they could get on with the business of fighting each other.

Top spot: The Mighty River Congo, still the only proper way to travel through central Africa and anyone who tells you otherwise is a bounder. **[17]**

Historical low: The Unwritten Law (1996–2003): *Any African country may send its troops into the DRC to 'save it from itself'*. As we all know, this is best achieved by indiscriminate killing and the wholesale looting of natural resources ('Diamonds as a preference, but send back what you can and we'll see if we can find a use for it', seems to have been the word from most presidents-cum-saviours). **[0]**

Pub fact: Half of the woodland in Africa can be found in the DRC, so if you've lost any recently there's a good chance it's here. **[16]**

Made in the DRC: The okapi, a cross between a gazelle and a zebra. Imagine a giraffe without the neck (but don't just leave the head floating six feet above the body – have some sense). **[13]**

National anthem in which soloists alone get to sing the words in brackets, an honour only slightly diminished by the fact that most of them are the word 'Congo': 'Blessed gift (Congo) of our forefathers (Congo)/Oh (Congo) beloved country (Congo)', etc., etc.' **[12]**

SCORE: 48 **WORLD RANKING: 146**

Costa Rica

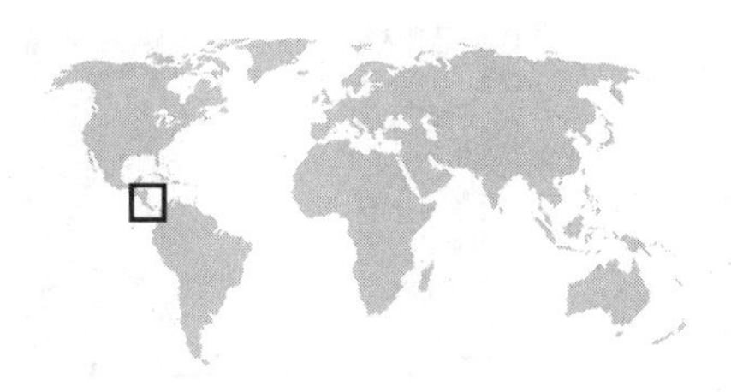

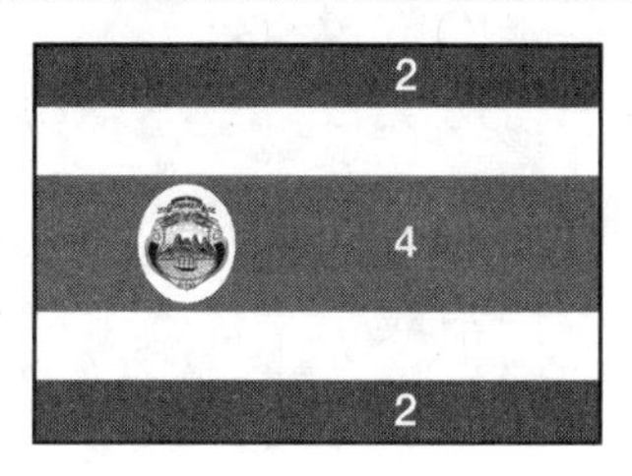

Name on driving licence: Republic of Costa Rica
Capital: San José
Population: 4 million
Dosh: Costa Rican colón = 100 céntimos
Size: 51,100 km^2 (2.5 Wales)

Complete history: There's Columbus, of course, annoying the living spleen out of everyone by turning up like a bad peseta and pretending to be nice. The Spaniards spent the next few hundred years lithping like anything and generally being thcurulouth. Independenthe (all right, stop it now) in 1838.

Where to avoid 500 years ago: The entire country. The name Costa Rica (lit. 'rich coast') was given spuriously by Columbus to lure his countrymen there in search of non-existent treasure while he bombed off to where the real action was. **[4]**

Gifts to the world: Franklin Chang-Díaz, astronaut; Clodomiro Picado Twight, toxicologist. **[13]**

Customs to treasure: Pacificity. The Costa Rican constitution is unique in banning the existence of a national army, meaning they don't have to waste untold millions paying a lot of suspect hoodlums who would much sooner stage a coup than fight off invaders anyway. Thus, while the rest of Central America scratches around trying to make ends meet or stares glumly into its canal, Costa Ricans enjoy an average standard of living comparable with that of Oman and (very nearly) Croatia. **[19]**

Made in Costa Rica: Butterflies. The locals can't move for them. However, since tourism is the mainstay of the economy, they just have to smile and pretend that nothing's amiss whenever some members of one of the friskier species evict them from their homes and empty their drinks cabinets. **[15]**

National anthem that highlights the peculiarly flappy nature of Costa Rican life: 'Noble homeland/Your beautiful flag gives us an expression of your life.' **[12]**

SCORE: 63 **WORLD RANKING:** 61

Cote D'Ivoire AFRICA

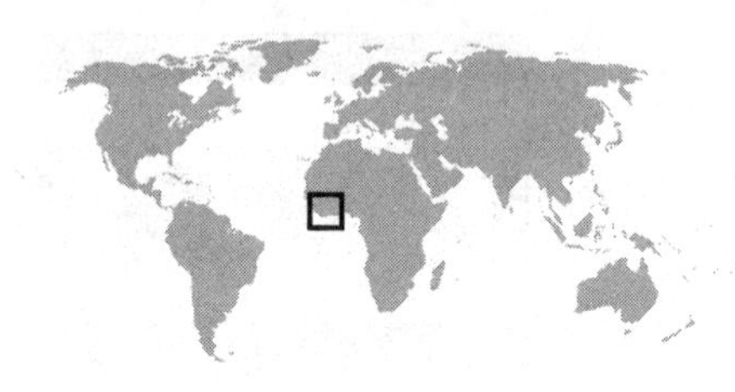

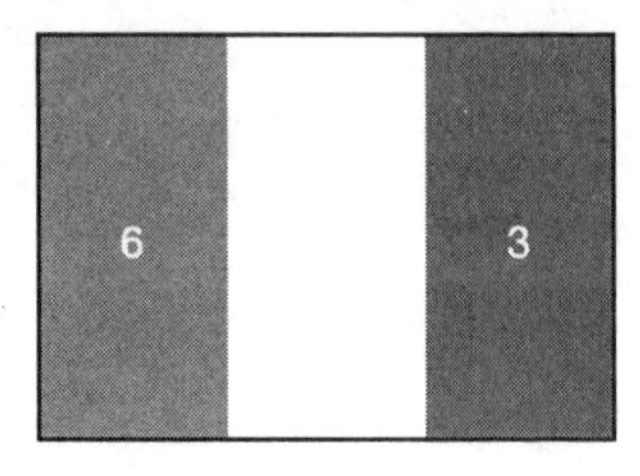

Name on driving licence: Republic of Côte d'Ivoire
Capital: Yamoussoukro
Population: 17 million
Dosh: CFA franc = 100 centimes
Size: 124,500 km² (16 Wales)

Complete history: A miserable slide from freedom into slavery into enforced Frenchness. Now independent, Côte d'Ivoire produces more cocoa than anyone else in the world. But is it really happy?

Top spot: The Basilique de Notre Dame de la Paix (The Basilica of Our Lady of the Peas) in Yamoussoukrou (built 1985–9) is purported to be the world's tallest place of Christian worship and to contain more stained glass than exists in all of France. Pope John Paul II was so aghast at the US$300 million it cost that he only agreed to consecrate it if a hospital was constructed nearby. The cornerstone he laid is still the only bit of it built. **[3]**

Sad fact: Deforestation – Côte d'Ivoire holds world records, you know. From 1977 to 1987, the country lost a jaw-dropping 42 per cent of all its wooded areas. **[0]**

Crowning achievement: Somehow getting people to believe that Abidjan is the 'Paris of Africa' on the mere basis that some French people once built some houses there. **[16]**

Popular misconception: '"Côte d'Ivoire" is merely the French rendering of Ivory Coast.' Well, it is, yes, obviously, but the country did officially change from Ivory Coast to Côte d'Ivoire two decades ago, so by rights we should have stopped calling it Ivory Coast by now in the same way we no longer refer to Bangladesh as East Pakistan, Holland as the Batavian Republic or the USA as 'that bunch of revolting troublesome land-grabbers'. Ah, I see, you've got a point. **[17]**

Opening lines of national anthem: 'We salute you, O land of hope/Country of hospitality.' **[12]**

SCORE: 38 **WORLD RANKING: 179**

EUROPE

Croatia

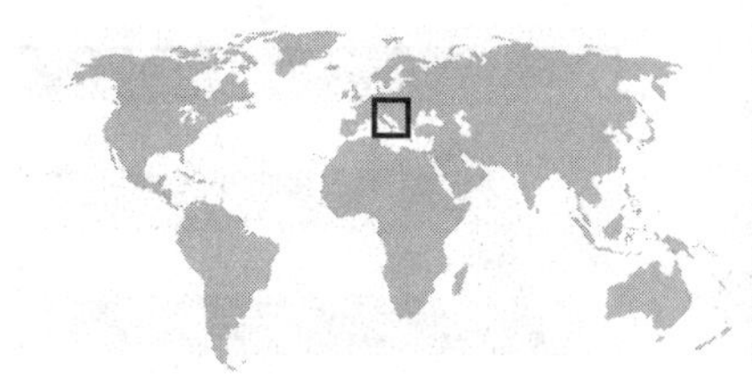

Name on driving licence: Republic of Croatia
Capital: Zagreb
Population: 4.5 million
Dosh: Kuna = 100 lipas
Size: 56,500 km² (2.7 Wales)

Complete history: You probably couldn't get away with it nowadays but in 1102 King Coloman of Hungary also became king of Croatia. It took 800 years and a world war to dissolve the union and for Croatia to be tempted into joining the joyously named Kingdom of the Serbs, Croats and Slovenes. Once they were in, however, the dastardly Serbs changed the name to drabsville old Yugoslavia. The Croats got their own back in 1934 by assassinating the Yugoslav king (a Serb), and the two peoples haven't been able to play a nice friendly game of Monopoly together since. Croats declared that they had stopped being Yugoslav in 1991. The Serbs grudgingly agreed a few years later.

Crowning achievement: Who can deny the glory that is the red-and-white chessboard worn by Croatia's national sides? (May not include chess team.) **[18]**

Top spot: The Dalmatian Coast. Recurrent rumours that there are in fact 101 Dalmatian Coasts are probably the result of a misunderstanding rather than due to any malicious intent. **[10]**

Customs to treasure: On St John's Day (24 June) it is seen as poor form if one does not jump over a bonfire. **[15]**

Gift to the world: The city of Dubrovnik is a World Heritage Site. This means that whenever Earth is visited by aliens, Dubrovnik is one of the places we take them to in order to give our inter-planetary tourism industry a much-needed boost. **[17]**

Opening lines of national anthem: 'Our beautiful homeland/Our dear, heroic land.' Difficult to decide whether this is more ho or hum. **[4]**

SCORE: 64 WORLD RANKING: 54

Cuba

AMERICAS

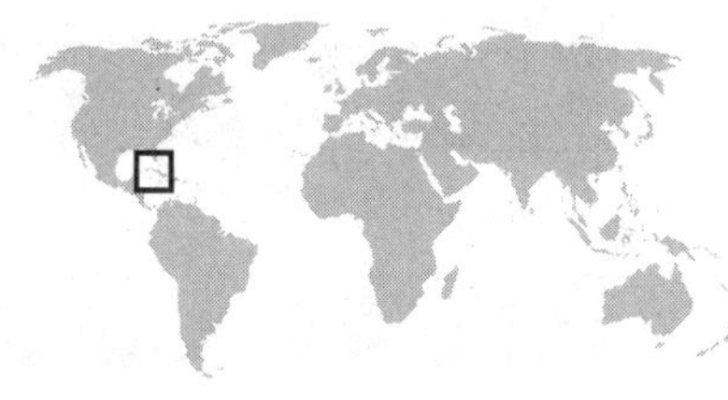

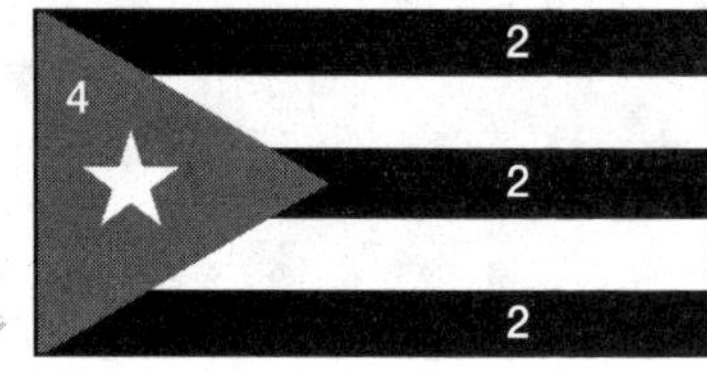

colour key: 1 = turquoise 2 = blue 3 = green 4 = red 5 = yellow 6 = orange 7 = pink 8 = purple 9 = brown

Name on driving licence: Republic of Cuba
Capital: La Habana
Population: 11 million
Dosh: Cuban peso = 100 centavos
Size: 110,900 km^2 (5.4 Wales)

Complete history: From 1898 to 1902 the USA ran Cuba, having (ahem) *bested* the Spaniards in the Spanish-American War. They then magnanimously handed the nation its independence, a decision they have lived to regret and presumably the reason why they see no inconsistencies between imposing a blockade on Communist Cuba while awarding Communist China 'Most Favoured Nation' status. Cuba is now famous for its cigars, its missile crises and its bays, which are full of pigs.

Crowning achievement: Cuba boasts the world's longest-serving leader (Fidel Castro), and the longest-serving defence minister (his brother Raúl), though don't mention this to anyone in Miami or you'll never get away. **[8]**

Popular misconception: 'Cuba is the home of Cubism.' *¡No, señor!* Cubism is so called because it was invented by early members of the Cub Scout movement. **[13]**

Pub fact: Cuba's Isla de la Juventud – once favoured by pirates as a place to hide behind before leaping out jovially to surprise ships and slaughter their crews – was the inspiration for Robert Louis Stevenson's *Treasure Island*. **[12]**

Gift to the world: *Son* (pronounced to rhyme with 'scone', assuming you're a person who pronounces 'scone' to rhyme with 'John' rather than 'Joan', which many modern at-ease-with-themselves English-speakers now prefer), one of the many rhythms to which only Latin Americans can move without leaving their dignity in tatters on the dance floor. **[14]**

National anthem pregnant with assumptions: 'Hasten to battle, men of Bayamo/ ... /You do not fear a glorious death since to die for the country is to live.' **[6]**

SCORE: 53 **WORLD RANKING: 119**

EUROPE Cyprus

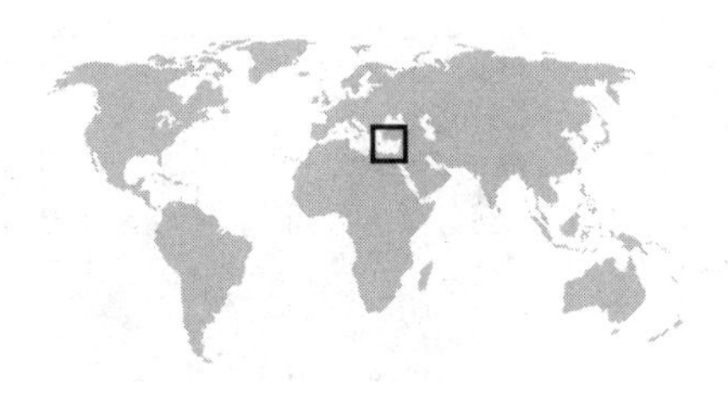

Name on driving licence: Republic of Cyprus
Capital: Nicosia
Population: 776,000
Dosh: Cypriot pound = 100 cents
Size: 9,300 km^2 (0.5 Wales)

Complete history: Cypriots' affiliation to other people's clubs, or 'empires' as they're sometimes known, has chopped and changed over the years as the distempered winds of fortune have blown about the Mediterranean. Thus it is that they have held membership cards at one time or another bearing the imprimatur of Greece, Egypt, Persia, Alexander the Great, the Ptolemies, Rome, Byzantium, the Arabs, Richard the Lionheart, the Knights Templar, Guy de Lusignan, Venice, Ottoman–Turkey and Great Britain. Independence was gained in 1960, but a 1974 coup staged by Greek-led Cypriots prompted a partial invasion by Turkey and the creation of the Turkish Republic of Northern Cyprus. However, the only country in the world to recognise this state is Turkey itself, so a bit of 'hearts and minds' work still required there.

Historical low: Difficult to say which is more embarrassing: being sold by Richard the Lionheart to the Knights Templar, or being leased to Britain by the Turks. **[4]**

Flag fact: Lest we forget, the official Cypriot flag – now used by no one but the editors of 'flags of the world' books – is emblazoned with two of the most underused olive branches in history. **[2]**

Pub fact: The Greeks named the country 'Kypros' meaning 'copper', a reference to the island's then excellent police force. **[15]**

Made in Cyprus: Independence leader Archbishop Makarios, a man not unfamiliar with the skills required to hide an illicit Kalashnikov in a beard. **[12]**

National anthem encouraging the disparagement of another's sword: 'I shall always recognise you/By the dreadful sword you hold.' **[18]**

SCORE: 51 **WORLD RANKING: 130**

Czech Republic EUROPE

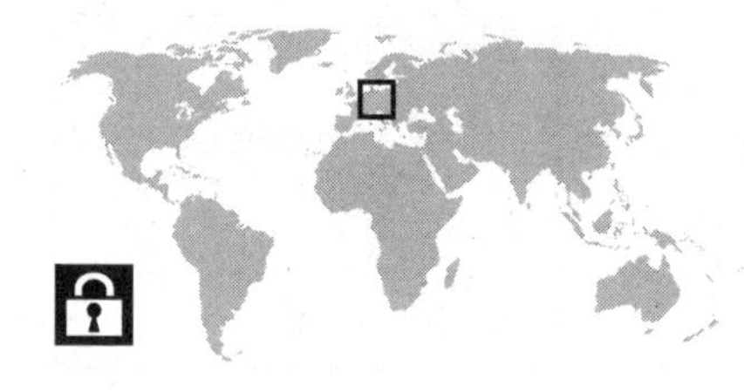

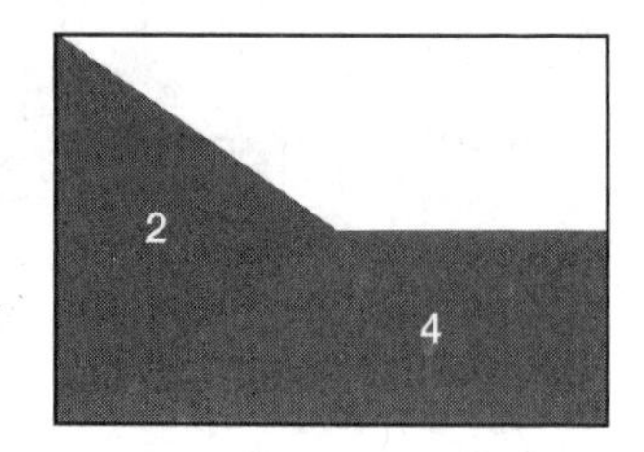

Name on driving licence: Czech Republic
Capital: Prague
Population: 10 million
Dosh: Czech koruna = 100 haleru
Size: 78,900 km^2 (3.8 Wales)

Complete history: It was a rebellion by Protestant Czechs in 1618 that sparked the Thirty Years' War. This was in reality an intermittent affair and thus reinvigorated the trend for conflicts that didn't quite do what they said on the tin (see also the 116-year Hundred Years' War). The split with Slovakia in 1993 reaped immediate dividends, with the Czechs managing to stay out of the Eurovision Song Contest while their Slovak cousins seriously embarrassed themselves.

Top spot: Prague is the definitive holder of the title the 'Paris of the East', despite whatever counterclaims you may hear from Bratislavans, Minskovites, and those inhabitants of eastern Paris who didn't quite understand the question. **[12]**

Customs to treasure: The annual commemoration of the death of Jan Palach (19 January), a 20-year-old student who self-immolated in 1969 in protest at the Soviet invasion. **[18]**

Made in the Czech Republic: Painted scenes in Bučovice manor of the 'world upside down', in which hares wreak a terrible revenge on men and dogs. Huzzah. **[19]**

Popular misconception: 'The Czechs can't get enough of fun-loving English-speaking tourists entering their shops to enquire, between guffaws: "I haven't got any cash, will you accept a Czech?"' (Also note that they probably won't, although they might accept a traveller's cheque if you wiped that idiot grin off your face and got your passport out.) **[7]**

Proof positive from the opening lines of the national anthem that the Czechs are a forgetful people, or at the very least a careless one: 'Where is my home?/Where is my home?' **[19]**

SCORE: 75 **WORLD RANKING:** 8

EUROPE

Denmark

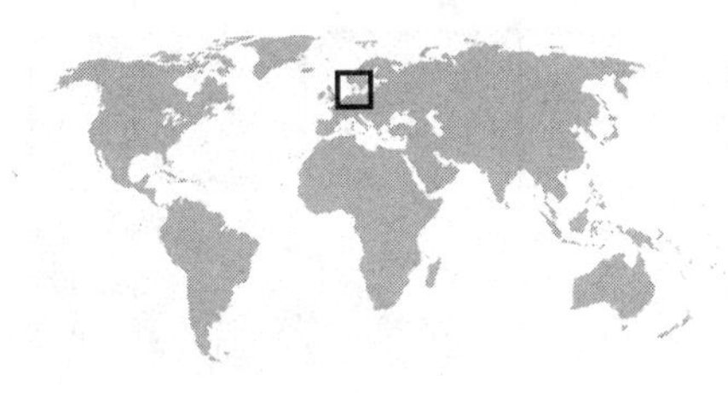

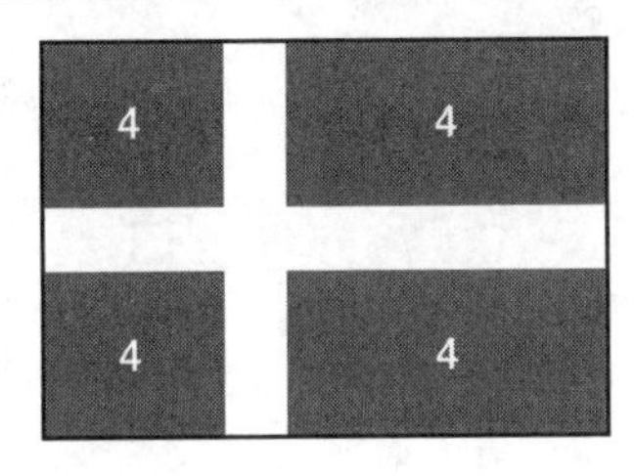

Name on driving licence: Kingdom of Denmark
Capital: København
Population: 5.4 million
Dosh: Danish krone = 100 øre
Size: 43,100 km^2 (2.1 Wales)

Complete history: Søren Kierkegaard, everyone's favourite loopy romantic and the father of existentialism, lived from 1813 to 1855. Nothing else has happened in Denmark, before or since. This tragic dearth of history is lamented each year on National Disappointment Day, during which Danes look at each other blankly for a split second before carrying on as if nothing had happened.

Historical high: The lack of notable events on the Jutland peninsula had become so upsetting by the time of Queen Elizabeth (no one is sure who, if anyone, was on the Danish throne) that William Shakespeare gallantly invented some history for the benighted nation in the form of *Hamlet*, the title arising from the Bard's misapprehension that all Danes lived in tiny villages. **[14]**

Gift to the world: Danish pastries. **[12]**

Made in Denmark: The Great Dane. Just a dog, so not really all that great, as great things go. **[5]**

Flag fact: The *Dannebrog* is possibly the oldest national flag in continuous use. It apparently fell from the heavens into the hand of King Waldemar II, an event that – had it happened in reality – would have made 1219 a red-letter year in Danish history. **[7]**

Opening lines of national anthem: The Danes boast two – a royal one and one for hoi polloi. The proletarian anthem is full of whimsy and expansive adumbral deciduana: 'I know a lovely land/With spreading, shady beeches'; whereas the royal version possesses indisputably the most stirring opening couplet of any national anthem anywhere: 'King Christian stood by the lofty mast/In mist and smoke.' **[20]**

SCORE: 58 **WORLD RANKING:** 91

Djibouti

AFRICA

colour key: 1 = turquoise 2 = blue 3 = green 4 = red 5 = yellow 6 = orange 7 = pink 8 = purple 9 = brown

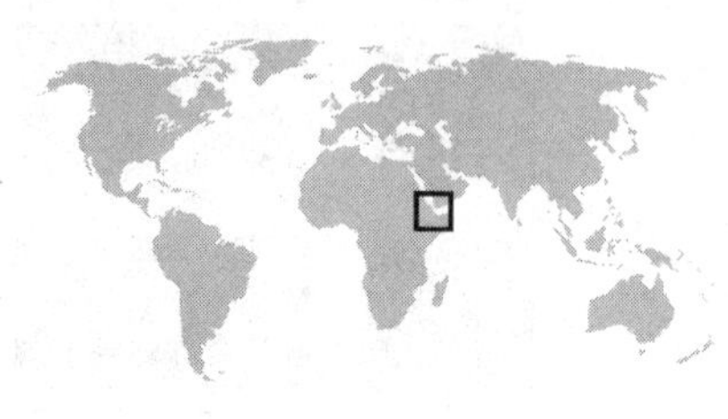

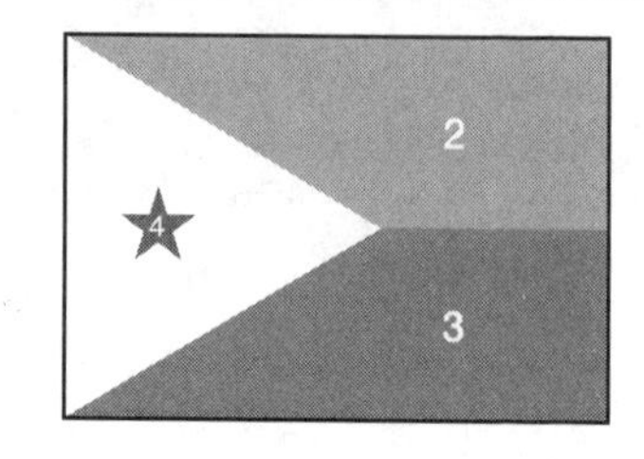

Name on driving licence: Republic of Djibouti
Capital: Djibouti
Population: 652,000
Dosh: Djiboutian franc = 100 centimes
Size: 22,000 km^2 (1.1 Wales)

Complete history: The Issas live in the south, around the capital, the nomadic Afars live, it must be said, further afar. Aside from the Afars' conversion to Islam in the 9th century, not a great deal can be said to have happened in Djibouti, which probably accounts for the country's friendly rivalry with Denmark.

Crowning achievement: Recovering from being landed with the technically accurate but fundamentally useless name The French Territory of the Afars and Issas (1967–77), one of the least successful attempts at naming a country since Columbus reckoned the Caribbean islands were off the coast of Asia and called them the West Indies. **[16]**

Low point: Djibouti possesses the lowest point on the African continent. The surface of Lake Assal is an ear-popping 155 metres below sea level. Even more impressively, it's not all that far from the Mabla Mountains whose peaks rise well above sea level, even when the tide's in. **[16]**

Customs to treasure: The chewing of qat, a mildly narcotic shrub. Apparently, qat chewing can provoke gum disease, heart attacks, oesophageal cancer, impotency and psychological disorders. In short, it would be safer to chew cats but for their endearing habit of making pre-emptive attacks on human facial tissue. **[7]**

Pub fact: Djibouti is the only country in sub-Saharan Africa to host a US military base, something which might have seemed an asset once upon a time, but now we're not quite so sure. **[6]**

National anthem bemoaning the price of the Djiboutian flag: 'We have raised our flag/The flag which has cost us dear.' **[14]**

SCORE: 59 WORLD RANKING: 85

Dominica

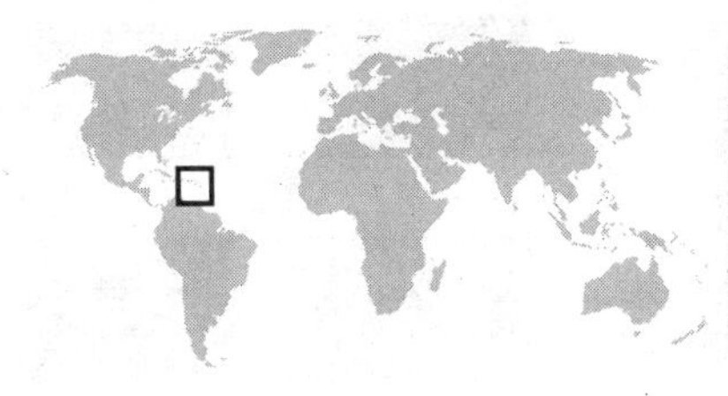

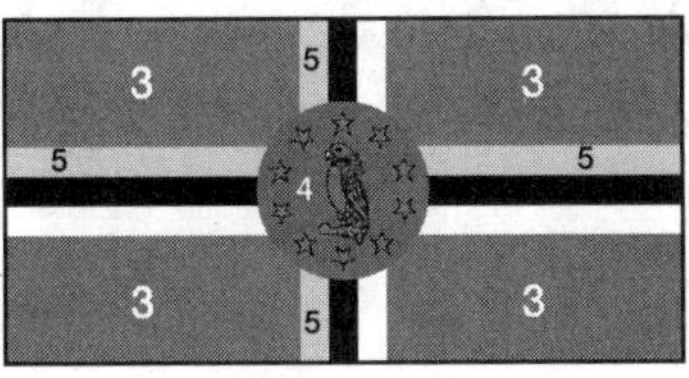

Name on driving licence: Commonwealth of Dominica
Capital: Roseau
Population: 70,000
Dosh: East Caribbean dollar = 100 cents
Size: 750 km² (0.09 Wales)

Complete history: Although it was Christopher (never just 'Chris' apparently) Columbus who inevitably biffed into Dominica one Sunday in 1493, it wasn't until the 17th century that European diseases got on with the hard task of decimation, which the local population fatally mistook for the mere conversion of weights and measures from imperial terms. By the time they realised their error, most of them were already dead. The French and English, both more or less inoculated against smallpox and the like by having survived several years in school classrooms, mooched around for a while, the latter even building something they called Fort Shirley, for reasons that were probably clear at the time.

Crowning achievement: Astonishingly, unlike the vast majority of countries in the Americas who went down like nine pins, the indigenous Carib population resisted colonisation by the Spanish, French and British until 1805. **[18]**

Top spot: The Boiling Lake in the Morne Trois Pitons National Park is the second largest boiling lake in the world, behind the one at Rotorua, New Zealand. **[10]**

Made in Dominica: The Sisserou Parrot, Dominica's national bird, is the largest of all the Amazon parrots. This somewhat makes up for the disappointment over the boiling lake, without entirely assuaging it. **[11]**

Pub fact: The Carib Territory (pop. 3,000) on the north-east of the island is the only indigenous community left in the Caribbean. **[17]**

National anthem cleverly sidestepping intellectual copyright laws: 'We must prosper! Sound the call, in which everyone rejoices, "All for Each and Each for All."' **[14]**

SCORE: 70 **WORLD RANKING:** 22

Dominican Republic AMERICAS

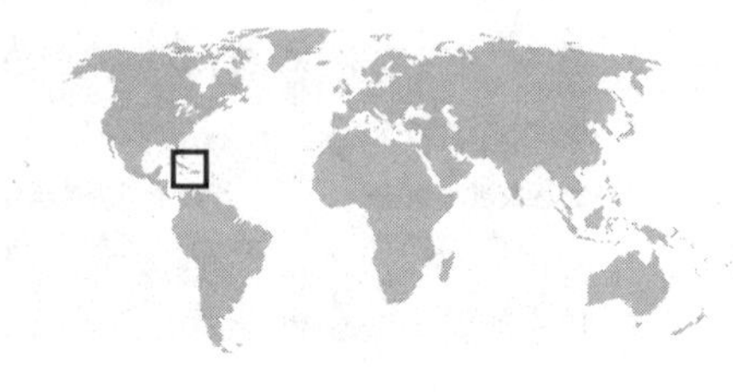

Name on driving licence: Dominican Republic
Capital: Santo Domingo
Population: 8.8 million
Dosh: Dominican peso = 100 centavos
Size: 48,400 km^2 (2.3 Wales)

Complete history: Although it was Christopher Columbus (never just 'Colomby' apparently) who inevitably biffed into Hispaniola (lit. '[this belongs to the] Spanish. *¡Hola!*'), it was his brother Bartholomew who was put in charge of killing the locals and establishing Santo Domingo (named after St Domingo, the patron saint of both dominoes and flamingos). The island was the darling of the Spanish empire until it came to their Imperial Majesties' attention that there wasn't much there. No one told the French, who garnered it in 1795 but handed it back as soon as the penny dropped. Next up were the Haitians who, bless them, took 22 years to cotton on. Astonishingly, no word of any of this got out to the Americans, who duly invaded in 1916 and again in 1965, just for luck.

Crowning achievement: Being wealthier than their island-mates, the Haitians, who are as poor as church mice. **[6]**

Top spot: Pico Duarte, at 3,098 metres, is the top spot in the whole of the Caribbean. **[16]**

Made in the Dominican Republic: Gold, though only bits and pieces that the conquistadores missed; and sugar, which was hidden from the Spaniards and exported secretly in the form of pink mice. **[14]**

Pub fact: Santo Domingo (founded 1496) is the Americas' oldest surviving European settlement. Even less happily, from 1936 to 1961 it was known as Ciudad Trujillo, after the country's self-effacing and not-at-all-psychotic dictator, Rafael 'Ciudad' Trujillo. **[3]**

National anthem that could lighten up a bit: 'No nation deserves freedom/If it is an indolent and servile slave.' **[3]**

SCORE: 42 **WORLD RANKING:** 171

East Timor

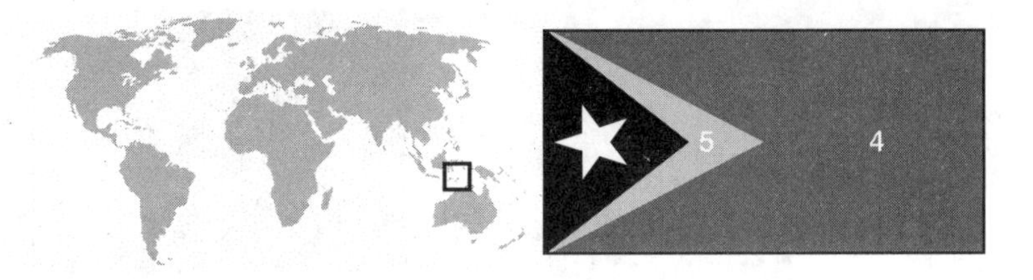

Name on driving licence: Democratic Republic of Timor Leste
Capital: Dili
Population: 1 million
Dosh: US dollar = 100 cents
Size: 14,900 km² (0.7 Wales)

Complete history: Timor was first squabbled over by the Dutch and Portuguese without reference to the Papuans who were living there at the time and who didn't care a fig for either. It wasn't until 1859 that the Portuguese prevailed, largely because the Dutch made themselves unpopular by wearing too much orange and clashing horribly with everything. The Japanese nosed in during World War II in their friendly devil-take-the-hindmost way. The Portuguese came back once they'd gone, but forgot they were European for a moment in 1974 and accidentally had a coup. East Timor declared independence, the signal for the Indonesians to march in and slaughter a third of the population.

Crowning achievement: Uniquely, not only is East Timor just part of an island (ditto Papua New Guinea, Dominican Republic and Haiti), it also has a stray bit of itself (Pante Makasar) marooned in another country (see Russia, Angola, Azerbaijan etc). Scorchio! **[12]**

Customs to treasure: Over-egging the pudding. Since '*timor*' just means 'east', East Timor means 'East East'. **[14]**

Pub fact: If it weren't for the Montenegro/Serbia divorce (2006), East Timor (2002) would be the world's newest country. **[17]**

Flag fact: The black triangle in the East Timorese flag represents the dark forces that the nation must still overcome. Visitors to the country should note that it's a criminal offence to come across a dark force and not report it to the relevant authorities. **[14]**

National anthem betraying a slightly over-optimistic estimate of the value of shouting: 'We are victorious over colonialism, shouting: Down with imperialism!' **[14]**

SCORE: 71 **WORLD RANKING:** 16

Eastern Gabon

AFRICA

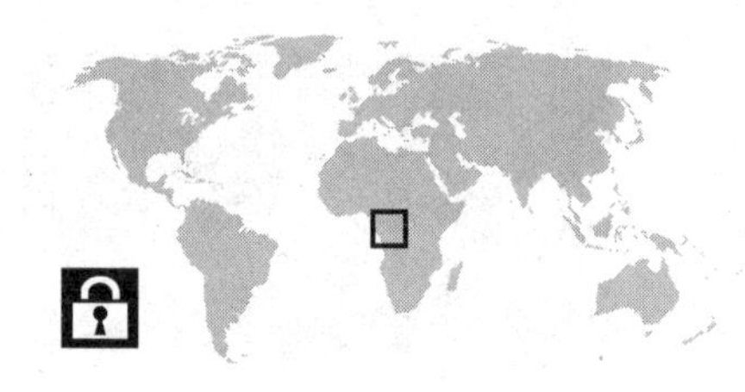

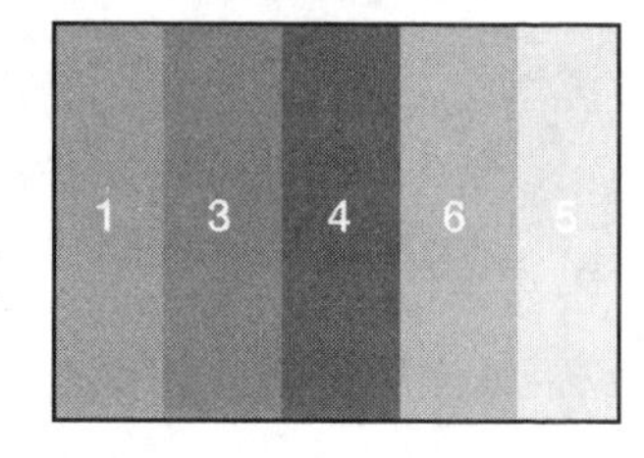

Name on driving licence: The Consecrated Republic of Eastern Gabon
Capital: Kumbayah
Population: 1.3 million
Dosh: Eastern Gabonese deco = 100 centaurs
Size: 35,300 km^2 (1.7 Wales)

Complete history: Lodged between Gabon and Congo, Eastern Gabon takes its name from the S'cat people, who have lived there for ages. The country was known as Portuguese Northern Angola and then Belgian Eastern Gabon before the infamous Kelvin Campaign in which a handful of British soldiers and 10,000 S'cat warriors under 'Bloody' Colonel Kelvin defeated well-trained but ultimately lacklustre Belgian forces. Rather than regain their homeland, however, the S'cat discovered that overnight their territory had become British West Congo. The country was only to gain independence in 1973, while the British were distracted by the wild sounds of David Bowie and Wizzard.

Crowning achievement: The Eastern Gabonese used to have a reputation for extreme querulousness, but they've more or less got it under control now. **[9]**

Top spot: The Lychee Lodge, the world's sole lychee-only restaurant. **[11]**

Customs to treasure: The S'cat people have a fear of white bread, which they call *lizi* (lit. 'too short to eat'). **[17]**

Made in Eastern Gabon: S'cat is the only known language to include a range of words which lie somewhere between adjectives and adverbs, and are intimated by the speaker's facial expression rather than his voice. **[12]**

National anthem: 'Eastern Gabonese, lovers of Peace/Come, Death, spill the blood/Of our eternal enemies.' Under the terms of the trilateral peace treaty signed in 1994, the second/third lines have been altered from the original: 'Come, Death, spill the blood/Of our eternal *Angolan and Congolese* enemies.' Since the melody has remained the same, the Eastern Gabonese now cheerfully hum the missing syllables. **[4]**

SCORE: 53 **WORLD RANKING:** 119

colour key: **1** = turquoise **2** = blue **3** = green **4** = red **5** = yellow **6** = orange **7** = pink **8** = purple **9** = brown

Ecuador

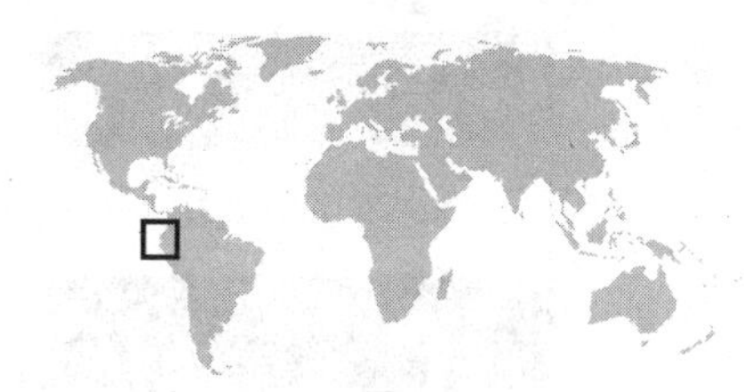

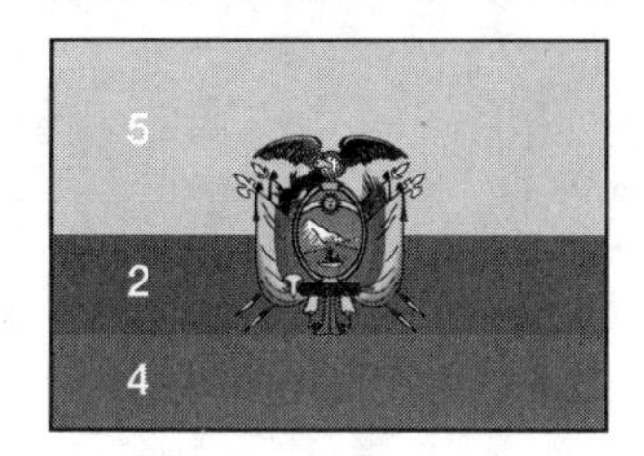

Name on driving licence: Republic of Ecuador
Capital: Quito
Population: 13 million
Dosh: US dollar = 100 cents
Size: 283,600 km^2 (13 Wales)

Complete history: Incas, Spaniards, Peruvians. A nightmare, really.

Crowning achievement: The Galapagos Islands. However, as anyone who has fetched up at an Ecuadorian port hoping to take a skiff over to them will tell you, it is a tad disappointing to learn that they are nearly 1,000 km away. Still, a couple of weeks' paddling should see you face to face with tortoises who actually met Darwin and will ask you for news of 'Mr Charlie'. **[17]**

Top spot: Cotopaxi (5,897 metres) is one of the world's highest active volcanoes. Refreshingly, few of the inhabitants of nearby Quito (the world's second-highest capital) seem all that bothered about it, claiming that the occasional dousing with lava 'flushes out the system a treat'. **[16]**

Pub fact: Ecuador is one of only two countries whose names suggest that they straddle the equator and, of those two, the only one that does. Strange times. **[10]**

Gifts to the world: Can be purchased at the Saturday market at Otavalo, which has been held since pre-Inca times when, of course, they didn't have Saturdays or indeed a seven-day week, and so had to guess when Saturdays might occur and hope for the best. **[14]**

Opening line of national anthem: 'Hello, oh homeland, a thousand times', followed by six interminable verses that cleverly give the effect of the singer actually having gone to the bother of greeting the homeland a thousand times. Instances of pubescent boys having begun the anthem as altos only to finish it as basses are frequent. **[2]**

SCORE: 59 **WORLD RANKING:** 85

Egypt

MIDDLE EAST

colour key: 1 = turquoise 2 = blue 3 = green 4 = red 5 = yellow 6 = orange 7 = pink 8 = purple 9 = brown

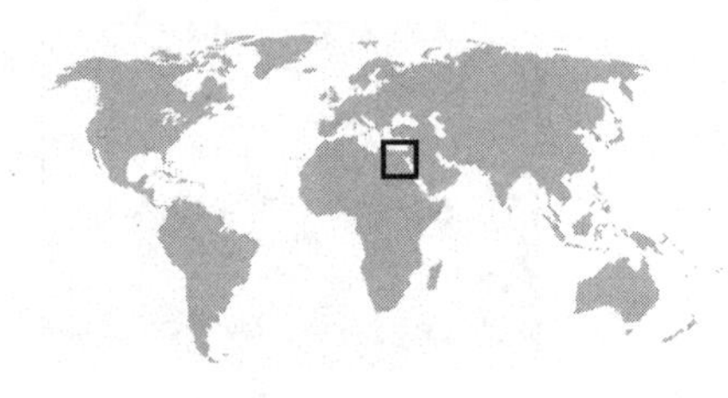

Name on driving licence: Arab Republic of Egypt
Capital: Cairo
Population: 76 million
Dosh: Egyptian pound = 100 piastres
Size: 1,001,500 km^2 (48 Wales)

Complete history: Egypt, or Ancient Egypt to give it its full name, is a country entirely comprising pyramids, sphinxes and mummified boy kings. This came about through 5,000 years of ancient civilisation, a period in which all ancient people were born male, built pyramids and sphinxes, succeeded to the throne, died young and were then mummified. After that came the Suez War, Nasser, the commendably brief Six-Day War, and the Aswan High Dam, which has brought irrigation to the desert, electricity to the masses and joy to generations of geography teachers.

Gift to the world: Egyptology. No other nation has a word in English relating specifically to the study of itself, an impressive achievement for a country with only two topographical features (the Nile and the Qattara Depression, since you ask). **[13]**

Made in Egypt: Omar Sharif. Actor, bridge player, polyglot and gambler. Sharif has also had time to earn a one-month suspended sentence for headbutting a policeman in a Parisian casino, and lose heaps of money trying to turn the world back on to the joys of papyrus. **[15]**

Historical high: The Pharos Lighthouse, one of the Seven Wonders of the World that no one can remember in an emergency (see also the Mausoleum of Halicarnassus). **[13]**

Opening lines of national anthem (since the 1979 peace accord with Israel): 'My homeland, my homeland, my homeland/My love and my heart are for thee.' Aah, sweet. **[20]**

Opening lines of national anthem (before the 1979 peace accord with Israel): 'O! My weapon!/How I long to clutch thee!' Erm, sorry? **[0]**

SCORE: 61 **WORLD RANKING: 70**

El Salvador

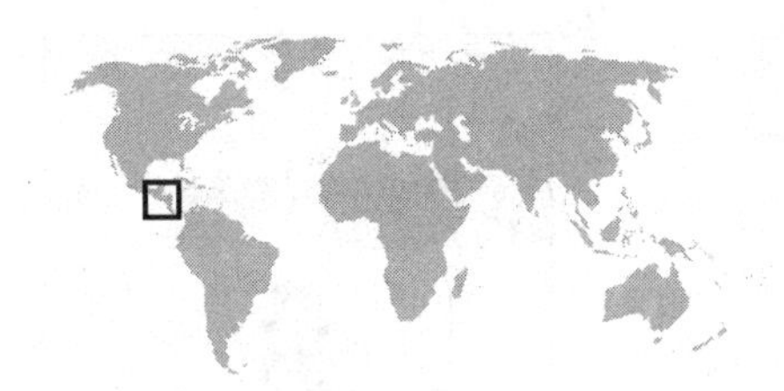

Name on driving licence: Republic of El Salvador
Capital: San Salvador
Population: 6.6 million
Dosh: US dollar = 100 cents
Size: 21,000 km² (1 Wale)

Complete history: Right up there with the 'War of Jenkins' Ear' (England vs. Spain), El Salvador's 'Soccer War' (1969) broke out after a World Cup qualifying match got a bit out of hand and ended with Salvadoran troops in possession of a good chunk of Honduras. This is unusual for a football match, even in Latin America, and the Salvadorans were prevailed upon to withdraw. El Salvador went on to qualify for the World Cup, but only thirteen years later when the country was at the height of another war, this time a civil one, if such a word can be used of a conflict in which the vast majority of the 75,000 casualties were unarmed non-combatants, many of them slaughtered by their own army.

Crowning achievement: El Salvador ('The Saviour') is the sole nation to be named after the founder of a major religion (reports that the people of Mauritania and Mauritius follow the teachings of a divine prophet known simply as Uncle Maurice are thought to be scurrilous). **[15]**

Top spot: The Izalco Volcano, aka the 'Lighthouse of the Pacific' because it used to erupt every five minutes, thus alerting passing ships to the presence of El Salvador every five minutes, like a lighthouse would. **[11]**

Made in El Salvador: Archbishop Oscar Romero. **[18]**

Gift to the World: The Montecristo Cloud Forest, the home of spiders, monkeys and spider monkeys. Is there something we're not being told? **[14]**

National anthem keen not to get everybody too down: 'Sad and bloody is her history/But sublime and brilliant as well.' **[13]**

SCORE: 71 **WORLD RANKING: 16**

England (*see also* United Kingdom) EUROPE

colour key: 1 = turquoise 2 = blue 3 = green 4 = red 5 = yellow 6 = orange 7 = pink 8 = purple 9 = brown

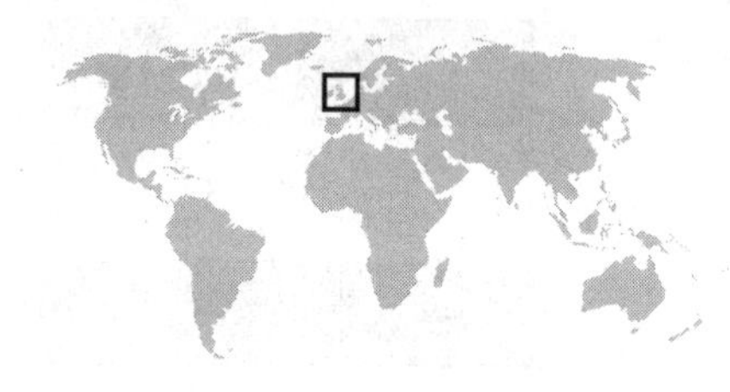

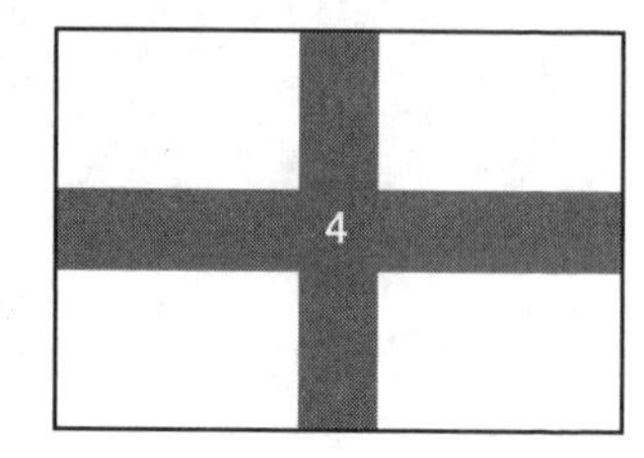

Name on driving licence: England
Capital: London
Population: 50 million
Dosh: Pound sterling = 100 pence
Size: 130,400 km^2 (6.3 Wales)

Complete history: Like most other places, England is a mishmash of immigrant peoples from the Angles onwards. The only two dates worth remembering, however, remain 1066 and 1966.

Made in England: Bit of a mixed bag: the wind-up radio (Trevor Baylis, 1991), the underground railway (1863), shrapnel (Henry Shrapnel, 1784). **[10]**

Gifts to the world: Football, rugby, cricket. The English display an enduring ability to feel aggrieved when losing at these sports to so-called lesser nations mixed with a resignation that they will never really be any good ever again combined with a madly optimistic hope that success is just around the corner topped off with the knowledge that any victory is sure to be followed by a good trouncing. **[14]**

Customs to treasure: Change ringing, the age-old practice of playing church bells according to mathematical formulae rather than any attempt at melody. Curiously, it seems to work. **[15]**

Popular misconception: 'The English are arrogant swine.' No they're not, they are genuinely superior. **[10]**

Anthem: What would England's anthem be if it were to have one of its own? In the running are 'Land of Hope and Glory', even though no one in England knows any of the words beyond the second line; 'Jerusalem', with its retelling of the ever-popular legend of a visit to Glastonbury by the teenage Jesus with his great uncle Joseph of Arimathea; and 'Three Lions', albeit that the 'thirty years of hurt' has to be updated every ten years. Or perhaps the English would prefer to stand and chant 'Ing-er-land' 40 times and sit down again. **[8]**

SCORE: 57 **WORLD RANKING:** 97

Equatorial Guinea

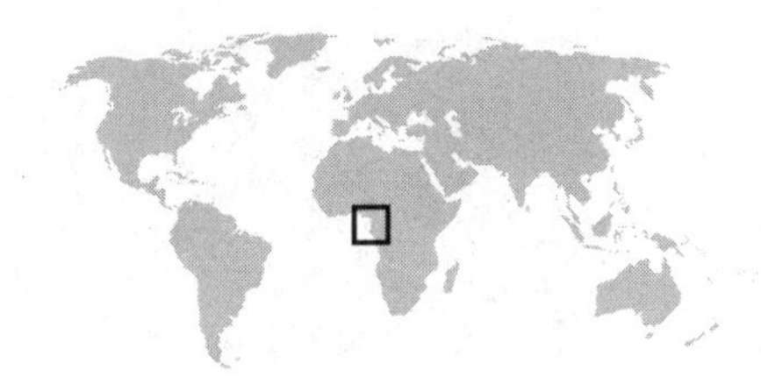

Name on driving licence: Republic of Equatorial Guinea
Capital: Malabo
Population: 523,000
Dosh: CFA franc = 100 centimes
Size: 28,050 km² (1.4 Wales)

Complete history: In the Southern Europe of 1472 there was a chronic shortage of proper names and people had to do the best they could with odd letters and bits of stray punctuation that served as hats and accents for vowels. So it was that the first European to sight Equatorial Guinea was Portuguese navigator Fernão do Pó. To be more precise, he clocked the island of Bioko, the largest of the five joined politically to mainland EG, as it is known to the cognoscenti (and thus often hilariously confused with thrusting West Sussex überville East Grinstead). Swapped by the Portuguese with the Spaniards for some of Brazil, then part-leased to the English, the five islands went on to become the Overseas Provinces in the Gulf of Guinea and finally, together with the mainland, the entirely independent EG. Then came Mark Thatcher.

Gift to the world: Eric 'The Eel' Moussambani who, had he been able to practise beforehand, might have shaved minutes off his 100-metre freestyle time at the Sydney Olympics. **[18]**

Made in Equatorial Guinea: Rent-a-loonyrightwinger Frederick Forsyth wrote *The Dogs of War* while in the capital, Malabo. **[3]**

Pub fact: Equatorial Guinea is the only Spanish-speaking nation in Africa (or 'Africa' to give it its Spanish name – see, it's not such a difficult language after all). **[11]**

Popular misconception: 'Equatorial Guinea is on the equator.' It isn't, it's around two degrees north. Perhaps people thought that Two Degrees North Guinea sounded ridiculous, and they may have a point. **[17]**

Opening lines of national anthem: 'Lets tread the paths/Of our immense happiness.' **[13]**

SCORE: 62 **WORLD RANKING:** 68

Eritrea

AFRICA

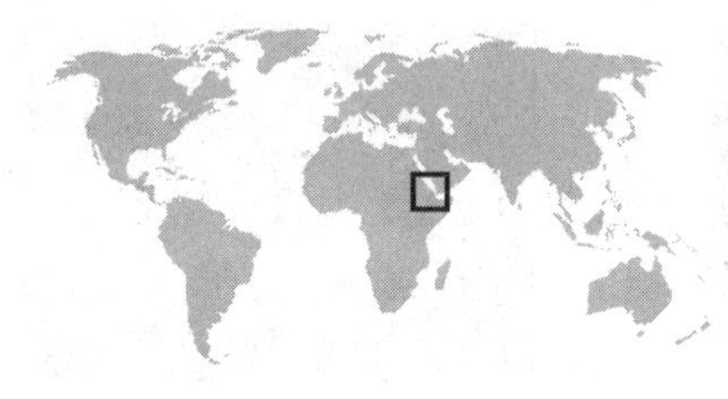

colour key: 1 = turquoise 2 = blue 3 = green 4 = red 5 = yellow 6 = orange 7 = pink 8 = purple 9 = brown

Name on driving licence: State of Eritrea
Capital: Asmara
Population: 4.4 million
Dosh: Nakfa = 100 cents
Size: 121,300 km^2 (5.7 Wales)

Complete history: Whisper it in Asmara, but Eritrea has been part of Ethiopia since antiquity was young. For over 500 years it was in Axum, one of the few Ethiopian kingdoms to double as a range of Avon toiletries for men. After the fall of the kingdom (known to the oppressed locals as the Axum of Evil), everyone just hung around, occasionally changing religion, until the Ottomans pitched up with their unlikely pluralisms and low upholstered stools. Then came the 1890s, a decade in which the Italians unusually roused themselves from their stupor to attack random bits of Africa. It was left to Britain to pretend to be the hero of the hour by driving them out during World War II. There followed some almighty shenanigans before Eritrea finally seceded from Ethiopia in 1993, though both would still rather see the other fall under a bus. The United Nations has wisely banned international sales of oversize buses to either nation.

Crowning achievement: The Eritrean railway system: narrow gauge, 1930s Italian-built carriages, steam engines. **[15]**

Top spot: The Danakil Depression is claimed to be the hottest place on Earth, if not the most depressed. **[18]**

Sad fact: In September 2001, when the world was otherwise distracted, the Eritrean authorities 'disappeared' all the independent journalists they could lay their hands on, while telling everyone that 'they're just away doing their national service'. **[0]**

Pub fact: Eritrea is named after the Latin for Red Sea (*Mare Erythraeum*) without the Sea bit. **[8]**

Opening lines of national anthem beloved of Ethiopians everywhere: 'Eritrea, Eritrea, Eritrea/Her enemy decimated.' **[5]**

SCORE: 46 WORLD RANKING: 157

EUROPE

Estonia

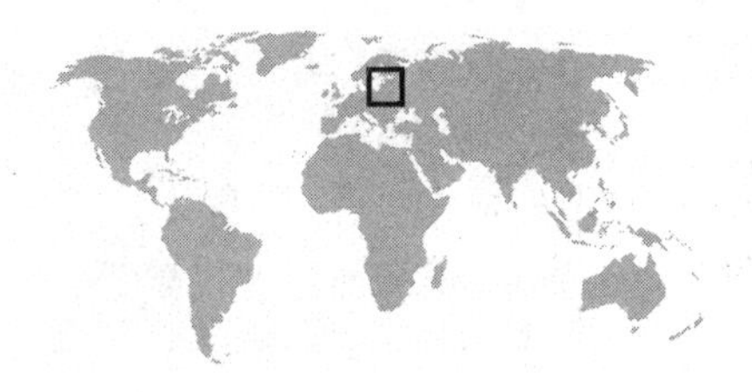

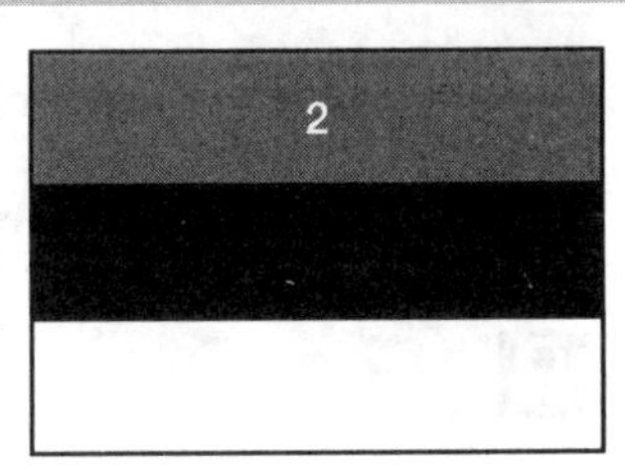

Name on driving licence: Republic of Estonia
Capital: Tallinn
Population: 1.3 million
Dosh: Estonian kroon = 100 senti
Size: 45,200 km² (2.2 Wales)

Complete history: Estonians used to keep watch on the coast for Vikings but forgot to look behind them and the south got invaded by Teutonic Knights. As soon as they turned round to look at the nice big horses, the Danes took over the north. By 1561, it was the Swedes in the north and the Poles in the south. Then it was the Russians (north and south), and finally, just Estonians.

Crowning achievement: Immortalised in the Morrissey song 'America is not the World': 'And don't you wonder why, in Estonia, they say "Hey you, you big fat pig, you fat pig, you fat pig"'? **[16]**

Top spot: Although Estonia is 47 per cent forest, it still has room for 1,400 lakes, including Lake Kavadi (not to be confused with the Hindu festival Thai Poosam Kavady, which is a different thing). **[12]**

Sad fact: Estonia is currently ranked first out of the 159 countries in The State of World Liberty Index produced by The State of World Liberty Project. In a nutshell, this means the government doesn't spend much on public services so, if visiting, take a hospital bed. **[6]**

Flag fact: The white bar does not represent 'peace' or 'harmony' as it does on the flags of other flouncier nations, but snow. In spring, when the snow melts, the flag presumably reverts to blue and black. **[13]**

National anthem of the twee: 'My native land, my joy, delight/How fair thou art and bright!/And nowhere in the world all round/Can ever such a place be found ... My little cradle stood on thy soil ...', etc., etc. **[3]**

SCORE: 50 WORLD RANKING: 136

Ethiopia

AFRICA

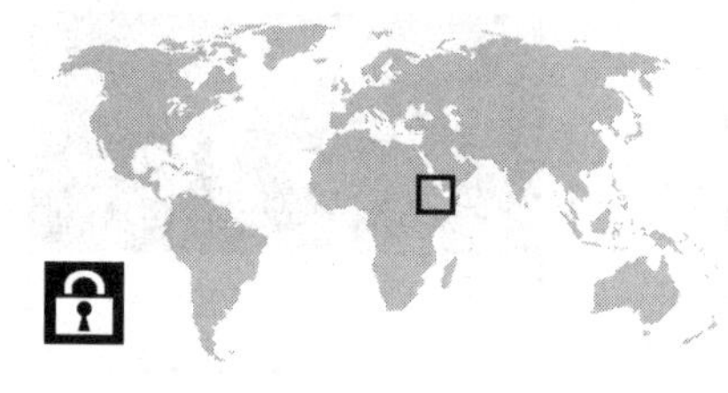

Name on driving licence: Ethiopia
Capital: Addis Ababa
Population: 68 million
Dosh: Birr = 100 cents
Size: 1,127,100 km^2 (53 Wales)

Complete history: If your country is going to be founded by a royal couple, it might as well be King Solomon and the Queen of Sheba. Unfortunately, even legend doesn't claim it was them but their alleged son Menelik I in the alleged year 1000 BC. In a largely failed bid to make himself as memorable as his parents, he visited Jerusalem and brought back the Ark of the Covenant, which just goes to show that souvenir shops of those times were of a much higher order than today's crop. The country fell into a state of desuetude until its inhabitants were called upon to stifle an Italian invasion in 1896.

Crowning achievement: Does beating the Italians count? **[11]**

Top spot: Lalibela, whose eleven rock-hewn churches were made with help from some angels, which was handy. **[13]**

Gift to the world: Emperor Haile Selassie (full title: His Imperial Majesty Haile Selassie I, Conquering Lion of the Tribe of Judah, King of Kings and Elect of God), whom Rastafarians believe was God incarnate. Selassie himself only appears to have been politely interested in the religion he inadvertently founded. **[12]**

Made in Ethiopia: The similar sounding Haile Gebreselassie (full title: World and Olympic champion, multiple world record holder, Athlete of Athletes, etc.), who, though a jolly fast runner, has not yet displayed any characteristics singling him out as God in human form. **[15]**

Opening lines of national anthem: 'Respect for citizenship is strong in our Ethiopia/National pride is seen, shining from one side to another.' Almost as inspiring as a list of hydroelectricity-generating facilities. **[3]**

SCORE: 54 WORLD RANKING: 114

colour key: 1 = turquoise 2 = blue 3 = green 4 = red 5 = yellow 6 = orange 7 = pink 8 = purple 9 = brown

Fiji

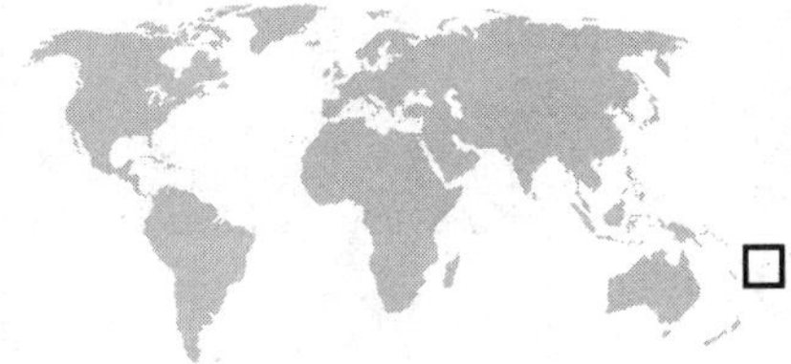

Name on driving licence: Republic of Fiji Islands
Capital: Suva
Population: 881,000
Dosh: Fiji dollar = 100 cents
Size: 18,300 km^2 (0.9 Wales)

Complete history: Fiji's 800 or so Melanesian islands have been visited by history's perennial bit-part players such as Captain Tasman (of Tasmania fame) and Captain Cook (of Cook Islands fame), but not, as far as can be established, anyone called Captain Fiji.

Crowning achievement: Fire-walking. Fijians do it on hot stones rather than the embers preferred in India. The trick, if you're trying it at home, is to go without coconut for two weeks beforehand. **[18]**

Made in Fiji: Vijay Singh. Ranked number one golfer in the world for 32 weeks (about the time it takes a human to go from conception to a stage in which he or she begins to inhale amniotic fluid) in 2004/05. **[17]**

Sad fact: Mel 'how do I hate the English, let me count the ways' Gibson owns the Fijian island of Mago, having reportedly paid US$15 million to a Japanese corporation for it in 2004. However, some of the locals want it back, citing the fact that their ancestors were cheated out of it in the early 1800s by the English (perhaps Mel's right after all), who gave them 2,000 coconut plants before evicting them at gunpoint. You feel they might have a point. **[3]**

Pub fact: Mago is apparently the largest privately owned island in the South Pacific, which doubtless only fuels the Fijians' desire to wash that man right out of their hair and send him on his way. **[3]**

Opening lines of national anthem: 'Blessing grant, o God, on nations on the isles of Fiji/As we stand united under noble banner blue.' **[7]**

SCORE: 48 **WORLD RANKING:** 146

Finland

EUROPE

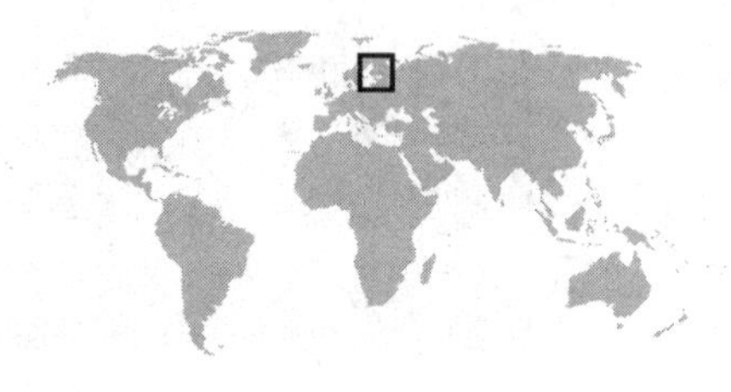

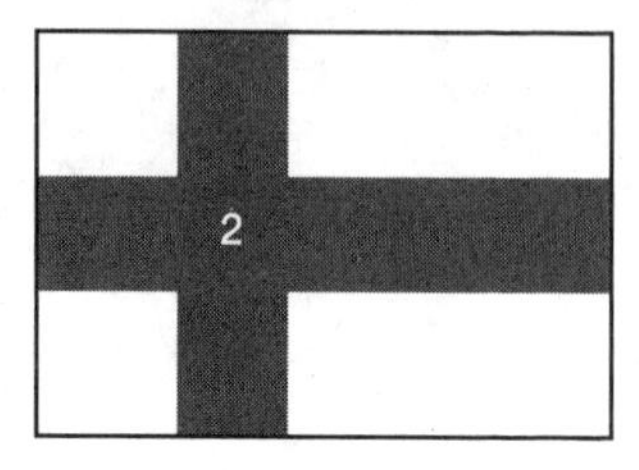

Name on driving licence: Republic of Finland
Capital: Helsinki
Population: 5.2 million
Dosh: Euro = 100 cents
Size: 337,000 km² (16 Wales)

Complete history: Don't talk about the Swedes. Unlike the Russians, who only managed to occupy the country for just over a hundred years on and off, Finland's western neighbours pretended they owned the place for more than six centuries. This has given rise to the famous Finnish joke: 'Swedes – can't love 'em, can't cook 'em.'

Crowning achievement: Having its own Monty Python song. As Michael Palin so adroitly observed: 'Eating breakfast or dinner/Or snack lunch in the hall/Finland, Finland, Finland/Finland has it all.' **[15]**

Sad fact: Just over a quarter of Finland is bog. Three-quarters of the country is tree. Of lakes, there are 187,000. Where liveth the people? **[8]**

Made in Finland: There is an unwritten law among media types that a Finnish person proceeding at anything more energetic than a swift walk must be referred to a 'Flying Finn'. This is largely the fault of Lasse Virén, who won the Olympic 5,000-metre and 10,000-metre events in 1972 and 1976 by 'flying', which seems a bit out of order, but there appears to have been some confusion over IOC rules in those days. **[16]**

Customs to treasure: The Finns rejoice in roasting saunas followed by dips in lakes through holes cut in the ice. This has given rise to the famous Swedish joke: 'A hot Finn is soon a cold Finn.' You had to be there, I guess. **[12]**

Opening lines of national anthem: 'Our land, our land, our native land/Oh, let her name ring clear!' (cf. the Estonian national anthem, which has the same tune but worse words). **[5]**

SCORE: 56 **WORLD RANKING:** 101

EUROPE France

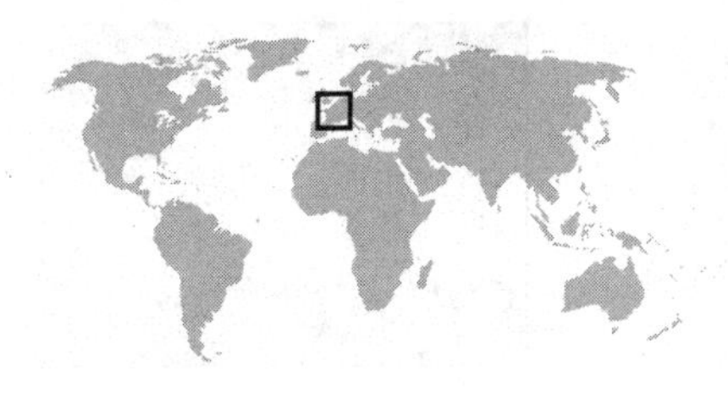

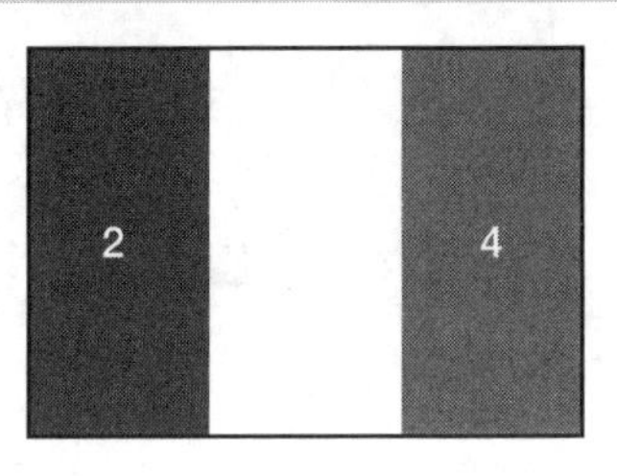

Name on driving licence: Republic of France
Capital: Paris
Population: 60 million
Dosh: Euro = 100 cents
Size: 547,000 km² (27 Wales)

Complete history: As is well known, the French like nothing more than to be reminded that they were saved from defeat in the last two shows by the British. They will claim, of course, that they were 'getting along just fine' and 'would 'ave won without the aid of you Rosbifs, sankyouverymuch', but I think we all know differently. Other topics not to bring up with the barman at the Folies-Bergère include Crécy, Poitiers, Agincourt, The Plains of Abraham, Trafalgar, Waterloo, the 2012 Olympics and Johnny Hallyday. Anything else will be fine.

Crowning achievement: Despite evidently having failed dismally on the world stage to a degree that has famously made France a nation of philosophers, l'Hexagone is the world's most popular tourist destination. Over 76 million people go there each year to pat the locals on the head and mispronounce their cheeses. **[17]**

Where to avoid: Lyons (16 February 1980), focal point of the world's longest traffic jam: 175 km of 2CVs, Renault 5s and carts taking noblemen to the guillotine. **[3]**

Gift to the world: Pasteurisation, a means of slaying the pathogens in milk. The irony is that the producers of the best French cheeses eschew this process, safe in the knowledge that the average French *consommateur* is only happy when dying of some bacterial disease. The procedure is named after it inventor, Louis Pasteurisation. **[16]**

Motto: '*Je ne regrette rien*.' **[4]**

National anthem: '*Aux armes, citoyens!/Formez vos bataillons!/Marchons! Marchons!/Qu'un sang impur/Abreuve nos sillons!*' 'La Marseillaise', the only proper national anthem ever written: seven verses of blood, entrails and mayhem. **[20]**

SCORE: 60 **WORLD RANKING: 76**

Gabon

AFRICA

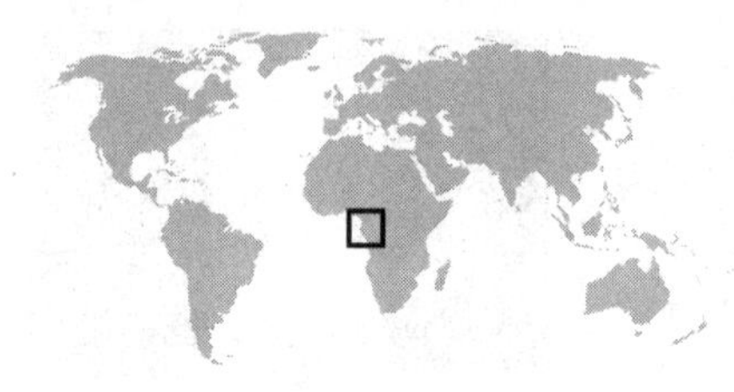

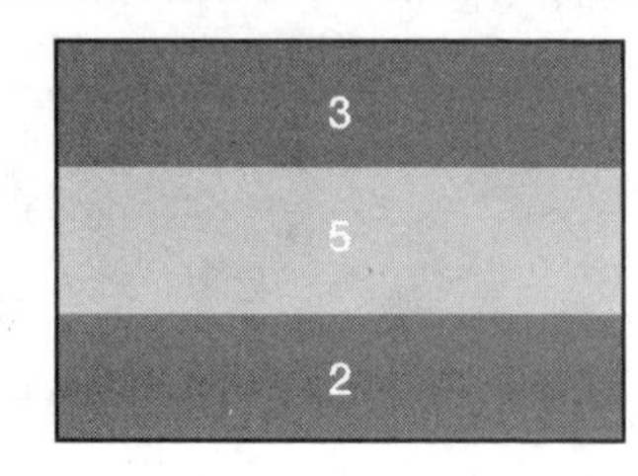

Name on driving licence: Gabonese Republic
Capital: Libreville
Population: 1.4 million
Dosh: CFA franc = 100 centimes
Size: 267,700 km² (13 Wales)

Complete history: The French arrived in 1839 and swanned around for a bit, but no one much noticed and they gradually drifted away, leaving only their language behind them, little knowing that that too faced global extinction by the year 2053. 'Gabon' comes from '*gabao*', the Portuguese for a hoodie.

Crowning achievement: The Fang. If you must have names for ethnic groups, they might as well be good ones. Also available as a language. **[16]**

Top spot: The Schweitzer Hospital, as founded in 1913 by organist, missionary doctor and Nobel Peace Prize laureate Albert Schweitzer. His cousin, Anne-Marie Schweitzer, gave birth to Jean-Paul Sartre, of course. **[17]**

Gift to the World: President Omar Bongo (1993–). If a novelist chose the name Mr Bongo for the leader of the fictitious country that supplies the backdrop to his searing story of jealousy, lust and betrayal under the harsh African sun, there'd barely be a straight face on the beach. Better to set the yarn in Gabon, make his basking readers aware that the head of state is actually *actually* called President Bongo, and let the libel suits flow. **[12]**

Sad fact: Gabon lays claim to the world's only natural nuclear reactor. Apparently, it all kicked off about 1.7 billion years ago when some uranium ore began reacting furiously at a place now called Oklo. It's all over now, however, just in case you were thinking of plugging your house into it. **[13]**

Opening lines of national anthem betraying an unhealthy predisposition to early rising: 'United in concord and brotherhood/Awake, Gabon, the dawn rises.' **[5]**

SCORE: 63 **WORLD RANKING:** 61

AFRICA (The) Gambia

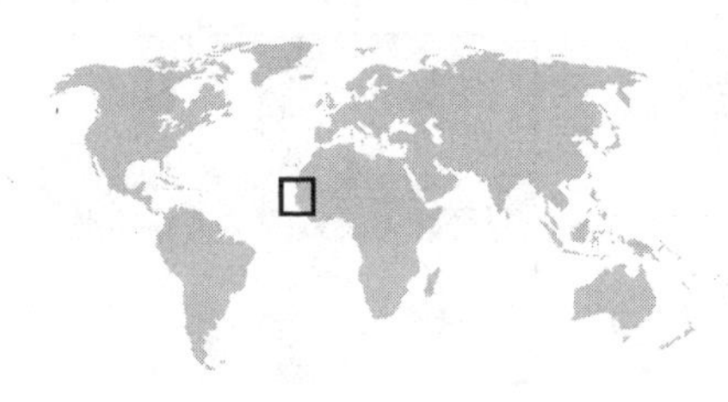

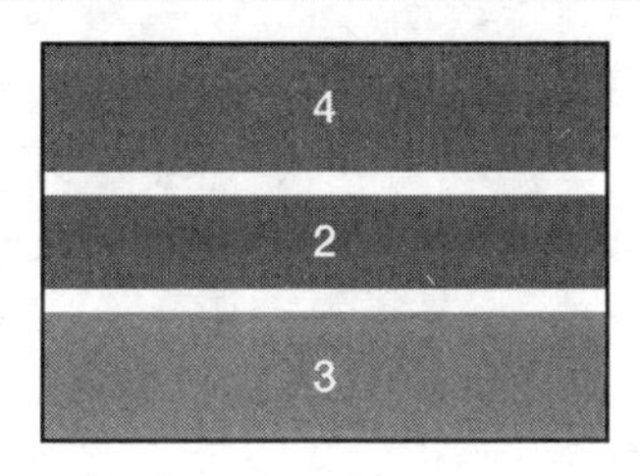

Name on driving licence: Republic of the Gambia
Capital: Banjul
Population: 1.5 million
Dosh: Dalasi = 100 butut
Size: 11,300 km^2 (0.6 Wales)

Complete history: There was a good deal of to-ing and fro-ing of Portuguese and British ne'er-do-wells for most of the country's history until 1783, when the latter handed it over to the French, possibly to make some point or other, before taking it back a century later, presumably to make some other point.

Crowning achievement: How many other countries enjoy (The) Gambia's ambivalent relationship with the definite article? Say 'Gambia' and it feels like something's missing. Call it 'The Gambia' and you sound like the sort of person who keeps the spirit of Empire alive by pronouncing Kenya as Keeenya and protesting surprise at the news that that nice Mr Smith is no longer running Rhodesia. **[15]**

Customs to treasure: The Wassau Stone Circles. Imagine Stonehenge built by a people whose religious beliefs centred on the importance of disappointing tourists. **[8]**

Popular misconception: '(The) Gambia is the gnarled finger of an oncologist poking into the side of Senegal and rummaging around in its gizzards.' Not so: this is merely the impression it gives to the casual map-glancerer. However, since the Gambians have been charging Senegal for this service ever since independence in 1965, it's probably best to keep this one under your hat. **[11]**

Pub fact: (The) Gambia is the smallest country on mainland Africa. It manages this feat by successfully labouring under the delusion that two riverbanks and a river constitute an actual country, rather than just a pleasant day-trip. **[12]**

Opening lines of national anthem: 'For The Gambia, our homeland/We strive and work and pray.' **[6]**

SCORE: 52 **WORLD RANKING:** 125

Georgia

EUROPE

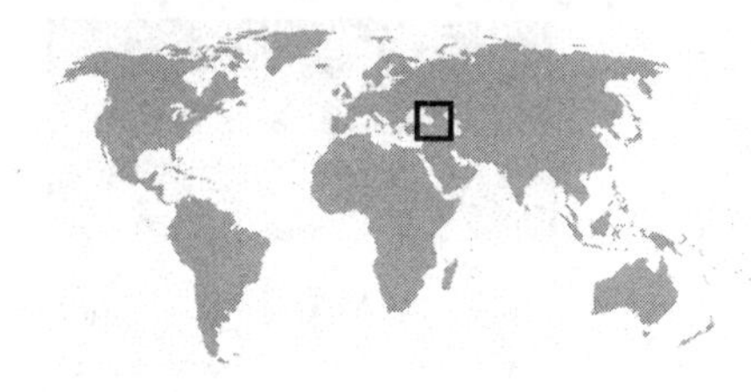

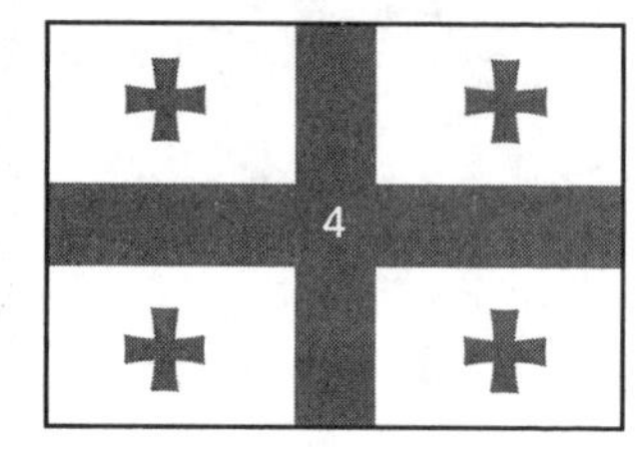

colour key: 1 = turquoise 2 = blue 3 = green 4 = red 5 = yellow 6 = orange 7 = pink 8 = purple 9 = brown

Name on driving licence: Republic of Georgia
Capital: Tbilisi
Population: 4.7 million
Dosh: Lari = 100 tetri
Size: 69,700 km^2 (3.4 Wales)

Complete history: The usual drab litany of Roman, Persian, Byzantine, Arab, Mongol, Ottoman Turk and Russian rulers is at least leavened by the spawning of no fewer than three separatist regions (Abkhazia, Ajaria and South Ossetia) in a country hardly large enough to swing (or indeed swig) a Molotov cocktail.

Crowning achievement: Producer of the world's cheapest 'champagne', which should prove interesting when Georgia attempts to join the EU. **[9]**

Top spots: Gori, the only place in the former Soviet Union with a public statue of Stalin (who was born there); and Mtskheta, where Christ's crucifixion robe is buried (there is some confusion over why one would go to the trouble of bringing it all the way from Jerusalem just to stick it in the ground, but then that's the joy of myths). **[8]**

Sad fact: The street cred vouchsafed unto Georgia by its inclusion in The Beatles' *Back in the USSR* ('Georgia's always on my mi-mi-mi-mi-mi-mi-mi-mi-mi-mind' or possibly 'Georgia's always on my my my my my my my my my my mind') is only marginally diminished by the song's troubling: 'Let me hear your balalaikas ringing out.' **[14]**

Customs to treasure: Georgian folk music may be the oldest surviving polyphonic tradition in the world. It uses a compressed major second and a stretched perfect fourth, notes sadly missing from Julie Andrews' much loved eulogy to the scale. **[16]**

National anthem not afraid to describe the rest of the planet as a sort of handy metal rack: 'My icon is my motherland/And the icon stand is the whole world.' **[15]**

SCORE: 62 **WORLD RANKING:** 68

EUROPE Germany

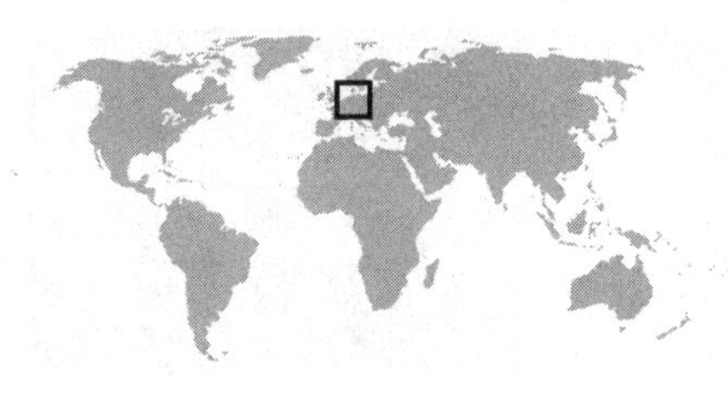

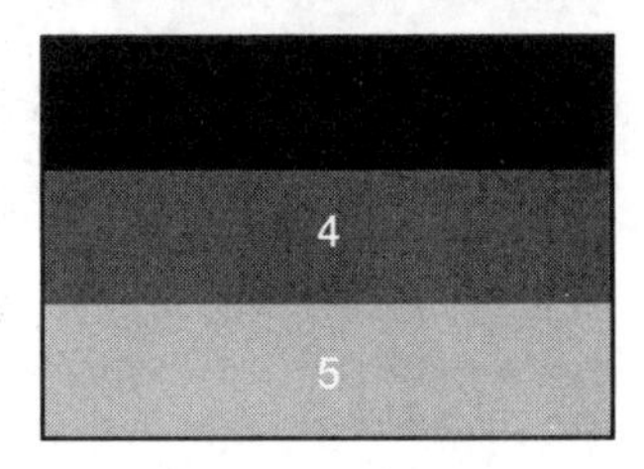

Name on driving licence: Federal Republic of Germany
Capital: Berlin
Population: 82 million
Dosh: Euro = 100 cents
Size: 357,000 km^2 (17 Wales)

Complete history: There are some folk who believe that German history can be summed up by the one word 'war'. This is a shame because the Germans haven't started one for over sixty years now, which is more than can be said for goody-goody Britain or her friend the United States. Even apparently happy-go-lucky countries such as El Salvador and India have invaded their neighbours in that time, while Israel makes a veritable virtue of it. Also, Germany produces over 300 types of bread which, you have to admit, is not the defining characteristic of a warlike nation.

Crowning achievement: Music. Germany has produced an absurdly long list of toe-tapping colossi including J.S. Bach, Handel, Beethoven, Mendelssohn, Schumann, Wagner, Brahms, Strauss and Nena, whose '99 Red Balloons' is universally regarded as representing the very pinnacle of German music. **[18]**

Top spot: Schloss Neuschwanstein, the fairy-tale castle built by King Ludwig II, a man who once declared: 'I want to remain an eternal mystery to myself and others.' Ripping. **[19]**

Customs to treasure: The Black Forest, often rashly believed to be good only for the odd gateau, is in fact also the spiritual home of the cuckoo clock (whatever the Swiss might tell you). And they say that Germans have no sense of humour. **[12]**

Gift to the world: Shouting. **[8]**

National anthem: Although it famously kicks off with the not-at-all-expansionist-oh-no-not-us-we-send-our-mothers-flowers-and-everything '*Deutschland, Deutschland über alles/Über alles in der Welt*', the only official verse is the third one, which begins with the heroically dull: 'Unity and Right and Freedom/For the German Fatherland.' **[4]**

SCORE: 61 **WORLD RANKING:** 70

Ghana

AFRICA

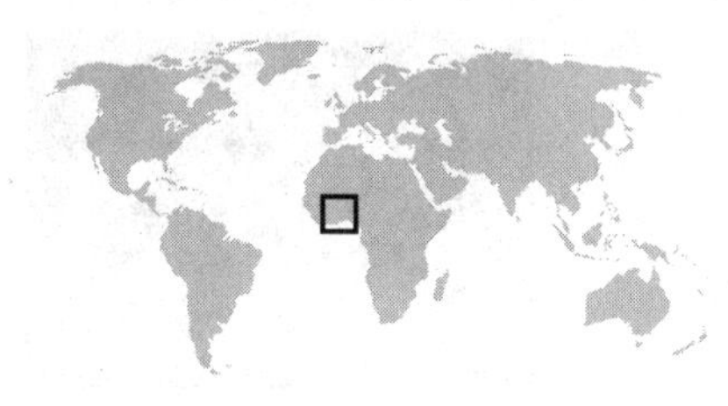

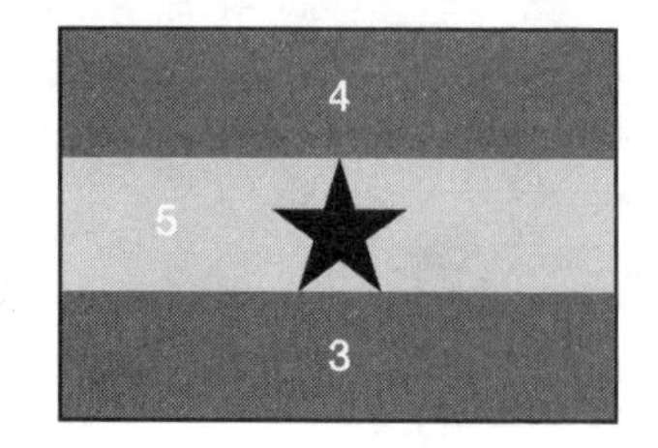

Name on driving licence: Republic of Ghana
Capital: Accra
Population: 21 million
Dosh: Cedi = 100 pesewas
Size: 238,500 km² (12 Wales)

Complete history: O, the happy days when the nation was merely a sleepy corner of a great empire stretching across West Africa. It was the Portuguese who ruined everything, arriving in 1471 with their floppy clothes and insistence that they weren't actually Spanish but *Portuguese*, thus starting an honourable tradition preserved today by Canadians and New Zealanders. Jerry Rawlings ruled from 1981 to 2001, initially on the grounds that he had the biggest gun, but later on the basis that he had the best ballot-box stuffers. Ghana was formerly known as the Gold Coast on account of the gold mined there and sent to London to be made into coasters.

Crowning achievement: Ghana was the first sub-Saharan nation out of the traps in the race to independence, just as Jack Kerouac was publishing *On the Road*, which is a coincidence because Ghana has roads too. **[14]**

Customs to treasure: Farmers' Day (first Friday of December). **[15]**

Pub fact: Ghana's Lake Volta, created by the Akosombo Dam, is the largest man-made lake on the planet. **[9]**

Popular misconception: 'The name Ghana is derived from the locals' ancient habit of putting their fingers in their ears and shouting "*Gaa Naa*" (lit. "Can't hear you/Not listening/Talk to the spear", etc.) every time a visitor with paella stains down his mantle claimed not to be Spanish.' **[10]**

National anthem in which the writer has a good stab at the name of the country, but is left regretting not having revised that bit for the exam: 'Arise, arise, O sons of Ghanaland/And under God march on for evermore!' **[8]**

SCORE: 56 **WORLD RANKING:** 101

Greece

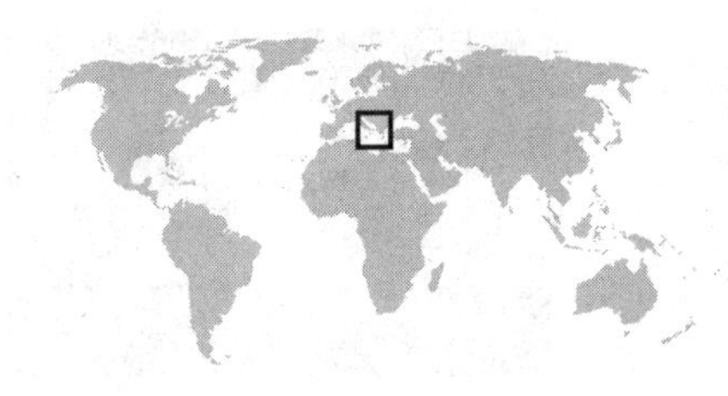

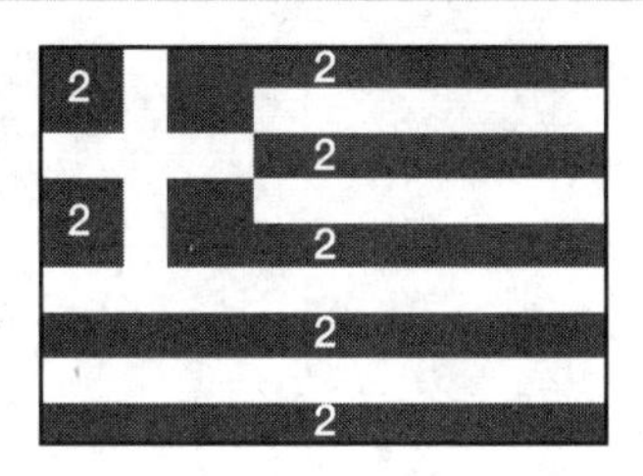

Name on driving licence: Hellenic Republic
Capital: Athens
Population: 11 million
Dosh: Euro = 100 cents
Size: 132,000 km² (6.4 Wales)

Complete history: If anything, Greece has a little too much history and it wouldn't do them any harm to give some of it away to places with hardly any at all (the people of Greenland wouldn't say no, I'm sure). On account of this surfeit, Greek history has had to be divided into two parts: Ancient Greece, which covers any event that has been recorded by someone inscribing it in stone and sticking it on the side of a temple; and Modern Greece, an era expected very shortly.

Crowning achievement: Being the birthplace of Western civilisation. The fact that Western civilisation now consists of American television series, fat (mostly) sweaty people, and arsenals of nuclear weapons is hardly their fault (except for that section of the Greek population who are fat and sweaty, which is their fault). **[19]**

Top spots: The island of Kefalonia, the setting for Louis de Bernières' literary classic *Captain Corelli's Mandolin*; and neighbouring Ithaca, the setting for the less well-known work *The Odyssey* by Homer, a Greek poet named after a character in an American cartoon. **[14]**

Popular misconception: 'Greece is named after the musical of the same name.' **[8]**

Gift to the world: The Olympic Games. The fact that the Games are now more an advertising opportunity for huge corporations than a festival of sporting prowess is hardly their fault (except at the 2004 Games in Athens, when it *was* their fault). **[13]**

Opening lines of national anthem sharing the Cypriot disparagement of their neighbour's sword: 'I shall always recognise you/By the dreadful sword you hold.' **[18]**

SCORE: **72** WORLD RANKING: **14**

Grenada

AMERICAS

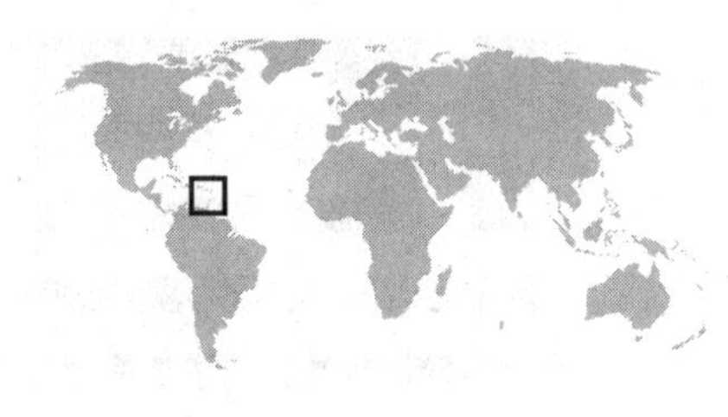

colour key: 1 = turquoise 2 = blue 3 = green 4 = red 5 = yellow 6 = orange 7 = pink 8 = purple 9 = brown

Name on driving licence: Grenada
Capital: St George's
Population: 89,000
Dosh: East Caribbean dollar = 100 cents
Size: 340 km² (0.02 Wales)

Complete history: Grenada wouldn't be with us at all had a volcano not erupted in the Caribbean. That was some while ago, of course, but it took until 1983 for the island to come to the attention of the Americans, who were naturally aghast when they discovered that they had not yet invaded it. Pointing to the expansion of the island's airport as clear evidence that the Marxist–Leninist government was planning to invade the US, the Americans launched Operation Urgent Fury, an act that immediately prompted a senate investigation into the identity of the anonymous twelve-year-old responsible for making up names for US military operations. A new government, much nicer than the previous one, was quickly installed, and the invasion of the US by the 1,500-strong Grenadan army was averted. Phew.

Pub fact: One in five of the world's nutmegs is grown in Grenada. **[10]**

Sad fact: When Hurricane Ivan hit the island in 2004 it partially destroyed a prison, allowing most of the inmates to escape, thus proving once again that it's an ill wind that blows nobody any good. **[8]**

Made in Grenada: Eric Gairy, former prime minister. Addressing the UN in October 1977, he called for the establishment of an 'Agency for Psychic Research into Unidentified Flying Objects and the Bermuda Triangle', and for 1978 to be designated the 'Year of the UFO'. **[17]**

Customs to treasure: It's illegal to build a wooden house in the capital, St George's. **[9]**

Opening lines of national anthem: 'Hail! Grenada, land of ours/We pledge ourselves to thee.' **[3]**

SCORE: 47 **WORLD RANKING:** 150

Guatemala

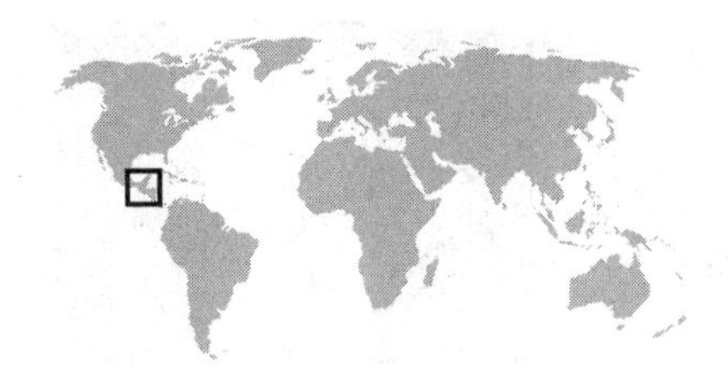

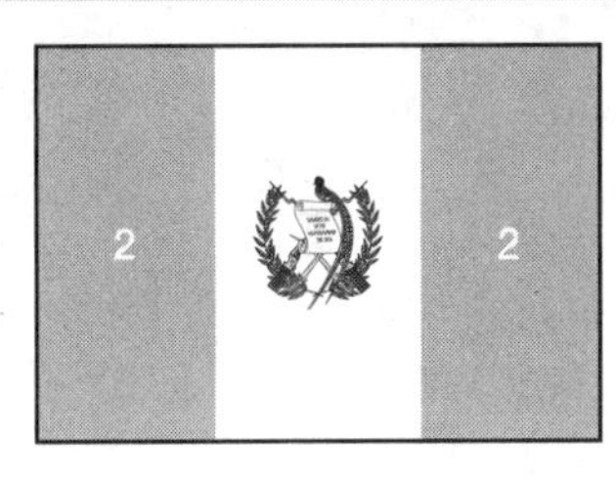

Name on driving licence: Republic of Guatemala
Capital: Guatemala City
Population: 12 million
Dosh: Quetzal = 100 centavos
Size: 108,900 km^2 (5.3 Wales)

Complete history: After the pleasingly baffling collapse of the Mayan empire in the 9th century, the history of Guatemala has been divided into Spanish People, Good People Who Were Democratically Elected and The Bad People With Whom The Americans Replaced Them By Force.

Crowning achievement: Despite there being 12 million Guatemalans, not one of them, or any of their forebears, has ever seen fit to win a medal at the Olympic Games. This is a feat of some magnitude given that athletes from places that aren't even countries – such as the US Virgin Islands and Bohemia – have been unable to resist climbing the Olympic podium. No doubt inspired by this, Guatemala's footballers have maintained an unblemished record of non-qualification for the World Cup. **[17]**

Top spot: Tajumulco Volcano (4,220 metres), the highest point in Central America. **[13]**

Made in Guatemala: No one you'll have heard of. **[20]**

Pub fact: The Mayan temples at Tikal were used for the exterior scenes on Yavin IV in *Star Wars IV* (i.e. the first one). None of the other five episodes featured the Mayan ruins and all consequently sold fewer tickets at the box office. Furthermore, *Heaven's Gate*, *Waterworld* and *Ishtar* all failed to make even a passing reference to Tikal and went on to run up some of the largest losses in the history of cinema. Yet you mention this to studio bosses in Hollywood and they laugh you out of their swimming pools. **[16]**

Opening lines of national anthem in which things go very wrong very early: 'Happy Guatemala! May your altars/Never be profaned by the executioner.' **[15]**

SCORE: 81 **WORLD RANKING: 2**

Guinea

AFRICA

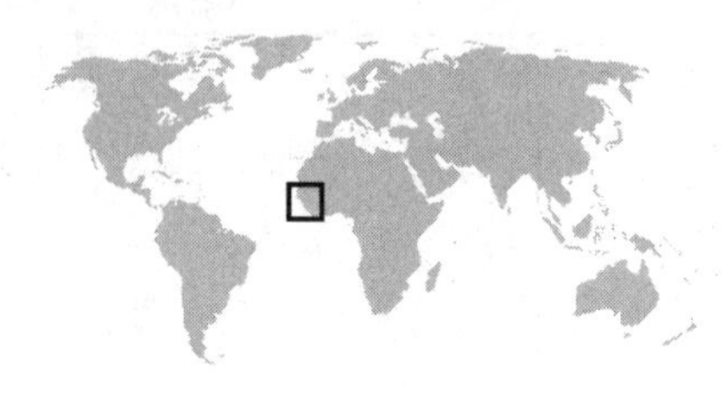

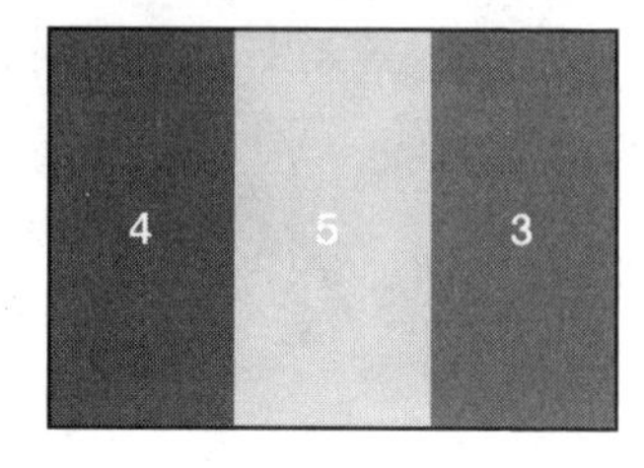

Name on driving licence: Republic of Guinea
Capital: Conakry
Population: 9.2 million
Dosh: Guinean franc = 100 centimes
Size: 245,900 km² (12 Wales)

Complete history: It should really have come as no surprise to the Portuguese in 1970 when their efforts to invade the country from Portuguese Guinea were crushed in a single day. Guineans had quite clearly had enough of being in other people's fiefdoms and were settling down to another fourteen years of misrule by their very own murderous and occasionally Marxist dictator Sékou Touré (1958–84). Prior to 1958, Guinea had been subjugated, in reverse order, by the French, the Fulani, the Moroccans, the Songhai empire, the Mali empire and the Sosso kingdom. Enough already.

Crowning achievement: No doubt as a result of spending eight centuries under the kosh, the Guineans were only too keen to vote themselves independence in 1958 and become the first French colony in Africa to gain the freedom to make their own hideous and awful mistakes (see Sékou Touré, above). **[15]**

Pub fact: Like a man with 84 shillings, the world has four Guineas. If they were to appear in alphabetical order after a colon they would look like this: Equatorial Guinea, Guinea, Guinea-Bissau and Papua New Guinea. **[12]**

Religious affiliations: The Kakimbon Caves in Conakry have great spiritual resonance for the Baga, a monotheistic people who believe in Kanu, the creator god. There's also a good deal of dead-ancestor appeasement involved, which must be trying. **[10]**

Popular misconception: 'The guinea pig comes from Guinea.' Nope. And while we're about it, it's not a pig. **[12]**

National anthem keen to introduce the African continent to a new past tense: 'People of Africa!/The historic past!' **[14]**

SCORE: 63 **WORLD RANKING:** 61

Guinea-Bissau

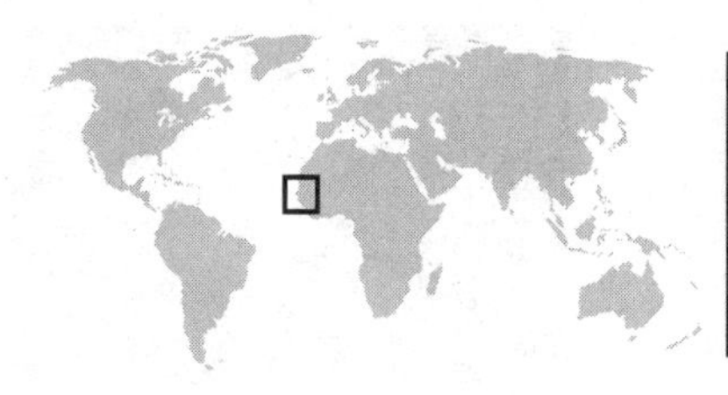

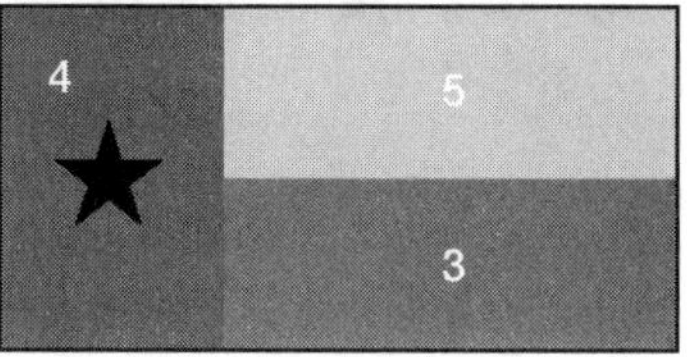

Name on driving licence: Republic of Guinea-Bissau
Capital: Bissau
Population: 1.4 million
Dosh: CFA franc = 100 centimes
Size: 36,100 km^2 (1.7 Wales)

Complete history: Belgians, eh? They look all butter-wouldn't-melty in their *Tintin* sweaters and their strings of Brussels sprouts, but what did they ever do for Guinea-Bissau? Very little, as it turns out, because the country was first enslaved, then colonised, by Portugal. Unlike in Guinea, where the French at least let the locals vote for independence, the Portuguese hung on to Guinea-Bissau as if their livers depended on it, which they did, in a sense, since the country's main crop, cashew nuts, is excellent for cleansing that particular organ. They finally conceded defeat and dragged their livers back to Lisbon in 1974. Nowadays, the country is plagued by invasions of locusts, which might conceivably be worse.

Crowning achievement: The hippos of the Bijagós Peninsula throw sand in the face of accepted wisdom, and indeed their accepted name, which is Greek for 'river horse', by wallowing around in the sea. No other hippos are known to do this. **[17]**

Pub fact: In 2003 Guinea-Bissau issued a set of stamps in honour of the New York Fire Department. **[10]**

Sad fact: The main attraction of Bolama, the first capital of Portuguese Guinea (as was), is a statue to a 1931 seaplane crash. The hotels are packed. **[5]**

Gift to the world: Gumbe, a polyrhythmic fusion of various local music styles. You might expect there to be a lot of drumming in it and you'd be right. **[10]**

Opening lines of national anthem that could almost be described as poetry: 'Sun, sweat, the greenery and sea/Centuries of pain and hope.' **[14]**

SCORE: 56 WORLD RANKING: 101

Guyana

AMERICAS

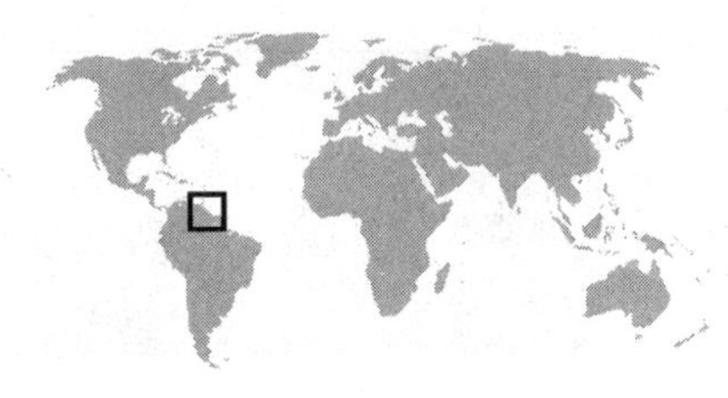

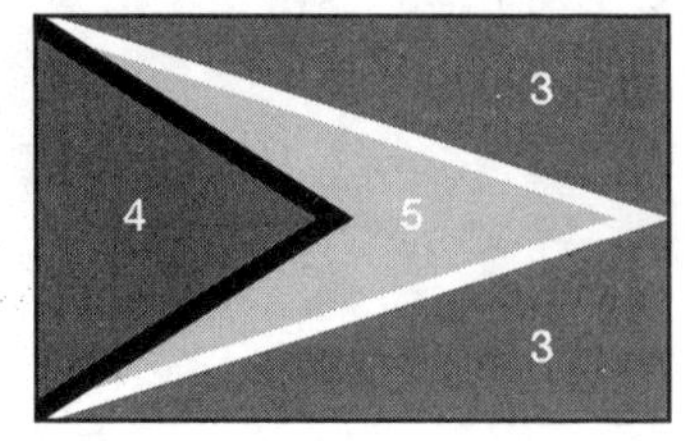

colour key: 1 = turquoise 2 = blue 3 = green 4 = red 5 = yellow 6 = orange 7 = pink 8 = purple 9 = brown

Name on driving licence: Co-operative Republic of Guyana
Capital: Georgetown
Population: 706,000
Dosh: Guyanese dollar = 100 cents
Size: 215,000 km^2 (10 Wales)

Complete history: ¡A co-operative republic! Just how trendy is that? Doubtless the original inhabitants, the notoriously collaborative Arawaks, would have been proud. Sadly, they were chased out long ago by the Caribs, who left to devote their attentions to naming the sea. The Dutch, French and British came next, the last of these clearing out in 1966 to devote their attentions to naming British Columbia, only to discover it was already called that.

Crowning achievement: Fooling half the world into thinking your country is in Africa, most likely in that bulgy bit in the north-west. **[15]**

Top spot: No. 63 Beach, which appeals to women (who like beaches) and men (who like numbers). There's also Shell Beach which, at 145 km long, is perfect for any children you might want to lose. **[14]**

Gift to the world: Nibbee, a sort of jungle vine leaf from which you (assuming you are a skilled Amerindian nibbee weaver) can make everything, from clothes to three-piece suites to boxes that could be mistaken in very poor light for hi-tech stuff like computers and plasma-screen televisions. **[13]**

Pub fact: The table-like Tepui Mountains inspired Sir Arthur Conan Doyle to write his 1912 novel *The Lost World*, the *Jurassic Park* of its day, though with fewer instances of people pointing up in terror and saying: 'Look out!' **[11]**

National anthem: 'Dear land of Guyana, of rivers and plains/Made rich by the sunshine, and lush by the rains/Set gem-like and fair, between mountains and sea/Your children salute you, dear land of the free.' Aah, sweet. **[18]**

SCORE: 71 **WORLD RANKING: 16**

Haiti

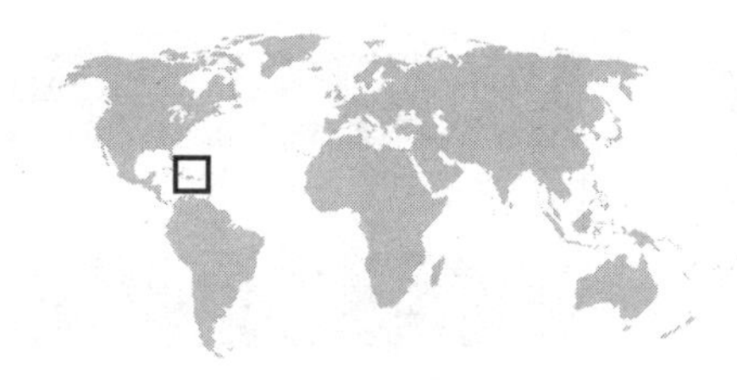

Name on driving licence: Republic of Haiti
Capital: Port-au-Prince
Population: 7.7 million
Dosh: Gourde = 100 centimes
Size: 27,750 km² (1.3 Wales)

Complete history: Oh dear. From the moment the Spaniards arrived and took just 25 years to exterminate the Arawak and Taino population of Hispaniola, it's all gone wrong. Realising that the western third of the island (aka Haiti) would never come to any good, the Spanish handed it over to the French in 1697, who then proceeded to make wads of cash out of it by making slaves grow sugar. By 1790, the slaves had had enough and within fourteen years had won independence. After a brief period of American occupation, it was left to 'Papa Doc' and 'Baby Doc' Duvalier and their Tonton Macoute enforcers to plunge the country into terror. Since then, it's all been floods, hurricanes, American invasions, corruption and drug-trafficking. Meanwhile, the wreck of Columbus' *Santa Maria* still lies mouldering off Cap-Haïtien like some sort of overworked allegory.

Crowning achievement: The hugely impressive L'Ouverture/Dessalines uprising is the first and, thus far, only instance of a successful revolt by slaves against an occupying power. **[19]**

Top spot: The mountaintop Citadelle Laferrière, built in fifteen years by 20,000 workers, helped dampen any French enthusiasm for returning. **[14]**

Sad fact: Haiti is the poorest country in the Western hemisphere by some margin, with an estimated 85 per cent of the population below the poverty line. **[0]**

Made in Haiti: Graham Greene's 'Papa Doc' novel *The Comedians*. **[15]**

National anthem aping the old '*Dulce et decorum est pro patria mori*' lie. Haven't they read any Wilfred Owen, for goodness' sake?: 'For the Flag, for the Homeland/To die is beautiful. To die is beautiful.' **[1]**

SCORE: 49 **WORLD RANKING: 140**

Honduras AMERICAS

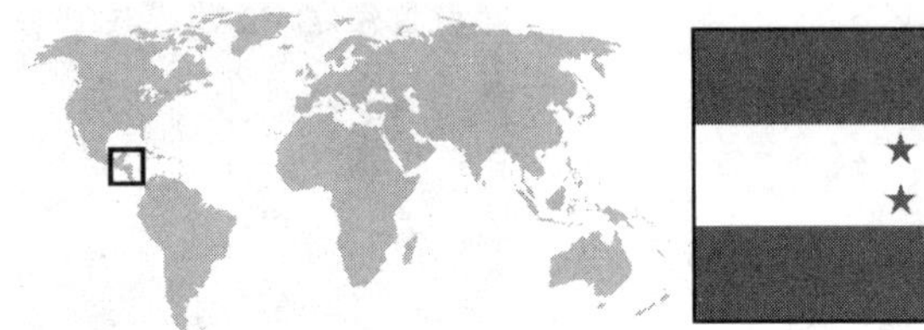

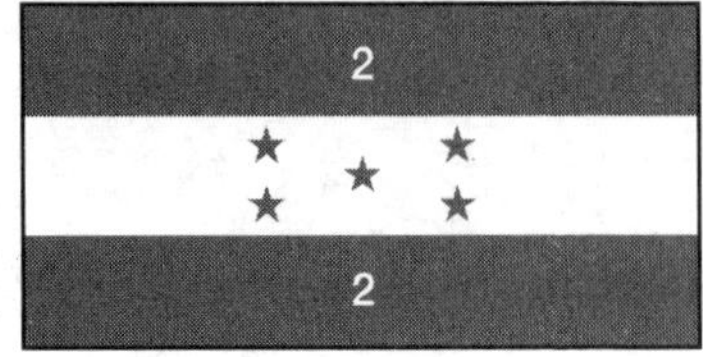

Name on driving licence: Republic of Honduras
Capital: Tegucigalpa
Population: 6.8 million
Dosh: Honduran lempira = 100 centavos
Size: 112,100 km^2 (5.4 Wales)

Complete history: Mayans. Spaniards. Americans. Bananas. Tourists.

Pub fact: In 1876, Honduras got through no fewer than five different presidents. **[19]**

Historical low: Hurricane Mitch (October/November 1998) killed 5,000 people and did for 70 per cent of the country's crops. **[0]**

Sad fact: Honduras was the original banana republic. In 1912, US businessman Sam Zemurray's banana company Cuyamel Fruit was in financial trouble, so he hit upon the novel idea of smuggling deposed president Manuel Bonilla back into Honduras, staging a violent coup, reinstalling him as president, and getting him to waive all Cuyamel's taxes for 25 years. Problem solved. He later took over the United Fruit Company (now Chiquita) chiefly to turn it into an incarnation of evil. **[0]**

Crowning achievement: Not just throwing in the towel. **[20]**

Not so much a national anthem as a day out: 'Your flag, your flag, is a shock of sky/Crossed by a block, by a block of snow/And one can see in its sacred depths/Five pale blue stars/In your emblem, which a murmuring sea/With its untamed waves shields/Behind the bare peak of a volcano, of a volcano/There is a star, there is a star of clear light.' And that's just the chorus. There are seven verses of jaw-dropping tediousness to go with it in which the entire history of Honduras is related and, indeed, marvelled at, while one verse is dedicated entirely to philosophising over the French revolution. No wonder the Salvadorans attacked them. I've half a mind to go there myself and give them all a good shaking. **[1]**

SCORE: 40 **WORLD RANKING: 176**

EUROPE Hungary

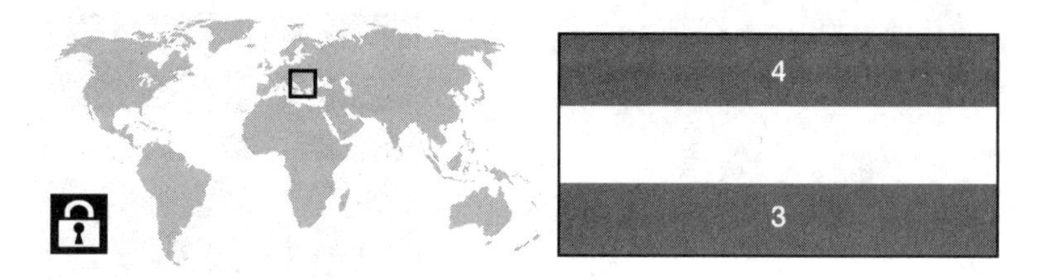

Name on driving licence: Republic of Hungary
Capital: Budapest
Population: 10 million
Dosh: Forint = 100 fillér
Size: 93,000 km^2 (4.5 Wales)

Complete history: For all their Magyar *this* and Magyar *that*, it turns out that the Magyars have been inhabiting Hungary for a piffling eleven centuries. Furthermore, they've since been overrun by Turks and Habsburgs. Then they accidentally started World War I by allowing a Bosniak student to bump off their Archduke Ferdinand. Then they joined the Nazis, only to be invaded by them, which is about as bad as ignominy gets. Finally, the Red Army invaded. Superstitious Magyars put all this down to something called the Curse of the Turan that has apparently been dogging them for years, which is bad luck really.

Crowning achievement: Handing the England football team its first ever home defeat (6-3, 1953) and opening the floodgates of misery. **[12]**

Top spot: At the fall of the Soviet Union, Hungary was littered with hundreds of statues of people nobody much cared for. The solution came in Budapest's Statue Park, which promises (and indeed delivers) 'Gigantic Monuments from the Age of Communist Dictatorship'. The Lenins are particularly good, as is the deeply affecting memorial to the Republic of Councils. **[17]**

Pub fact: Modern Hungarian is related to Finnish, which is quite a triumph because Finnish itself doesn't seem to be related to Finnish. **[13]**

Gifts to the world: The vinyl record, BASIC, the noiseless match. **[17]**

Opening line of national anthem: 'God bless the Hungarians with good cheer.' For any Hungarian embarking on the epic journey that is their national anthem, this is a most heartfelt plea. Eight verses and half a lifetime later, they're singing: 'Pity, God, the Magyar, then.' **[4]**

SCORE: 63 **WORLD RANKING: 61**

Iceland

EUROPE

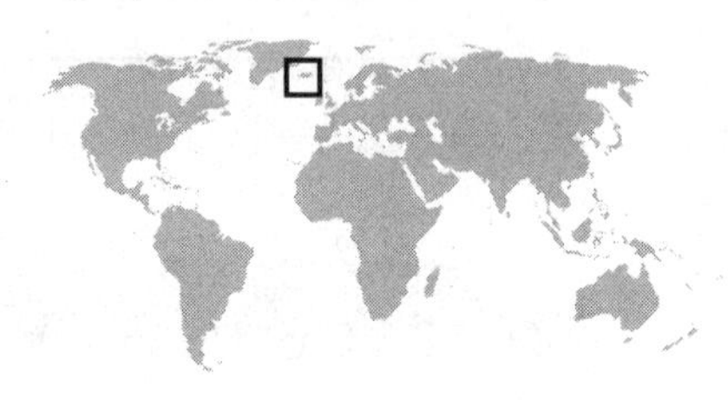

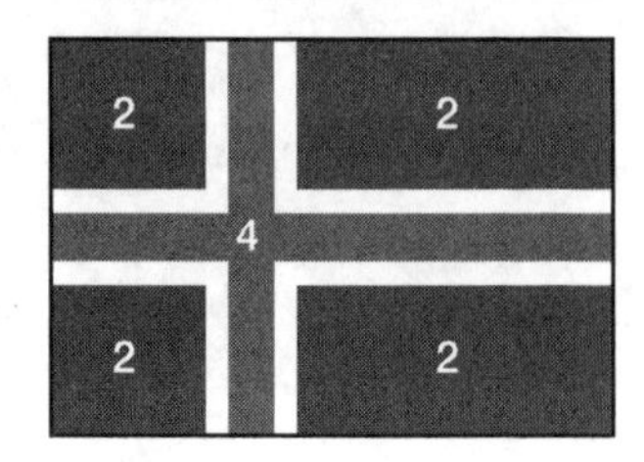

colour key: 1 = turquoise 2 = blue 3 = green 4 = red 5 = yellow 6 = orange 7 = pink 8 = purple 9 = brown

Name on driving licence: Republic of Iceland
Capital: Reykjavik
Population: 294,000
Dosh: Icelandic króna = 100 aurar
Size: 103,000 km² (5 Wales)

Complete history: Icelandic history is entirely composed of names that can only be pronounced properly whilst under sedation. Help to keep Iceland's famed oral tradition alive by contracting a disease for which you require minor surgery (an endoscopy is perfect, for example) and, once the little needle is safely in the back of your hand, proclaim: 'Naddoddr, Garthar Svavarsson, Flóki Vilgertharson, Ingólfur Arnarson Haraldur Harfagri, Althingi, Eiríkr rauthi, Thorgeirr Ljósvetningagothi, Ísleifr Gizurarson, Sturlungaöld, Snorri Sturluson, Sturla Sighvatsson, Gissurarsáttmáli, Geir Haarde, Eidur Gudjohnsen.' The nurse will understand.

Crowning achievement: Joining NATO without actually having any armed forces whatsoever. I bet the people at NATO were cross when they found out. **[16]**

Top spot: Reykjavik is not only the world's northernmost capital city, it also means 'Smoky Bay' (a shortening of 'Smoky Bacon', a popular crisp flavour). **[15]**

Pub fact: Up until 1984, Icelandic television didn't broadcast on Thursdays or at any time during July so that people could read books. As a consequence, Icelanders bought more books than anyone else in Europe. Nowadays, Icelandic television is on all the time, except for between 6 pm and 10 pm on Christmas Eve, so Icelanders only buy books for reading then. Many Icelandic authors are alcoholics. **[13]**

Made in Iceland: Magnús Magnússon started in Reykjavik but finished up in Scotland. This was the origin of his popular catchphrase: 'I've started so I'll finish.' **[12]**

Opening lines of national anthem: 'Our country's God! Our country's God!/We worship Thy name in its wonder sublime.' That's 'The God of our country' rather than 'Our country is God', we hope. **[12]**

SCORE: 68 **WORLD RANKING:** 31

ASIA India

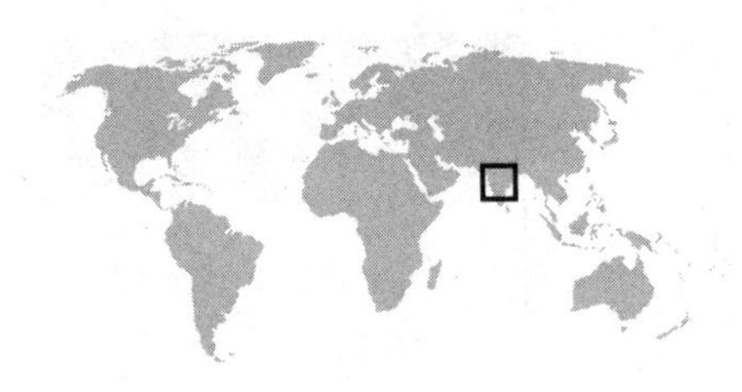

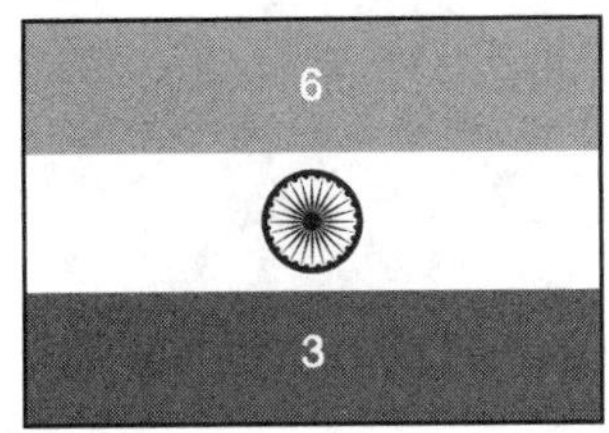

Name on driving licence: Republic of India
Capital: New Delhi
Population: 1,065 million
Dosh: Indian rupee = 100 paise
Size: 3,287,600 km^2 (159 Wales)

Complete history: Aryans, Dravidians, Mughals, Brits and Mahatma Gandhi (or 'Ben Kingsley' to give him his proper name).

Crowning achievement: With its 1.6 million workers, India's national railway is the largest employer on the planet. It's also the world's most dangerous railway, a crash or two a day accounting for around 800 people every year. However, even this pales in comparison with the mayhem on the roads, where a staggering 85,000 people per annum are killed in accidents. This may go some way to explaining the popularity in India of walking. **[12]**

Top spot: A trifle clichéd it may appear to our jaundiced 21st-century eyes, but the Taj Mahal is still a cracking building and, lest we forget, really just a tomb that got a bit out of control. **[18]**

Flag fact: India's flag is the only one on the globe to feature a wheel and, if that weren't shocking enough, also the only one to feature a band of colour representing 'renunciation'. As might be imagined, this noble aspiration doesn't quite stretch to nuclear weapons, which India holds a stock of just in case the good people of the Maldive Islands wake up one morning and decide they fancy their chances. **[8]**

Gifts to the world: Aside from Hinduism, Buddhism and Sikhism, there's Indian ink, Indian clubs, the Indian rope-trick, the Indian elephant, Indian summer, Indian file, etc., etc. **[19]**

Opening line of national anthem apparently ripped from the pages of *1984*: 'Thou art the ruler of the minds of all people.' **[3]**

SCORE: 60 **WORLD RANKING: 76**

Indonesia

ASIA

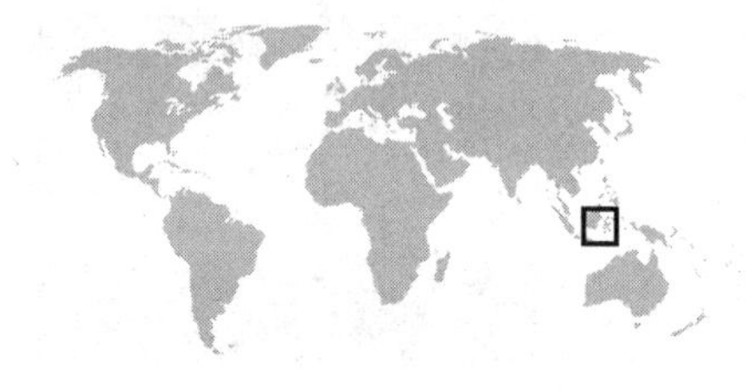

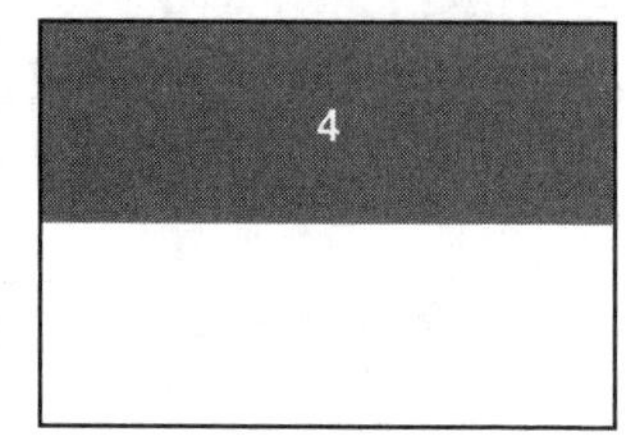

Name on driving licence: Republic of Indonesia
Capital: Jakarta
Population: 238 million
Dosh: Rupiah = 100 sen
Size: 1,919,400 km^2 (92 Wales)

Complete history: Full of power mongers about whom one normally hears so little. Take the Srivijaya empire, for instance, which was followed by the Sailendra and Mataram dynasties and the Majapahit kingdom. The Dutch East India Company (of whom admittedly we have heard) took over later, followed by all the rest of the Dutch, then all the Japanese, then all the Dutch again. Happily, independence in 1949 brought the deeply obscure Achmed Sukarno to the presidency. Like all good presidents, he promptly threw the country into chaos.

Crowning achievement: Being the world's largest archipelago. However, most people will swear blind that they've seen larger flocks of archipelagos on nature programmes. This is nonsense, of course: as anyone knows, archipelagos go round in herds. **[13]**

Top spot: Borobudur has 232 Buddhist temples, some of which got lost in the jungle but were found again by the ever helpful British in 1815, just in time for them to rush back to fight at Waterloo (the British, that is, not the temples). Visitors are advised to bring earplugs to guard against the almost constant background noise of one hand clapping. **[12]**

Pub fact: Kopi Luwak, the world's most expensive coffee, is made from beans that have passed through the digestive system of the palm civet, a sort of cat that lives in a tree. **[15]**

Made in Indonesia: The country's first proper rock band was called God Bless. It doesn't get more rock 'n' roll than that. **[16]**

National anthem: 'Indonesia, our native land/Our birthplace.' Naturalised Indonesians are well-versed in la-la-la-ing the opening lines. **[4]**

SCORE: 60 **WORLD RANKING:** 76

colour key: 1 = turquoise 2 = blue 3 = green 4 = red 5 = yellow 6 = orange 7 = pink 8 = purple 9 = brown

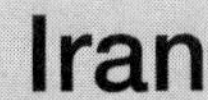

Iran

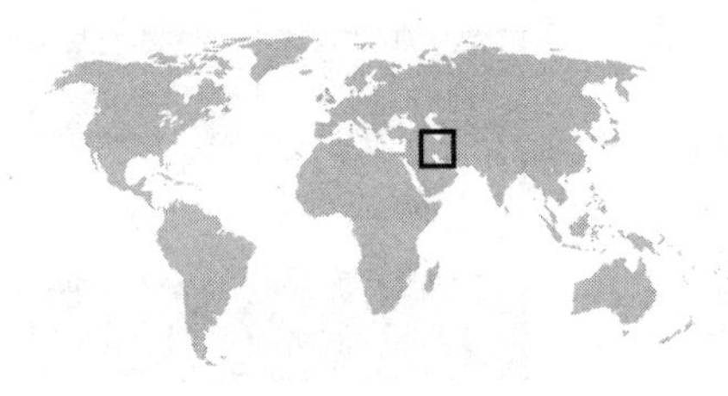

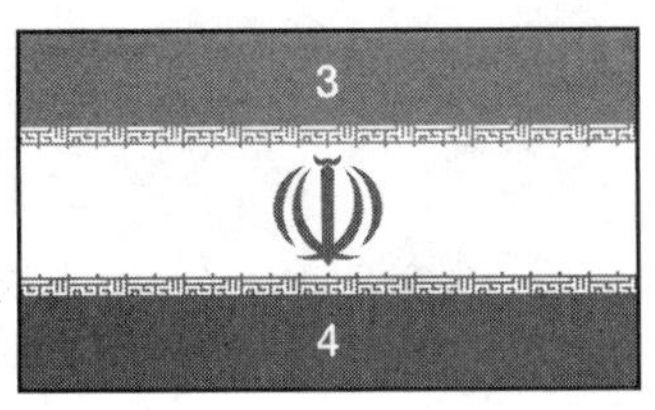

Name on driving licence: Islamic Republic of Iran
Capital: Tehran
Population: 69 million
Dosh: Iranian rial = 100 dinar
Size: 1,648,000 km^2 (80 Wales)

Complete history: There was Ancient Persia, of course: one of the most powerful empires of all time. Unfortunately, it happened to be around at the same time as Alexander the Great, who made rather short work of it, probably astride a unicorn or similar, just because he could. After that, everything – and here we mean *absolutely* everything – is the fault of the British. The British contend that it was also the fault of the Soviets and then the Americans, which is true but still doesn't absolve them. The only thing the Iranians like about the British is that everything that is wrong with Iraq is also their fault.

Crowning achievement: Ali Daei, with 109 international goals notched on his bedpost, is far and away the world's most successful striker ever. Natch, it does help if you get to play the Maldives (17-0) and Guam (19-0) but, as pundits everywhere will tell you, you can only beat who you get to play, or something. **[12]**

Sad fact: There are over 100 crimes, including 'apostasy' (converting from Islam), for which you can be executed. **[0]**

Made in Iran: Weapons-grade plutonium. Only kidding. **[10]**

Customs to treasure: The shaking minarets of Manar Jomban. Shake one of the pair of minarets atop this 14th-century tomb and the other will shake, as if by magic. Fun for all the family. **[15]**

National anthem that helpfully explains why it's so noisy in Iran: 'O Martyrs! The time of your cries of pain rings in our ears/Enduring, continuing, eternal.' **[13]**

SCORE: 50 **WORLD RANKING: 136**

Iraq

MIDDLE EAST

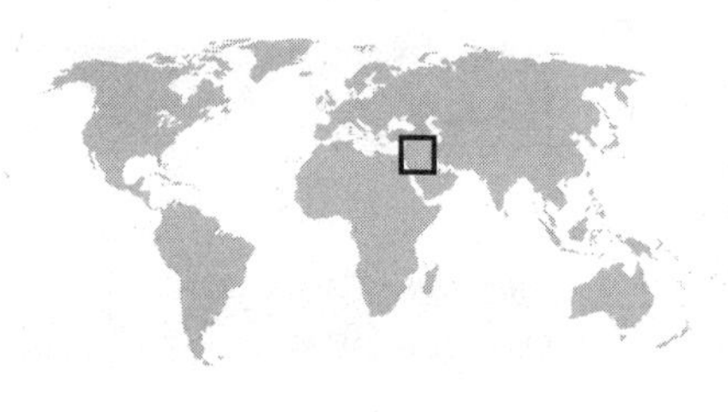

Name on driving licence: Republic of Iraq
Capital: Baghdad
Population: 25 million
Dosh: New Iraqi dinar = 1,000 fils
Size: 437,100 km^2 (21 Wales)

Complete history: Sumerians, Babylonians, Assyrians, Persians and Ottoman-Turks all inhabited Mesopotamia very happily. Then the British invaded, imposed a monarchy, renamed the country Iraq, and everyone became miserable. The end.

Crowning achievement: Mesopotamia is popularly believed to have been the cradle of civilisation, which is quite something. **[18]**

Top spot: You can't swing a history teacher in Iraq without hitting some heavyweight from the world of antiquity. There's the Hanging Gardens of Babylon (currently a military base, always a winner), the site of the (possibly mythical, admittedly) Tower of Babel, and the Garden of Eden, which you can find at Al Qurnah, just off the main road from Baghdad to Basra. **[19]**

Pub fact: The first minimum wage was introduced by the Babylonian authorities a mere 4,000 years ago, no doubt much to the chagrin of the Friedmanite faction in their midst. **[20]**

Flag fact: The three stars were added in 1963 when Iraq was about to join in holy alliance with Egypt and Syria. Sadly, this never came about. Sadder still, the Iraqis forgot to remove the other two stars. Saddest of all, you might feel, the Arabic words for 'God is great' on the flag's 1991–2004 incarnation are reputedly in Saddam Hussein's own handwriting. **[2]**

Opening lines of national anthem: 'My homeland/My homeland.' Granted, it takes a while to get into its stride but then it really cracks along. 'The youth will not get tired/Their goal is your independence/Or they die/We will drink from death' are sentiments which must really cheer the youth up before international sporting events. **[8]**

SCORE: 67 **WORLD RANKING:** 37

colour key: 1 = turquoise 2 = blue 3 = green 4 = red 5 = yellow 6 = orange 7 = pink 8 = purple 9 = brown

EUROPE Ireland

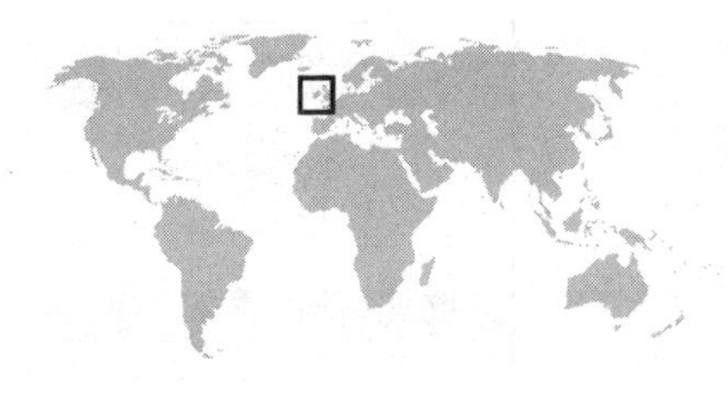

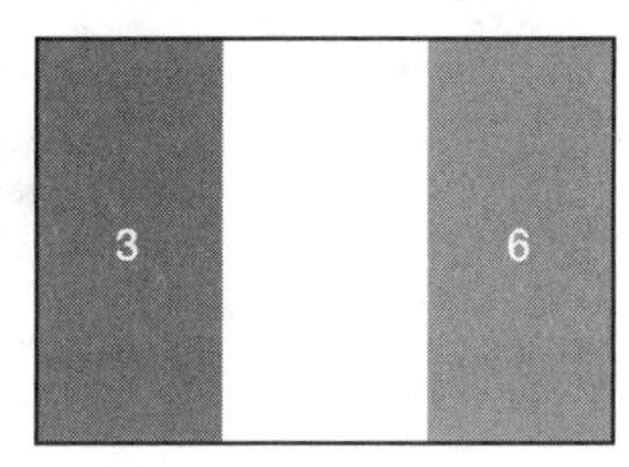

Name on driving licence: Republic of Ireland
Capital: Dublin
Population: 4 million
Dosh: Euro = 100 cents
Size: 70,300 km² (3.4 Wales)

Complete history: Irish history has forever been dogged by The Irish Question. This was first asked when the British realised they had spent centuries invading the cockatoo-shaped landmass, as Chris Morris so poignantly described it, without ever conquering it. Unfortunately, their answers to The Irish Question nearly always involved attempting to conquer the country 'but properly this time'. When this inevitably failed – due to the Irishman's indomitable spirit, and the unsporting tactics of leprechauns who would trip the English up before vanishing behind a ceilidh – the nature of the question changed to 'How do we get out of Ireland while still looking like we really won?' to 'How do we get out of Ireland?' and finally, to 'What should be done about The Corrs?'

Crowning achievement: Five victories in the European Song Contest have made Europe's third-largest island the competition's most successful nation and subsequently the one most often bankrupted by the 'winner is next year's host' rule. **[13]**

Top spot: The Twelve Bens of Connemara. Believed to be the only family in Europe in which all dozen sons have been given the same name. **[12]**

Made in Ireland: Fionn mac Cumhaill ('Finn McCool'), builder of the Giant's Causeway and husband to a deer. Brilliant. **[18]**

Gifts to the world: Craic, blarney, the colour green. **[14]**

Opening lines of national anthem: 'Soldiers are we whose lives are pledged to Ireland/Some have come from a land beyond the wave.' Written in the days of penury when the Irish Sea could afford but the single wave. **[8]**

SCORE: 65 **WORLD RANKING:** 50

Israel

MIDDLE EAST

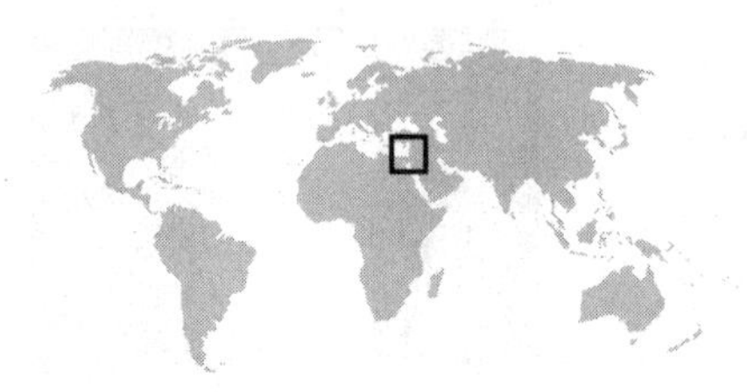

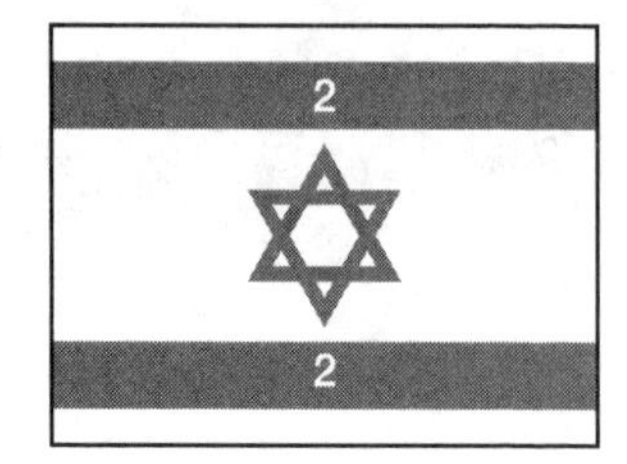

Name on driving licence: State of Israel
Capital: Jerusalem
Population: 7 million
Dosh: New Israeli shekel = 100 agorat
Size: 20,800 km^2 (1 Wale)

Complete history: The great thing about the history of Israel is that whatever you say about it will be contested in its entirety by everyone else, on the grounds that your perspective is 'too Jewish', 'too Palestinian', 'too Arabic', 'too conservative', 'too liberal', 'too imperialist', 'too mint imperialist', etc.; or conversely 'not Jewish enough', 'not Palestinian enough', etc., or in extreme cases 'not stupid enough'. Therefore it's safest to draw a veil over the whole thing and say nothing. If you cannot bring yourself to do this, you should declare to all and sundry that the most persuasive claims to Jerusalem are actually voiced by the people of the Cook Islands and run out of the room as fast as you can.

Top spot: Jerusalem, the only capital of a UN-recognised country not to enjoy UN recognition as a capital. Work that one out, o you mite. **[15]**

Pub fact: The Dead Sea is not only the lowest point on the planet but also the world's deepest hypersaline lake, a fact your audience will find less interesting than you do. **[16]**

Sad fact: Despite being ranked 99th in terms of population, Israel is in the top ten in terms of arms exports. **[0]**

Religious affiliations: Sundry. [4]

National anthem(s): Compare and contrast 'The soul of a Jew yearns ... to be a free people in our land' (the official anthem) with 'By my strong will and my inflaming rage, my volcanic revenge' (the Palestinian anthem), and it all starts to make sense. **[0]**

SCORE: 35 WORLD RANKING: 184

Italy

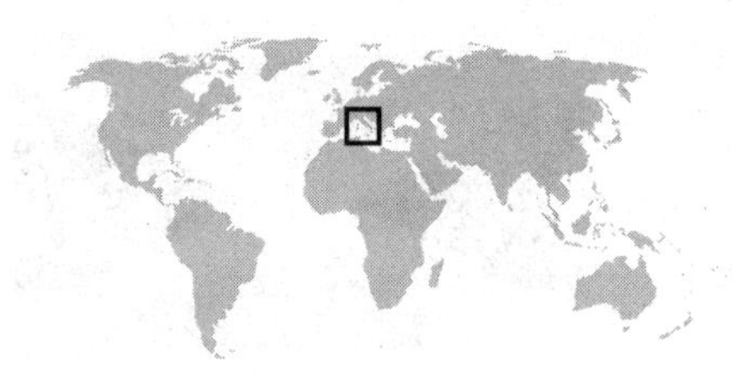

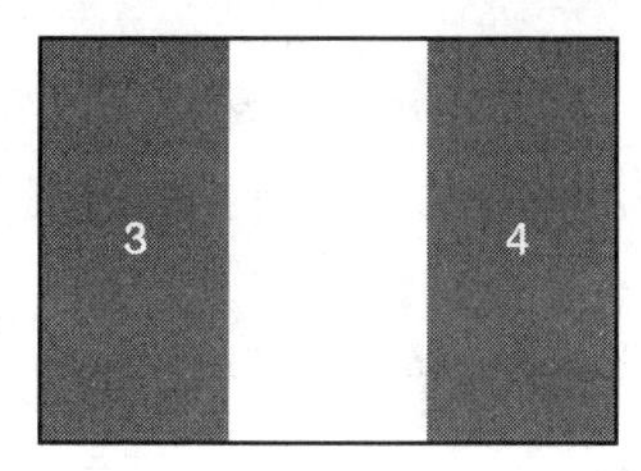

Name on driving licence: Republic of Italy
Capital: Rome
Population: 58 million
Dosh: Euro = 100 cents
Size: 301,200 km² (15 Wales)

Complete history: Italy has only been with us as an entity since 1861, or perhaps 1871, depending on your idea of what constitutes an entity, but whatever that is, nobody in Italy is likely to get unduly exercised about it. Garibaldi unified the country largely through the astute use of the biscuit that bears his name. Before that, all was chaotic city states and petty kingdoms and Renaissance men. Before *that* came the Romans, whose major contribution to society was to provide essential material for the joke about them never doing anything for Palestine.

Crowning achievement: Leonardo da Vinci, who drew detailed designs for helicopters, tanks, submarines and parachutes several centuries before any of these were invented. Also did some nice pictures. **[16]**

Top spot: Venice, which is famously sinking, although apparently not as swiftly as was once believed. This is rather disappointing for the Italians who like to have something they can get really worked up about before shrugging their shoulders and wandering away. **[13]**

Customs to treasure: In a freakish distortion of Oscar Wilde's dictum on tragedy, all Italian women become like their mothers (or, where possible, their grandmothers), as do Italian men. **[12]**

Gifts to the world: Gesticulating, scooters, anarchy, sometimes all at once. **[13]**

Chorus of national anthem: 'Let us band together/We are ready to die/We are ready to die/Italy has called us.' If this is the reaction every time Italy calls her people, it's probably just as well she doesn't do it all that often or the place would be an uninhabited desert. Get over yourselves. **[5]**

SCORE: 59 **WORLD RANKING:** 85

Jamaica

AMERICAS

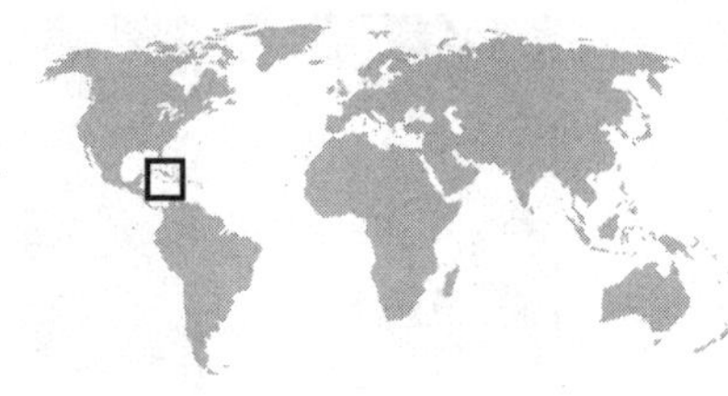

Name on driving licence: Jamaica
Capital: Kingston
Population: 2.7 million
Dosh: Jamaican dollar = 100 cents
Size: 11,000 km^2 (0.5 Wales)

Complete history: Christopher Columbus (never just 'CC' apparently), inevitably, landed in 1494 and decided that the best use Spain could put it to was as his family's private estate. This is the sort of thing that was quite acceptable back then, but it would be quite wrong if you tried it yourself today. There were people there, after all, and one suspects that some of them were not thoroughly consulted over the island's change of use. Perhaps this is why they all decided to die off within about twenty years. This selfish act provided something of a headache for Spain (Columbus having popped his clogs by then) and forced them to invite Africans to work there instead. The Africans enjoyed producing Jamaican sugar so much that they never left, even when the British took over.

Crowning achievement: Blue Mountain coffee, despite it not having gone through the digestive system of a tree-climbing Indonesian cat, is still claimed to be the best in the world. **[12]**

Pub fact: The father of English poet Elizabeth Barrett Browning owned lots of Jamaica. Try not to hold it against her. **[3]**

Gift to the world: 'My mother went to the West Indies.' '*Jamaica*?' 'No, she went of her own accord.' **[20]**

Made in Jamaica: Bob Marley who, despite dying at 36, still managed to father thirteen children in between gigs. Claims that he also gave his name to a West African nation have been strongly denied by the government of Mali. **[18]**

Opening lines of national anthem: 'Eternal Father, bless our land/Guide us with Thy Mighty Hand.' Breathe in those rhyming couplets. **[14]**

SCORE: 67 **WORLD RANKING:** 37

ASIA Japan

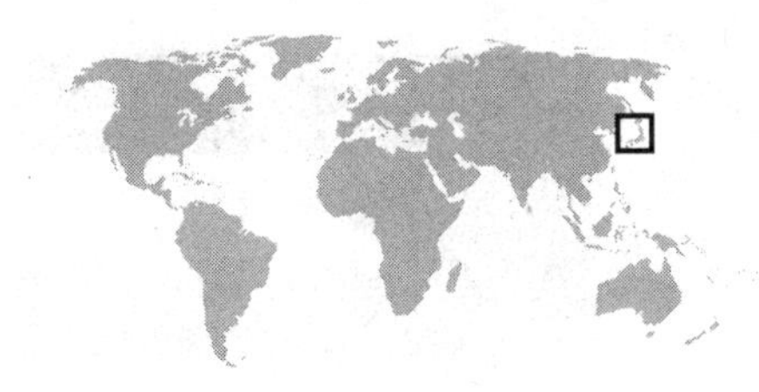

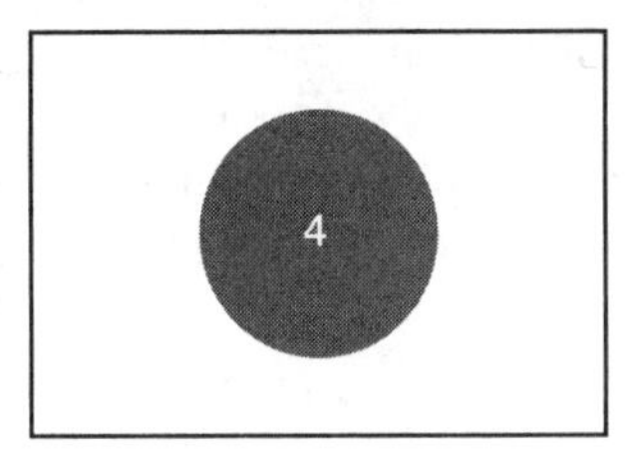

Name on driving licence: Japan
Capital: Tokyo
Population: 127 million
Dosh: Yen = 100 sen
Size: 377,800 km² (18 Wales)

Complete history: It is still a little bit extraordinary how an island nation, crushed by defeat in World War II, with hardly any natural resources or much spare land to speak of, and unused to dealing with the world at all except at a bayonet's length, came to be one of the world's economic superpowers. The six eras of Japan are: Jomon, Yayoi, Classical, Medieval, Edo and finally, inevitably, the Modern Era (as sponsored by Sony).

Crowning achievement: One of the few countries to be made into a verb, hence 'to japan something' is to cover it with a glossy black lacquer. Not outstanding perhaps, but a start. **[14]**

Gift to the world: Karaoke. **[12]**

Popular misconception: '"*Donkey Kong*" is so called due to a mistranslation into English from the original Japanese "*Monkey Kong*".' Disappointingly, *Donkey Kong*'s inventor, Shigeru Miyamoto, has stated that he chose the 'Donkey' bit to conjure up something stubborn while the 'Kong' bit refers to the game's gorilla. Hey ho. **[5]**

Made in Japan: Prince Akishino's moustache. Facial hair is not something you associate with the Japanese, so it's possible he grew his simply because he could. Combined with his devil-may-care centre parting, it makes him quite the swell of Japanese high society. **[19]**

Entire national anthem: 'May thy peaceful reign last long/May it last for thousands of years/Until this tiny stone becomes a massive rock/And the moss covers it all deep and thick.' Laudably brief and fantastically enigmatic. If only all national-anthem writers looked to Japan for inspiration rather than their own countries, we'd all be a lot happier. **[20]**

SCORE: 70 **WORLD RANKING: 22**

Jordan

MIDDLE EAST

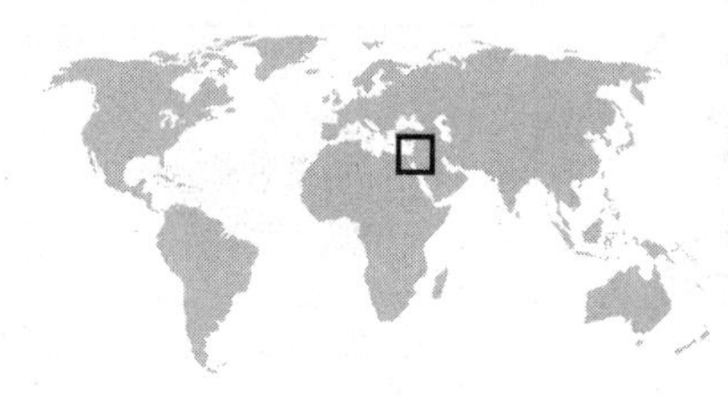

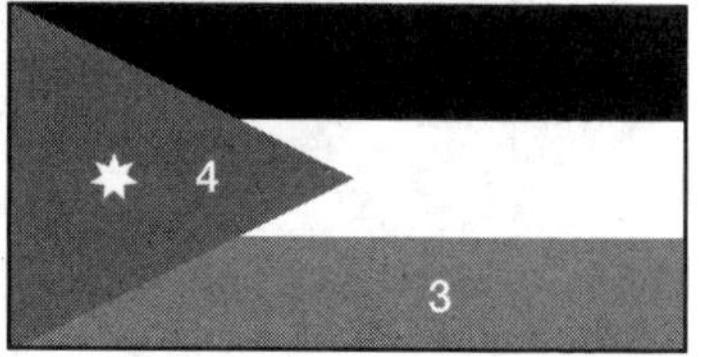

Name on driving licence: Hashemite Kingdom of Jordan
Capital: Amman
Population: 5.6 million
Dosh: Jordanian dinar = 1,000 fils
Size: 92,300 km^2 (4.3 Wales)

Complete history: The formation of Jordan is a troublesome thing to grasp on the best of days and even if you expect to live into your eighties, you still might not wish to bother with anything more than a cursory understanding. There are some royal families involved (in a tight corner, guess the Hashemites), the French and British dig their meddling oars in, and then it all goes horribly wrong for the PLO in something called Black September.

Crowning achievement: Jordan managed to negotiate its way out of being landlocked in 1965 by swapping loads of desert with Saudi Arabia in return for a few beaches around Aqaba. Switzerland take note. **[15]**

Top spot: Petra, not the quiet dog out of *Blue Peter* who was upstaged by Shep, but the ancient city carved out of rock, was so good the Romans actually added some bits and pieces rather than laying waste to it, which was their preferred method of interacting with things foreign. **[19]**

Pub fact: Some of the 250-fils coins minted in 1996 went into circulation even though they had planchets clipped so severely that they made the clipping error on the 1985 1-dinar coin look mild by comparison. **[11]**

Sad fact: Irritatingly for the country's leaders, despite huge public relations campaigns, visitors to Jordan are still apt to confuse it with Michael Jordan, Joe Jordan and Jordan. **[8]**

Entire national anthem: 'Long live the King!/Long live the King!/His position is sublime/His banners waving in glory supreme.' Even shorter than the Japanese anthem. Get them to a nunnery. **[20]**

SCORE: 73 WORLD RANKING: **11**

Kazakhstan

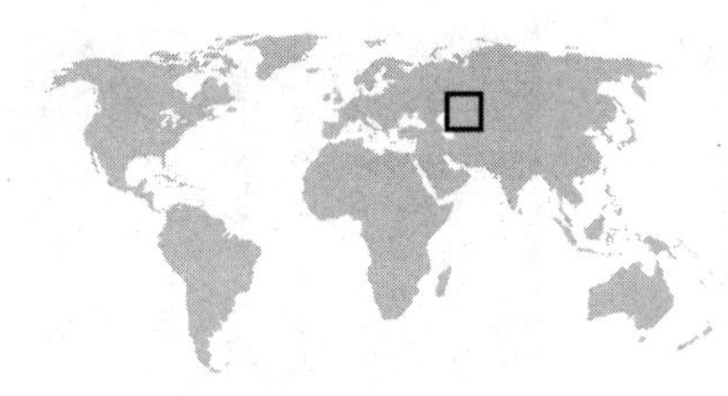

Name on driving licence: Republic of Kazakhstan
Capital: Astana
Population: 15 million
Dosh: Tenge = 100 tiyn
Size: 2,717,300 km^2 (132 Wales)

Complete history: Kazakhstan is huge, one might even venture to call it Brobdingnagian, and yet its people are sparse, so most events in its history have had to be conducted through loudhailers. The first people to develop nodes on their larynxes were Stone Age folk who, if horse historians are right, were the first humans to domesticate horses rather than eat them. A few years later the Soviets took over and the Kazakhs were back to eating the horses again.

Crowning achievement: It's true what they say, any country sporting the suffix 'stan' is ridiculously rich in the sorts of minerals any other country would feel a bit embarrassed about, like uranium (second-largest reserves in the world), chromium (ditto), lead (ditto ditto), zinc (ditto ditto ditto) and manganese (only third-largest – life has bitter pills for us all). **[14]**

Made in Kazakhstan: Former capital Almaty hosts the largest speed-skating ice rink known to humankind. **[13]**

Popular misconception: 'All Kazakhs are like Borat, the Sacha Baron Cohen creation.' Not so. At least half of them are women. **[10]**

Customs to treasure: Sending things into space. The Soviet-built Baikonur Cosmodrome still rockets assorted bleeping flashing objects up into the stratosphere for all our good. **[12]**

Opening lines of national anthem: 'Golden sun in heaven/Golden corn on the steppe.' The anthem's engagingly brisk, some might even say preppy, tune comes with words revised and updated in 2006 by none other than current president, the 'autocratic, *moi*?' Nursultan Nazarbayev. **[18]**

SCORE: 67 **WORLD RANKING: 37**

Kenya

AFRICA

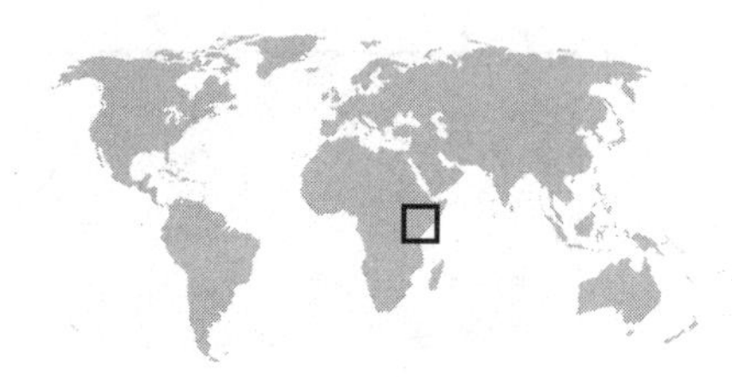

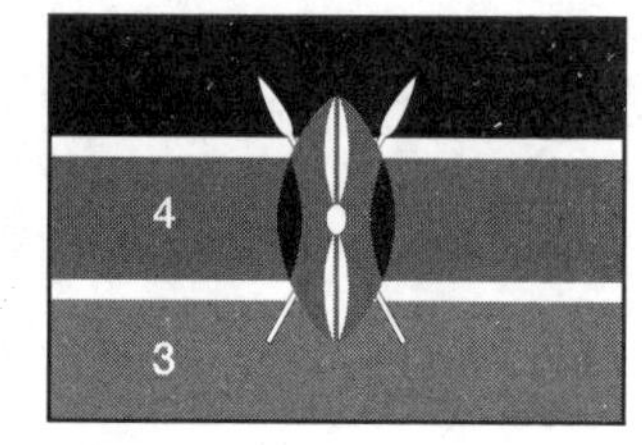

colour key: 1 = turquoise 2 = blue 3 = green 4 = red 5 = yellow 6 = orange 7 = pink 8 = purple 9 = brown

Name on driving licence: Republic of Kenya
Capital: Nairobi
Population: 32 million
Dosh: Kenyan shilling = 100 cents
Size: 582,700 km^2 (28 Wales)

Complete history: It was two millennia ago that the Kenyan people breathed a collective sigh of relief on discovering that the sailors who had come to visit them weren't Spanish, Portuguese or even Dutch but Arabs from Eastern Asia. The Arabs showed no interest in claiming the locals' land for their own, selling them into slavery, or butchering them to a man 'just in case'. Instead they traded objects with them and went away again. Not so the Portuguese. Or the British.

Crowning achievement: Providing the setting for the immortal line: 'I had a farm in Africa ...' which continues '... at the foot of the Ngong Hills', but keep this to yourself because nobody likes a show-off. **[12]**

Sad fact: Lucy Kibaki, Kenya's epically unhinged first lady, was so incensed at reports in *The Nation* [the *Daily Nation*?] of her hissy fits that she staged a midnight sit-in at the newspaper and proceeded to scweam and scweam and scweam for the next five hours. **[3]**

Made in Kenya: Jomo Kenyatta, a president whose brave attempt in 1963 to set a trend for heads of state to have more or less the same name as the country over which they preside started promisingly with the election of India's Indira Gandhi (1966), before rather running out of steam. **[10]**

Top spot: Lake Nakuru, still the only park in Africa to have been created specifically to protect the lesser flamingo. **[14]**

National anthem with a rhyming scheme cleverly devised to force the old school 'Keeenya' pronouncers into a corner: 'And our homeland of Kenya/Heritage of splendour.' **[15]**

SCORE: 54 **WORLD RANKING:** 114

Kiribati

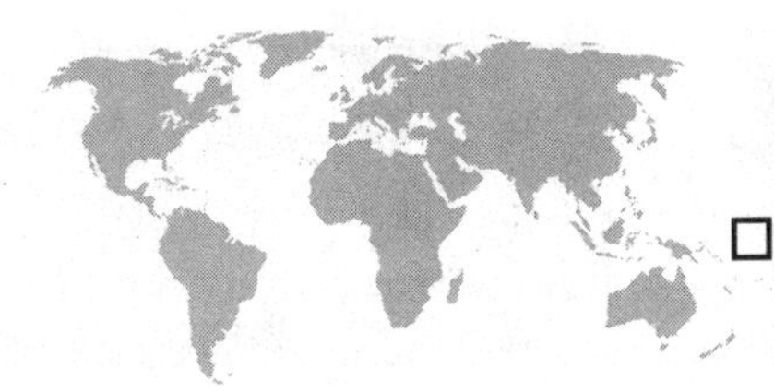

Name on driving licence: Republic of Kiribati
Capital: Bairiki
Population: 101,000
Dosh: Australian dollar = 100 cents
Size: 720 km² (0.04 Wales)

Complete history: The former Gilbert Islands comprise three groups of coral atolls spread across over 5 million km² of sea, and are all so low-lying that it took the British until 1837 to notice them. Sensing that the rising oceans would soon swallow them up, the colonists left in 1979. The islanders sought revenge by casting a spell over the British electorate that took nearly twenty years to wear off.

Top spot: Christmas Island, essential for those collecting the set of islands named after major dates in the Christian calendar. See also Easter, Christmas (the other, better-known one), Trinity (Antarctica), All Saints (Ireland) and Ascension. **[16]**

Crowning achievement: Christmas Island is also the world's largest coral atoll and includes towns named London, Paris and Banana, making it indisputably the bestest place on the entire planet. **[21]**

Pub fact: Kiribati is pronounced Kee-ree-bass. No, really. **[9]**

Sad fact: Up until 1995, the International Date Line split Kiribati in twain. This was a source of perpetual misunderstandings between islanders on either side. Today all the islands are in tomorrow, a most vexing state of affairs for those who like to put things off until then, because they find they're already there and so have to defer their actions until the day after tomorrow, which begins to get confusing. **[12]**

National anthem distressingly bereft of exhortations to murder one's enemies for the sake of the homeland and/or die in the attempt: 'The attainment of contentment/And peace by our people/Will be achieved when all/Our hearts beat as one/Love one another!/Promote happiness and unity!' **[18]**

SCORE: 76 **WORLD RANKING: 6**

Korea, North

ASIA

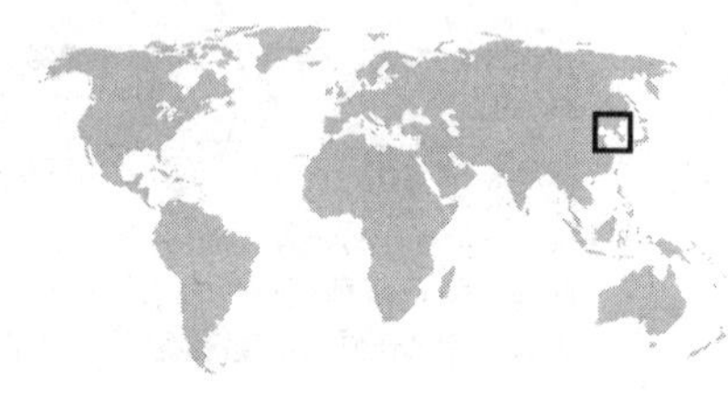

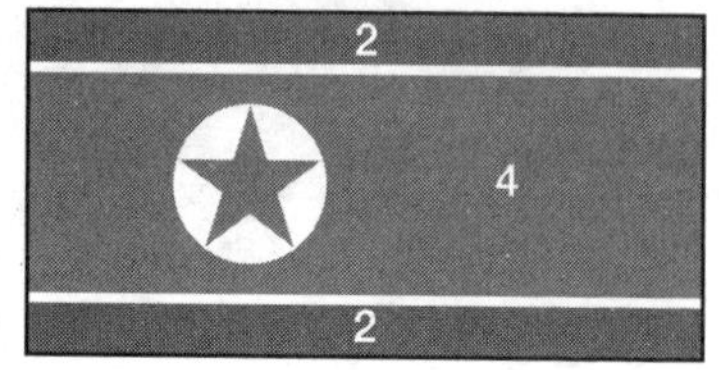

Name on driving licence: Democratic People's Republic of Korea
Capital: Pyongyang
Population: 23 million
Dosh: North Korean won = 100 chon
Size: 120,500 km^2 (5.8 Wales)

Complete history: The planet's most heavily defended border keeps the dirty Yankee-loving capitalist southerners from attacking the hard-working honest peace-loving northerners. This is probably just as well because North Korea may not have a technologically advanced army but it is apparently the fifth largest in the world.

Crowning achievement: Banning both the private car and mobile phones. This ensures that citizens do not waste time idly cruising around ringing their friends, and can thus devote themselves to more worthwhile pursuits such as making flags to wave enthusiastically at passing motorcades. **[15]**

Pub fact: The Mass Games celebrates, in dance and gymnastics, the history of the Workers' Party Revolution. It takes two months to do this. Book early. **[6]**

Gift to the world: The North Koreans perform nuclear tests out of mere curiosity and so that they can share their findings with other nations. They wouldn't dream of actually making a nuclear weapon themselves. **[5]**

Customs to treasure: Not letting death get in the way of a successful presidency. The 1998 constitution proclaimed the completely and utterly dead Kim Il Sung as the 'Eternal President of the Republic'. His son Kim Jong Il has to make do with being Chairman of the National Defence Commission, now the highest post in the land open to someone who is not yet actively decomposing. **[13]**

Opening lines of a national anthem that are not as mad as you might previously have suspected: 'Let morning shine on the silver and gold of this land/Three thousand leagues packed with natural wealth.' **[14]**

SCORE: 53 **WORLD RANKING: 119**

ASIA

Korea, South

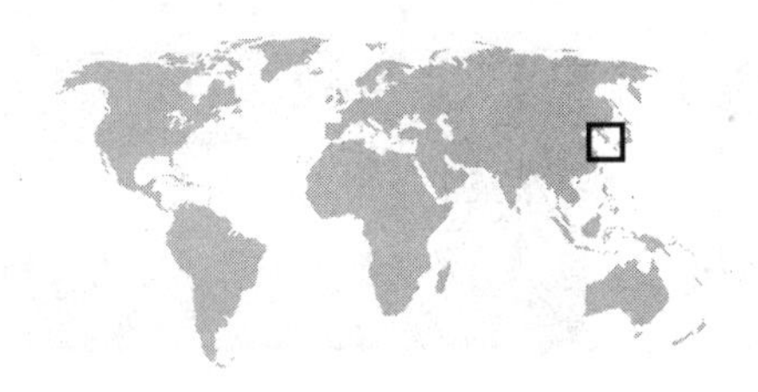

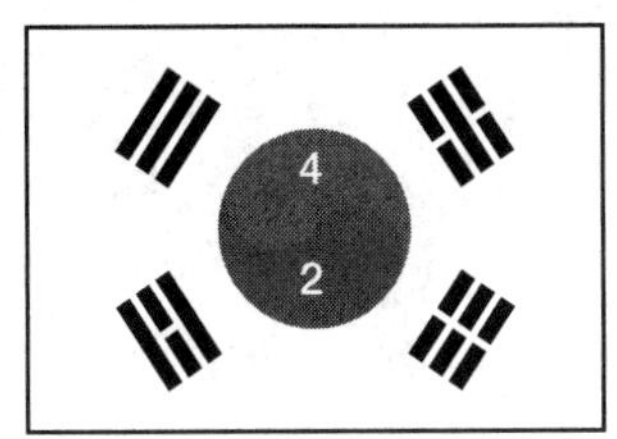

Name on driving licence: Republic of Korea
Capital: Seoul
Population: 49 million
Dosh: South Korean won = 100 chon
Size: 98,500 km^2 (4.8 Wales)

Complete history: The planet's most heavily defended border keeps the scary brainwashed freedom-hating doctrinaire northerners from attacking the hard-working honest peace-loving southerners. This is probably just as well because it doesn't take a great deal of imagination to picture it all spiralling out of control and ending with tactical nuclear weapons buzzing about the Asian skies. Which also makes one consider whether the word 'tactical' has ever found itself in a context in which it has been used quite so loosely.

Top spot: The Demilitarised Zone, under which the North Koreans tunnelled furiously in order to effect an invasion. The South Koreans rumbled them when they became suspicious about a wooden horse the northerners were using for gym practice. **[12]**

Crowning achievement: South Korea so resembles itself that it is known as the 'South Korea of Asia'. **[10]**

Gift to the world: Tae kwon do – aka 'the way of the foot, the way of the fist, and the way of life'. Also the way of death, since tales abound of South Korean soldiers using the technique to wipe out hordes of Vietcong fighters. Now an Olympic sport, although competitors are encouraged not to fight to the death if it can be avoided. **[5]**

Pub fact: The world's most couply wedding involved 35,000 Moonies getting hitched, inspiring 17,500 best men to crack jokes about honey-moonies. **[6]**

National anthem that shares more or less the same tune as that of North Korea's, which may come in handy one day: 'Until the East Sea's waves are dry/And Mount Baekdusan's worn away.' **[12]**

SCORE: **45** WORLD RANKING: **163**

Kuwait

MIDDLE EAST

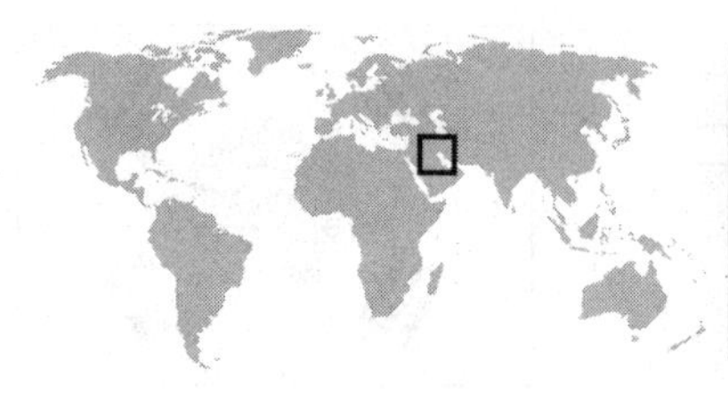

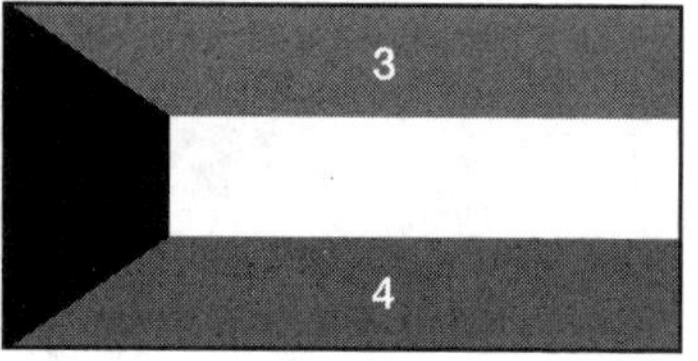

colour key: 1 = turquoise 2 = blue 3 = green 4 = red 5 = yellow 6 = orange 7 = pink 8 = purple 9 = brown

Name on driving licence: State of Kuwait
Capital: Kuwait City
Population: 2.9 million
Dosh: Kuwaiti dinar = 1,000 fils
Size: 17,800 km^2 (0.9 Wales)

Complete history: Contrary to press reports, loads of stuff happened to Kuwait before the first Gulf War. There was independence from Britain in 1961; the discovery of oil in 1936; the British swooping in to protect Kuwait from those swinish devils the Turks in 1899; and the founding of Kuwait City in 1710. Before that, admittedly, Kuwait's history can largely be told in terms of sand and camels, but who's to say that's a bad thing?

Top spot: Al Jahrah, Kuwait's highest point at 250 metres. Thrill-seekers should be aware that at that altitude there is a whole 3 per cent less oxygen in the air than at sea level, so it might be wise to wait until a fully equipped hospital has been established on the peak before attempting to conquer it. **[8]**

Sad fact: Kuwaiti citizens are a minority in Kuwait. Of nearly 3 million people milling about in the country, only 1 million are Kuwaiti. The rest are not, as popularly supposed, American soldiers, but stateless Arabs and assorted ex-pats. **[10]**

Popular misconception: 'Kuwaiti society is deeply conservative.' Couldn't be more wrong. Why, women have had the vote for well over a year now. **[5]**

Pub fact: Kuwait is unique in not having a single natural lake, river or reservoir to its name. Water is thus obtained from desalination plants, and by asking for it on planes and then secreting the bottles in hand luggage while the stewardess isn't looking. **[13]**

National anthem apparently inspired by a scratched record: 'Kuwait/Kuwait/Kuwait/My country ... Your face bright/Your face bright/Your face bright with majesty/Kuwait/Kuwait/Kuwait/My country.' **[14]**

SCORE: 50 **WORLD RANKING:** 136

ASIA

Kyrgyzstan

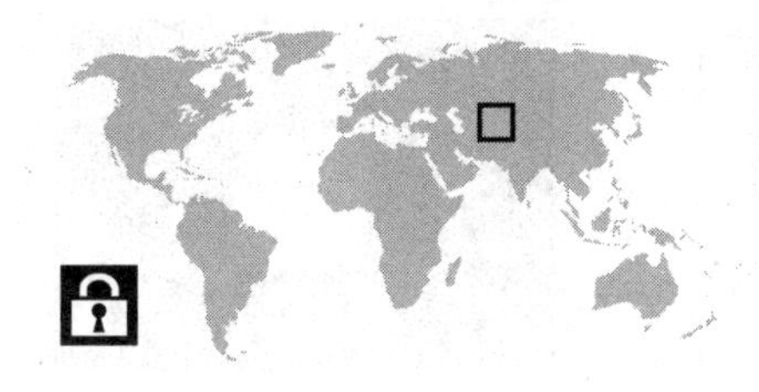

Name on driving licence: Kyrgyz Republic
Capital: Bishkek
Population: 5.1 million
Dosh: Som = 100 tyiyn
Size: 198,500 km^2 (9.7 Wales)

Complete history: Not much is known about the country's history except that its ancient Soviet-built capital, Bishkek, is the Kyrgyz word for a churn used to make fermented mare's milk. However, the nation is set to burst onto the world scene by dint of the faux travel-based game show *Refresh My Memory, Which One's Kyrgyzstan Again?* in which minor British comedians point at pictures of foreign people and laugh.

Top spot: Issyk-Kul, the second largest Alpine lake in the world (Titicaca is número uno, but I expect you knew that). **[13]**

Customs to treasure: Horses used to be the principal currency. If you wanted to buy something that wasn't worth a whole horse, you got your change in lambskins. **[11]**

Made in Kyrgyzstan: *The Epic of Manas*, a poem recalling the deeds of a chap called Manas as he defended the nation against the Uyghurs about a thousand years ago. With nearly half a million verses, this is not a poem for the faint-hearted, or indeed for anyone with anything less than a week to live. **[19]**

Pub fact: Kyrgyzstan is a veritable pot-pourri of ex/enclaves. There's the exclave of Barak (population 627), which is in Uzbekistan, and six enclaves: two of them biggish (the Uzbeki towns of Sokh and Shakhrimardan), two of them mediumy (Vorukh and Kairagach, both Tajikistani), and two of them piddly beyond all reason (the Uzbeki Chuy-Kara and Zhangail). **[15]**

National anthem intimating that life for the Kyrgyz forefathers was somewhat tiresome: 'Our fathers lived amidst the Ala-Toe/Continually saving their motherland.' **[12]**

SCORE: 70 WORLD RANKING: 22

Laos

ASIA

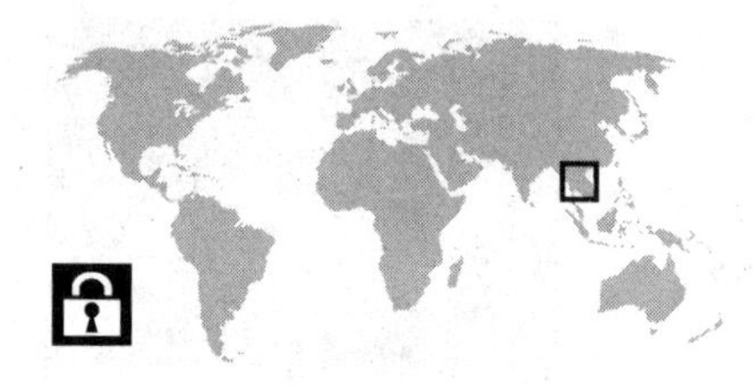

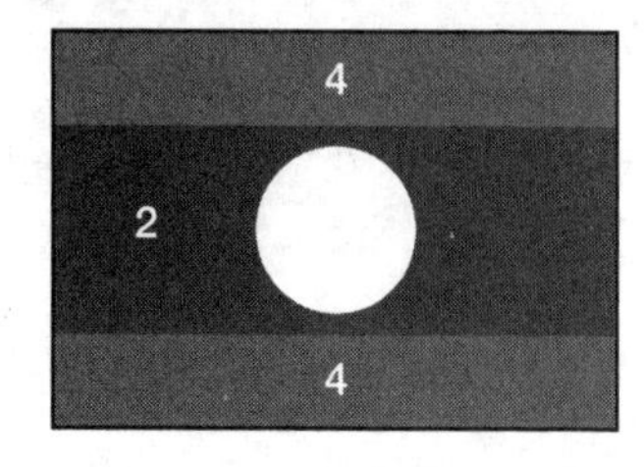

Name on driving licence: Lao People's Democratic Republic
Capital: Vientiane
Population: 6.1 million
Dosh: Kip = 100 at
Size: 236,800 km² (11 Wales)

Complete history: Compile a list of the world's communist countries and the chances are you'll overlook dear old Laos. However, it would be an unwise man who declared that Laos has always been a communist country, for it has not. Indeed, before 1975 it had never been a communist country. Instead there were French people, smoking Gauloises and introducing the locals to the concept of suave. Before them, there were many kingdoms and splittings and rejoinings thereof and Burmese invaders. It was really quite exciting.

Crowning achievement: By the time the Vietnam War had lurched to a halt, Laos had become the most heavily bombed country ever. This is quite something for a nation whose name didn't even get into the title of the war it was in. **[12]**

Popular misconception: 'Laos is pronounced "Laos"'. No, mister. Unlike in 'Slough', the 's' is silent. To remember this while visiting the country, use the phrase 'Come, friendly bombs and fall on Laos' every time you meet someone new until your visa is finally rescinded. **[14]**

Sad fact: Naga fireballs, often observed in the Mekong River, are believed to be produced by mythical serpent Phaya Naga, or by Laotian soldiers loosing off tracer bullets, which is just as romantic. **[15]**

Pub fact: Laos used to be part of Lan Xang, the 'Land of a Million Elephants'. This is approximately 998,500 more than Laos' current elephant population. **[2]**

Opening lines of national anthem: 'For all time the Lao people have glorified their Fatherland/United in heart, spirit and vigour as one.' **[8]**

SCORE: 51 **WORLD RANKING:** 130

colour key: 1 = turquoise 2 = blue 3 = green 4 = red 5 = yellow 6 = orange 7 = pink 8 = purple 9 = brown

EUROPE Latvia

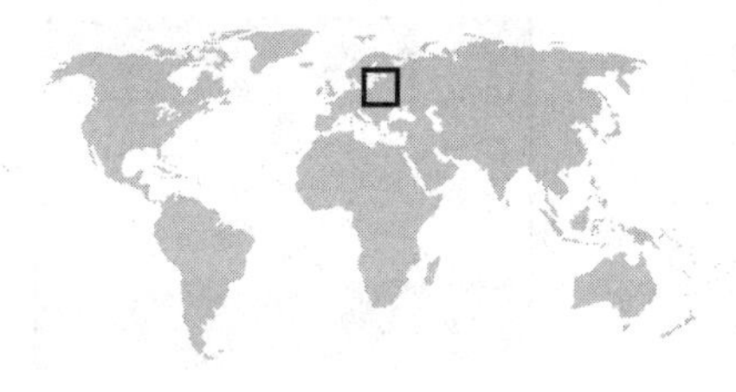

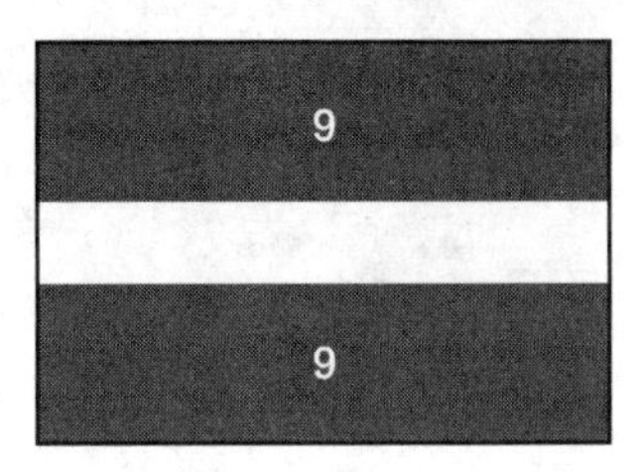

Name on driving licence: Republic of Latvia
Capital: Riga
Population: 2.3 million
Dosh: Latvian lat = 100 santimi
Size: 64,600 km² (3.1 Wales)

Complete history: Poor Latvia. The proto-Baltic state spent centuries under the Germans, got invaded by everyone else, and is now full of Russians.

Crowning achievement: Some people – I shan't name names – claim that having the world's fourth-largest pipe organ can in no wise be described as an achievement. Your mission is to find these knockers and jade-be-goods and explain to them, using force if necessary, that if they were the country with the world's fifth-biggest organ, having the fourth biggest would not only seem an achievement but an aspiration. **[20]**

Top spot: The mountain resort of Sigulda which, with an inevitability bordering on predestination, is known as the 'Switzerland of Latvia'. **[10]**

Customs to treasure: 'Good evening' is rendered in Latvian as '*Labvakar*!' This is fine at first but, to the English-attuned ear, it soon gives way to the impression that Latvians are continually registering their approval of each other's Trabant. **[12]**

Flag fact: Maroon is indisputably a brave choice for a flag, especially if you've nothing to distract the eye from it but a slim sliver of white. Naturally enough, it turns out that the colour has been chosen to represent the blood of some great Latvian hero from antiquity as it poured out onto a conveniently placed sheet. This is all well and good, but going by the resultant hue, he would surely have died from blood poisoning had his enemies not got to him first. **[13]**

National anthem explicitly devised to sound fantastic in English: 'Bless Latvia, O God/Our verdant native sod/Where Baltic heroes trod.' **[18]**

SCORE: **73** WORLD RANKING: **11**

Lebanon

MIDDLE EAST

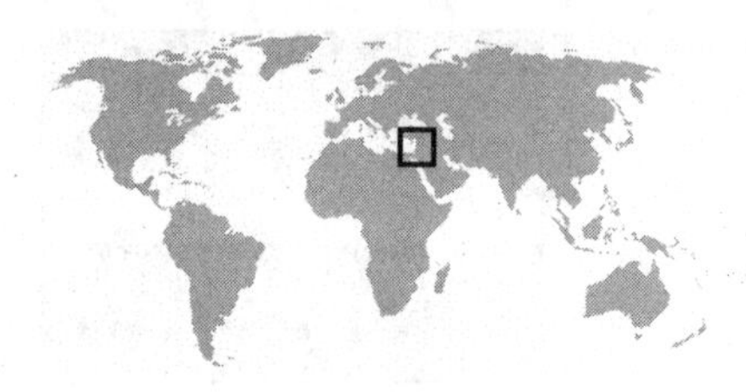

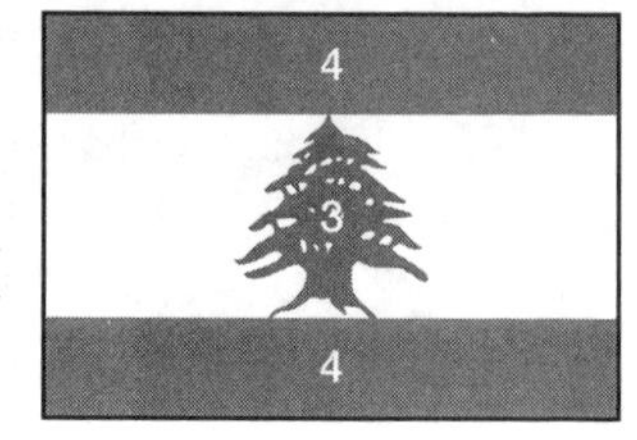

Name on driving licence: Republic of Lebanon
Capital: Beirut
Population: 3.8 million
Dosh: Lebanese pound = 100 piastres
Size: 10,400 km^2 (0.5 Wales)

Complete history: Those happy-go-lucky sea-faring Phoenicians were sadly trounced by eternal bully Alexander the Great, who burnt Tyre into the bargain. Next up were Persians, Romans, Byzantines, Arabs, Crusaders, Ottomans and Frenchies, all eager to possess what had been the Biblical 'land of milk and honey'. Lebanon has been independent since World War I, give or take three Israeli invasions and the not-so-brotherly attentions of Syria. Today, the country is never more than five minutes away from the next demonstration, with its unmistakable aroma of burning tyre, something of which Alexander would have approved, no doubt.

Sad fact: Lebanon's banking reputation caused it to be known as the 'Switzerland of the Middle East', while tourist-mecca Beirut became, somewhat confusingly, the 'Paris of the Middle East'. It's no wonder the country descended into civil war. **[13]**

Top spot: The Anti-Lebanon Mountains. Any nation sufficiently at ease with itself to acknowledge that its own mountains are against it can't be all bad. **[16]**

Gift to the world: The Cedar of Lebanon. **[14]**

Pub fact: Lebanese law is based on the Napoleonic Code, a sort of Da Vinci Code for grown-ups. **[9]**

National anthem: They've missed a trick here. Rather than going with The Human League's 1984 guitarfest chartbuster 'The Lebanon', which includes the timeless lines 'Before he leaves the camp he stops/He scans the world outside/And where there used to be some shops/Is where the snipers sometimes hide', they prefer something called 'Kalluna lil-watan lil 'ula lil-'alam' with its: 'All for the country, for the glory, for the flag.' All rather disappointing. **[4]**

SCORE: 56 **WORLD RANKING: 101**

AFRICA **Lesotho**

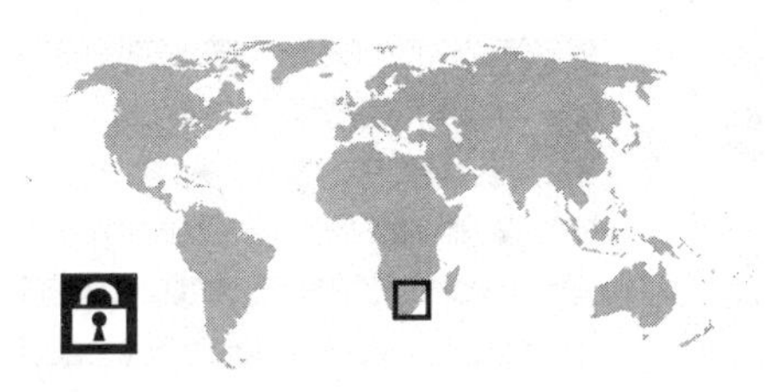

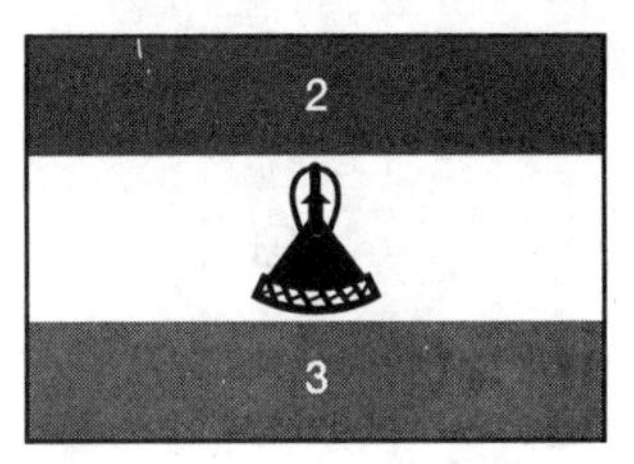

Name on driving licence: Kingdom of Lesotho
Capital: Maseru
Population: 1.9 million
Dosh: Loti = 100 lisente
Size: 30,400 km² (1.5 Wales)

Complete history: There's nothing like a bit of hunter-gathering to build up an appetite, so one can only imagine that the Khoisans were a hearty-eating bunch before they got shunted off the map by migrating Bantus, who enjoyed mountains more than almost anything, without necessarily feeling the need to eat them. However, it took ages before anyone thought to found a new country entirely within South Africa. Basutoland, as was, came about in 1822 and a mere 144 years later was free of British meddling. Drinks all round.

Crowning achievement: Large numbers of Lesotho's men travel to South Africa to work, leaving their womenfolk in charge of things. Lesotho has the highest female literacy rate in Africa (94 per cent). Read into this what you will. **[17]**

Pub fact: Lesotho is the only nation state above 1,000 metres in its entirety. Its lowest point is actually 1,400 metres, which smacks of showing off, I'm afraid. **[13]**

Made in Lesotho: The Basotho pony, a cross between a Javanese horse and a European full mount. There's probably some unicorn in there somewhere too, but no one talks about that. **[12]**

Actual motto: '*Khotso, Pula, Nala*' ('Peace, Rain, Prosperity'). It's not every country that includes a weather forecast with its national motto, but Lesotho, as their tourist board will happily inform you, is not every country. Indeed, sometimes it's barely any country and the South Africans have to send in their army to stop the rioting. **[14]**

Opening lines of national anthem: 'Lesotho, land of our Fathers/You are the most beautiful country of all.' **[5]**

SCORE: **61** WORLD RANKING: **70**

Liberia

AFRICA

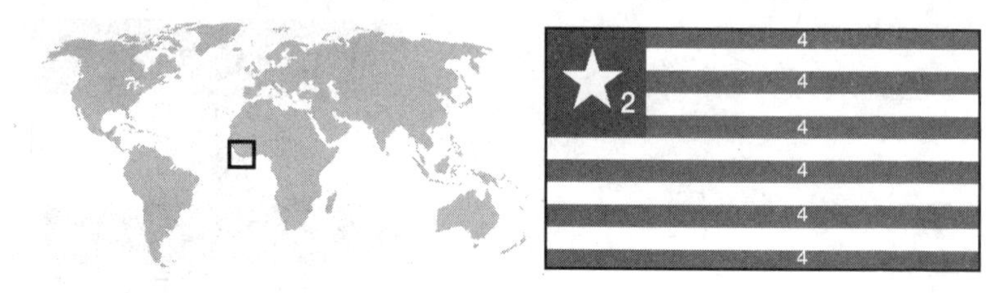

Name on driving licence: Republic of Liberia
Capital: Monrovia
Population: 3.4 million
Dosh: Liberian dollar = 100 cents
Size: 111,400 km² (5.4 Wales)

Complete history: The 'Land of the Free' was America's only proper stab at meddling in African affairs prior to the invention of the aircraft carrier. Freed black American slaves arrived in 1822 and promptly became slave-owners. More recently, Liberians have focused on civil war as a means of creating misery.

Crowning achievement: In November 2005, Liberia elected Africa's first-ever female head of state and the world's first black female president. If this were not enough, Ellen Johnson-Sirleaf also became the first African to have beaten a former Chelsea footballer (George Weah) in an election run-off. **[18]**

Top spot: Monrovia, twinned with Dayton (Ohio), home of Yugoslavian peace accords and the Incredible Hulk. **[12]**

Pub fact: There's around 1 billion tonnes of iron ore inside Mount Nimba, so when global warming really kicks in and we're reduced to some sort of subsistence living on a par with the Iron Age, Liberia will be in pole position to cash in. Until then, iron ore deposits will remain the Whitechapel or the Old Kent Road of natural resources – better than nothing, but you'd still rather come second in a beauty competition. **[10]**

Actual motto: 'The love of liberty brought us here.' With hindsight, they might have plumped for the rather more accurate: 'The love of poverty brought us here.' **[3]**

National anthem written by one untroubled by the daunting task of taking poetry to previously unscaled heights: 'This glorious land of liberty/Shall long be ours/Though new her name/Green be her fame/And mighty be her powers' **[17]**

SCORE: 60 **WORLD RANKING:** 76

colour key: 1 = turquoise 2 = blue 3 = green 4 = red 5 = yellow 6 = orange 7 = pink 8 = purple 9 = brown

Libya

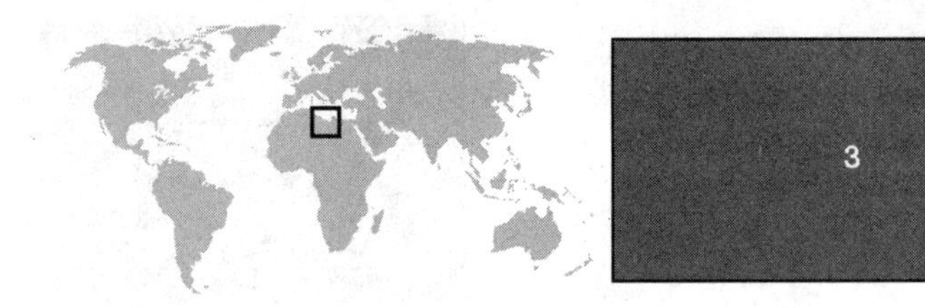

Name on driving licence: The Great Socialist People's Libyan Arab Jamahiriya
Capital: Tripoli
Population: 5.6 million
Dosh: Libyan dinar = 100 dirhams
Size: 1,759,500 km² (85 Wales)

Complete history: The Café Libya offers an unrivalled selection, including Berbers, Phoenicians, Carthaginians, Greeks, Romans, Vandals, Byzantines, Arabs, Ottoman-Turks, Italians (bless), Franco-Brits (disturbing but true) and Colonel Muammar Abu Minyar al-Gaddafi.

Crowning achievement: The Great Man-made River, which shunts water around Libya, is the world's largest underground network of pipes. With natural modesty, President Gaddafi has named it the 'Eighth Wonder of the World' (the Ninth Wonder being his continued hold on power). **[15]**

Top spot: The ruined city of Leptis Magna. Add the sound of a ballista bolt slicing through the rib cage of a Vandal, and you're back in the Roman Empire. **[16]**

Gift to the world: Couscous (actually, thank the downtrodden Berbers for this). **[13]**

Flag fact: It's difficult to imagine a flag greener than that of Libya. Indeed, to make a Libyan flag, all you need is some material in the shape of an oblong (a rectangle will do at a pinch) in the ratio 1:2. Now wave it. Everyone in Libya owns at least one green bedspread just in case President Gaddafi turns up unexpectedly at their door and demands some show of faithfulness to the cause. It is claimed that he has seen so many green bedspreads hung loyally out of windows that he no longer knows what the real flag looks like. **[20]**

National anthem: 'The enemy's army is coming/Wishing to destroy me/With truth and with my gun I shall repulse him/And should I be killed/I would kill him with me.' Steady on, chaps, they might just be tourists. **[2]**

SCORE: 66 **WORLD RANKING: 44**

Liechtenstein EUROPE

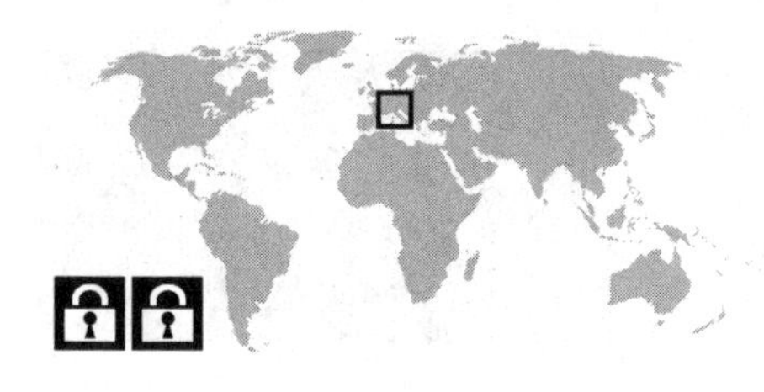

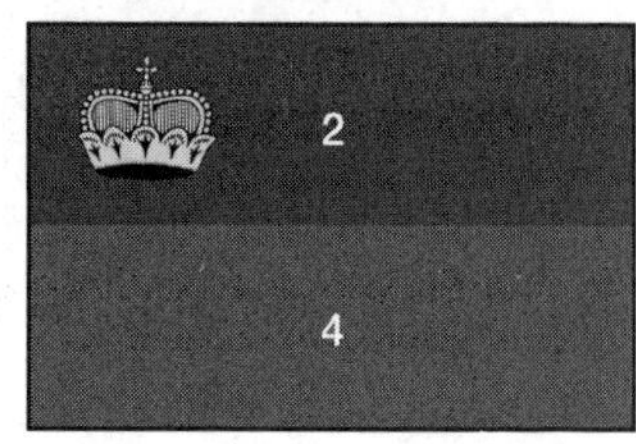

Name on driving licence: Principality of Liechtenstein
Capital: Vaduz
Population: 33,000
Dosh: Swiss franc = 100 centimes
Size: 160 km² (0.008 Wales)

Complete history: Liechtenstein is not so much a backwater of Europe as a minor spillage. Indeed, it has become so overlooked that no one outside Liechtenstein actually knows where it is, or whether it is in fact a fabled land, like Narnia, only with more frequent Christmases. Occasional news reports that the country has become extinct are only shown to be misleading when the Winter Olympics come round and the national team comes 22nd in the two-man bobsleigh. However, Liechtensteiners would have it no other way, gladly eschewing the modern lust for fifteen minutes of fame for a life of dignity, obscurity and, if at all possible, futility. (NB. Visitors to Germany should be aware that since 1992 it has been illegal for anyone to deny the existence of Liechtenstein.)

Crowning achievement: Liechtenstein owns a set of the world's strictest banking codes (viewing on request). **[10]**

Popular misconception: 'If Liechtenstein is anywhere, it's rammed up against Germany.' *Es ist nicht so*. It was last sighted squashed into an armpit formed by Switzerland and Austria. **[12]**

Sad fact: Women were given the vote as recently as 1984 (see also The Deeply Progressive State of Kuwait). However, since a hereditary royal family retains huge chunks of executive power, this is less of a treat than you might suppose. **[3]**

Pub fact: A rare example of a 'doubly landlocked' nation, in that it is bordered by countries that are themselves landlocked. **[10]**

Refreshingly indolent national anthem: 'High above the young Rhine/Lies Liechtenstein, resting.' The same tune is used by the British, which is nice. **[16]**

SCORE: 49 **WORLD RANKING: 140**

colour key: 1 = turquoise 2 = blue 3 = green 4 = red 5 = yellow 6 = orange 7 = pink 8 = purple 9 = brown

EUROPE Lithuania

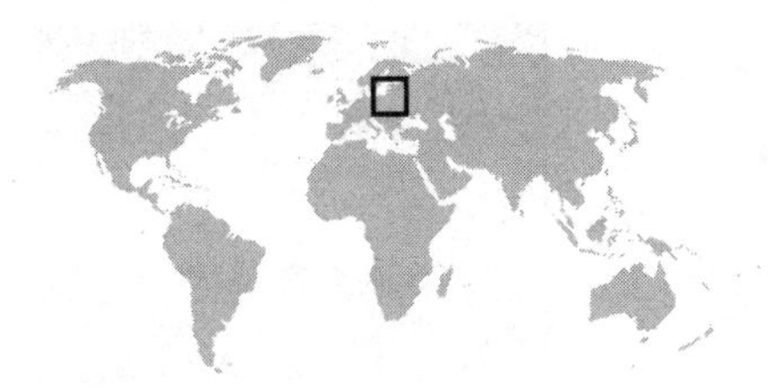

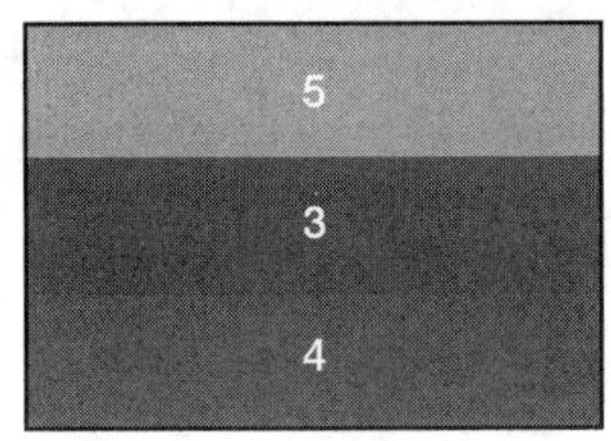

Name on driving licence: Republic of Lithuania
Capital: Vilnius
Population: 3.7 million
Dosh: Litas = 100 centu
Size: 65,200 km^2 (3.2 Wales)

Complete history: Lithuania may not look much now, but it used to stretch from the Black Sea to near Moscow. However, by 1915 it had become so diminutive that it was invaded by Germans, Bolsheviks, Bermontians, Poles, Soviets, Germans and Soviets, which must have been a shock.

Crowning achievement: Lest we forget, Lithuania was the first Soviet republic to declare its independence (go on, check your diary entry for 11 March 1990). **[16]**

Top spot: The Old Town of Vilnius is not only the largest Old Bit of a City in all of Lithuania but in the whole of Europe, even including really old manky cities of Europe, like Milan, where nothing gets torn down ever. **[15]**

Popular misconception: 'Amateur Radio Direction Finding is not practised in Lithuania.' Piffle and nonsense – the sport's winning combination of orienteering and the detection of signals from radio transmitters hidden in wooded areas is actually quite popular. **[13]**

Made in Lithuania: *The Manchurian Candidate* star Laurence Harvey (1928–73). His real name was Zvi Mosheh (Hirsh) Skikne, though confusingly he always claimed it was Laruschka Mischa Skikne. To make matters worse, he changed his name to Harry Skikne while growing up and then Larry Skikne when he moved to England, before finally settling somewhat ill-advisedly on Laurence Harvey. His daughter was Domino Harvey who, in the circumstances, had little choice but to start out life as a model and end it as a drug-addicted bounty hunter. **[8]**

Opening lines of national anthem: 'Lithuania, our homeland/You are a land of heroes.' But not, apparently, of humility. **[4]**

SCORE: 56 **WORLD RANKING: 101**

Luxembourg EUROPE

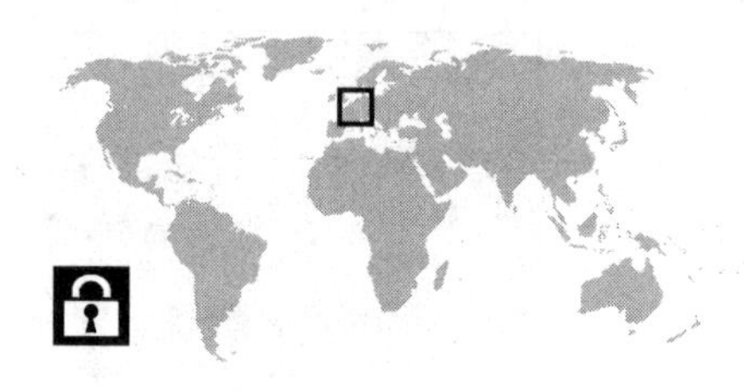

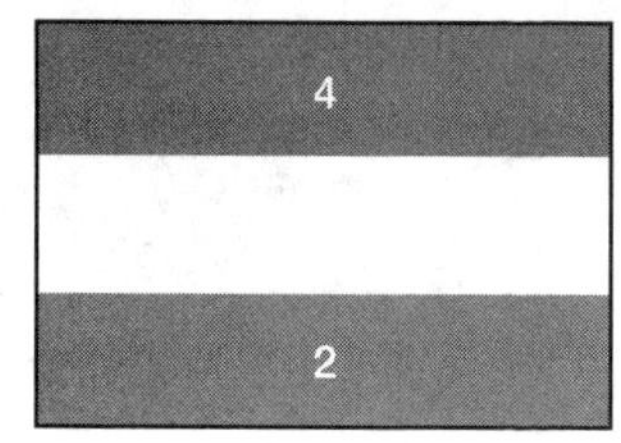

Name on driving licence: Grand Duchy of Luxembourg
Capital: Luxembourg
Population: 463,000
Dosh: Euro = 100 cents
Size: 2,600 km^2 (0.1 Wales)

Complete history: Often confused with the mythical nation of Liechtenstein, Luxembourg is in fact an almost entirely different country. Growing up somewhere near Belgium, the impressionable young state became a duchy in 1354, on the mistaken premise that duchyhood was what all the smart European countries were getting into. As it turned out, only Cornwall followed suit. Deflated, it succumbed to French rule, Habsburg rule and back to French rule. However, when the French carelessly lost the battle of Waterloo, the Netherlands seized the opportunity with both clogs and took over until 1890 when Queen Wilhemina succeeded to the Dutch throne. The wildly misogynous laws of Luxembourg did not allow for a female head of state and so the grand (by then) duchy felt obliged to break away. It's worth noting that since then only the Germans have bothered to invade.

Crowning achievement: Somehow providing most of the backdrop to the film *An American Werewolf in Paris*. **[8]**

Top spot: Müllerthal, a region popular with walkers, is known to locals as 'Little Switzerland', presumably because they haven't a single imaginative thought between them. **[7]**

Gift to the world: Radio Luxembourg. **[15]**

Pub fact: Luxembourgers enjoy far and away the world's highest gross national income per capita. It doesn't necessarily make them happy though. **[10]**

Opening lines of anthems: Who amongst us is worthy to choose between the competing merits of the royal anthem ('Two royal children truly in love/Became widely and deeply separated') and the national anthem ('Where the Alzette slowly flows/The Sura plays wild pranks')? **[18]**

SCORE: 58 **WORLD RANKING:** 91

EUROPE Macedonia

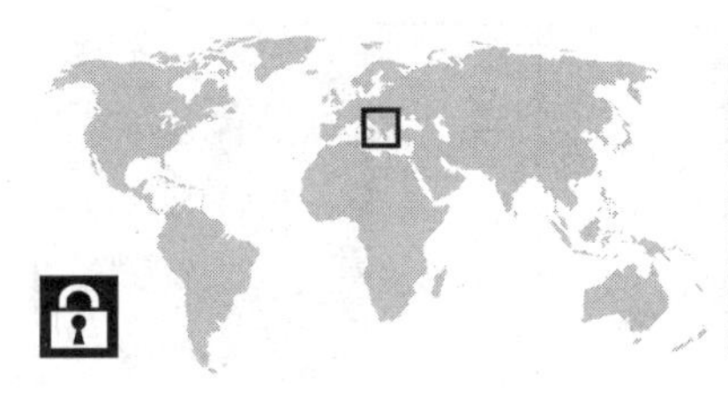

Name on driving licence: Republic of Macedonia
Capital: Skopje
Population: 2 million
Dosh: Macedonian denar = 100 paras
Size: 25,300 km² (1.2 Wales)

Complete history: Many years ago, in the days when a smart jerkin was considered essential attire for the upwardly mobile man, Bretagne (Britain) became Grande Bretagne (Great Britain) in order to delineate it from the other Bretagne (Britanny). In 1991, when Macedonia came into being, the inhabitants of the Greek region of Macedonia threw up their arms in horror. 'How will everyone know which Macedonia is which?' they cried (quite literally, this being Greece). The solution, of course, would have been to rename the (somewhat larger) Greek region Great Macedonia. Too simple. The Greeks and Macedonians would rather spend the next four centuries in a spiral of mutual hatred culminating in a spectacular bloodbath. Honour restored, the last surviving Greek and Macedonian embrace as brothers and everyone's happy again.

Top spot: Ohrid. Not at all 'orrid, whatever the hilarious English-speaking Greek comedians might tell you. **[13]**

Made in Macedonia: Alexander the Great (a name apparently not given to distinguish him from some other more mundane Alexander). **[12]**

Popular misconception: 'Mother Theresa came from Albania.' Get thee to a nunnery. She was born in Skopje. **[14]**

Flag fact: The Greeks (see above) managed to get the Macedonians to change their flag from one which featured the Vergina Sun (a sort of funky stylised star as used by Greek Macedonians) to a nice, big sunny design which is really much cooler, so that rather backfired on them. **[11]**

Opening lines of national anthem: 'Today over Macedonia/A new sun of freedom rises/Macedonians fight/For their rights/Macedonians fight/For their rights' to party. **[15]**

SCORE: 65 WORLD RANKING: 50

Madagascar AFRICA

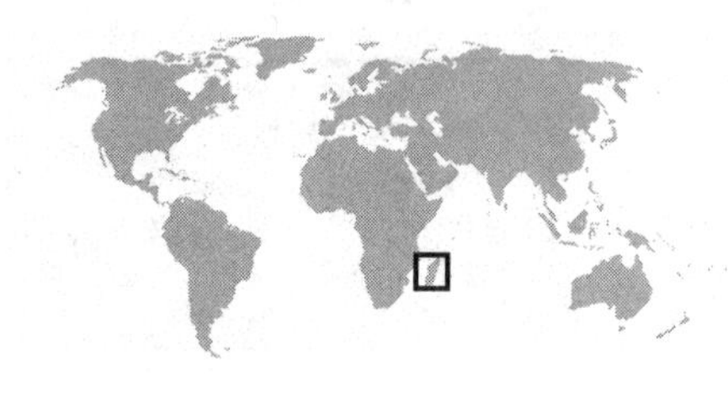

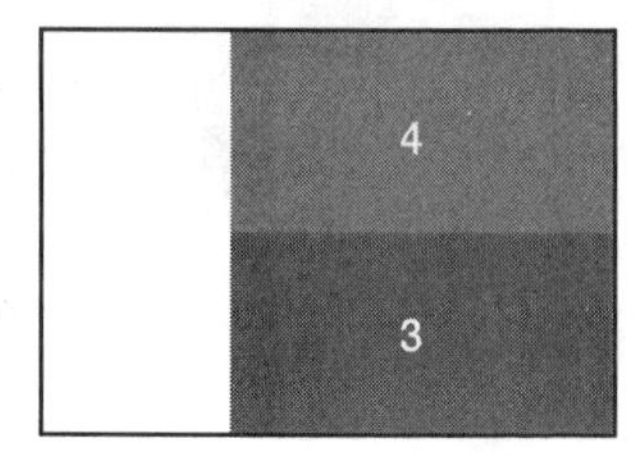

colour key: 1 = turquoise 2 = blue 3 = green 4 = red 5 = yellow 6 = orange 7 = pink 8 = purple 9 = brown

Name on driving licence: Republic of Madagascar
Capital: Antananarivo
Population: 17 million
Dosh: Malagasy franc = 100 centimes
Size: 587,000 km^2 (28 Wales)

Complete history: Madagascar, the world's fourth-largest island, was cast adrift from the rest of Africa some time ago. Don't bother asking even really ancient relatives about it because the whole rupture-rumble-shuffle-off-into-the-sea business would almost certainly have pre-dated them. They might remember the first Asian settlers, however, or the Europeans, pirates and army generals who followed them (most recent attempted coup November 2006).

Top spot: Stall holders at the Zuma Market in Antananarivo claim it's the second largest in the world which, in the circumstances, is remarkably constrained of them. After all, how many punters are likely to know what the largest market in the world is or even how big one's market has to be in order to be in with a chance of being the biggest? Not this young family coming along now with a baby-buggy full of vangavanga bracelets who are about to be kidnapped and sold into white slavery by an armed street gang, that's for sure. **[9]**

Made in Madagascar: Vangavanga bracelets, 100 per cent protection from bad luck guaranteed. **[10]**

Pub fact: Madagascar is known as the 'eighth continent' since an estimated 80 per cent of all its plant and animal life is unique to the island. Recent reports that such plant and animal life are simply better at hiding in other countries have been dismissed by government officials as 'nonsense'. **[18]**

Gift to the world: Vanilla. **[11]**

National anthem that allows no real hope for any development in the relationship between land and humanoid: 'Dear Madagascar/Our love for you will never change/But will be the same forever.' **[6]**

SCORE: 54 **WORLD RANKING: 114**

Malawi

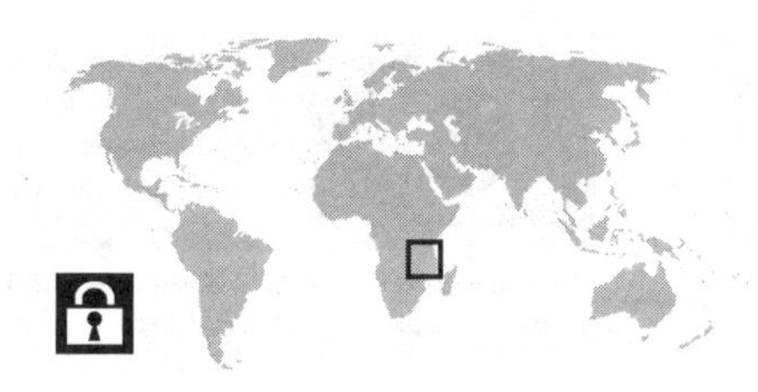

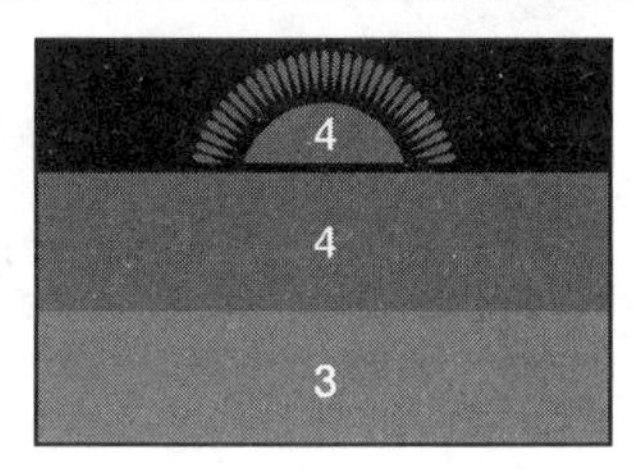

Name on driving licence: Republic of Malawi
Capital: Lilongwe
Population: 12 million
Dosh: Malawian kwacha = 100 tambala
Size: 118,500 km^2 (5.7 Wales)

Complete history: The former British Protectorate of Nyasaland became independent Malawi in 1964 and immediately found itself in thrall to Dr Hastings Banda, one of a select group of African presidents named after resort towns on the south coast of England (see also Swaziland's Bognor Msweko and Rwanda's Bournemouth Constitutional).

Crowning achievement: A fifth of Malawi is lake, which is impressive until you consider that a whole third of Emerson, Lake and Palmer was Lake and that they released more albums called *Brain Salad Surgery* than the whole of Malawi could muster. **[8]**

Made in Malawi: Tobacco, tea, sugar. Very nearly the breakfast of champions. **[11]**

Pub fact: The town of Livingstonia is named after missionary/explorer Dr David Livingstone, while Malawi's largest city Blantyre is a nod to his Scottish place of birth. Livingstone's faithful dog Chitane made the ultimate sacrifice, being eaten by crocodiles (or possibly just the one since poodles are quite small dogs) while crossing a lake and yet, perversely, there are no Malawian towns called Chitane, Poodle, or, as far as can be established, Snack. **[12]**

Top spot: Lake Malawi, which contains more types of fish than any other lake in the world (between 500 and 1,000 species, depending on your taxonomy and/or your desire to lure tourists). **[13]**

Opening lines of national anthem: 'O God bless our land of Malawi/Keep it a land of peace.' The anthem carries on in such a cheery manner that one can't help but feel that Malawians must be the most optimistic sort of folk whose jovial mien is seldom darkened by the grim shadows of gloomy dejection. **[15]**

SCORE: 59 **WORLD RANKING:** 85

Malaysia

ASIA

colour key: 1 = turquoise 2 = blue 3 = green 4 = red 5 = yellow 6 = orange 7 = pink 8 = purple 9 = brown

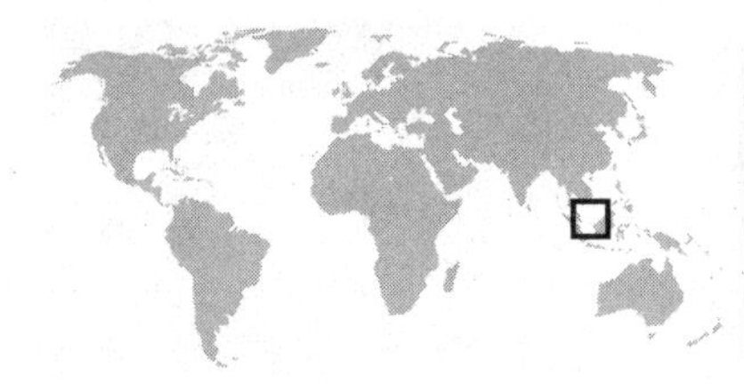

Name on driving licence: Malaysia
Capital: Kuala Lumpur
Population: 24 million
Dosh: Ringgit = 100 cents
Size: 329,800 km^2 (16 Wales)

Complete history: A sloppy country, Malaysia has draped itself across all manner of bits and pieces of South-East Asia without the slightest consideration for the finer points of aesthetics or the feelings of cartographers. Understandably, having blundered so comprehensively on the geographical front, it has not been trusted with any history.

Made in Malaysia: The orang-utan – the Malay for 'person of the forest' – has been shown to share 98 per cent of its DNA with humans, but only if they ask nicely. **[19]**

Sad fact: The (elected) King of Malaysia rejoices in the name Duli Yang Maha Mulia Al Wathiqu Billah, Al-Sultan Mizan Zainal Abidin Ibni Almarhum Al-Sultan Mahmud Al-Muktafi Billah Shah Al-Haj. Since this is a bit of a mouthful, particularly if you just want him to pass the cheese, those in his most intimate circle are apparently permitted to miss off the Al-Haj. **[12]**

Popular misconception: 'Malaysia is so called because those who live there for too long suffer a general malaise.' Not so. It is usually only a partial malaise which can often be put down to the over-consumption of satay. **[14]**

Pub fact: There's some controversy over whether Kuala Lumpur's Petronas Twin Towers (452 metres) was once the world's tallest building. Thankfully, the furore revolves around arcane arguments over whether spires and antennae count or not. It is, however, certain that if you drop a penny from the top floor, an urban myth cannot be far behind. **[10]**

Opening lines of national anthem: 'My country, my native land/The people living in unity and progress' (and malaise). **[8]**

SCORE: 53 WORLD RANKING: 119

ASIA Maldives

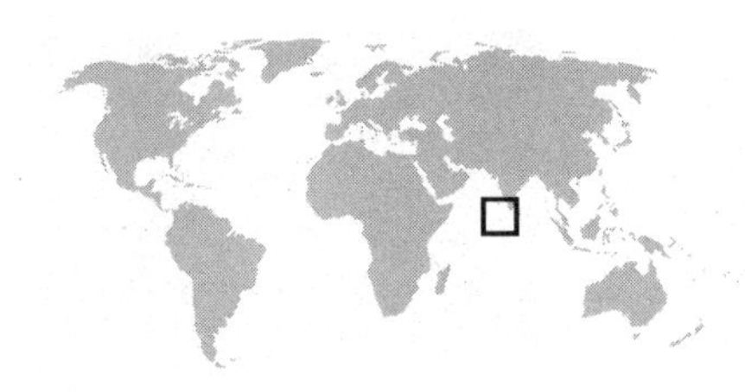

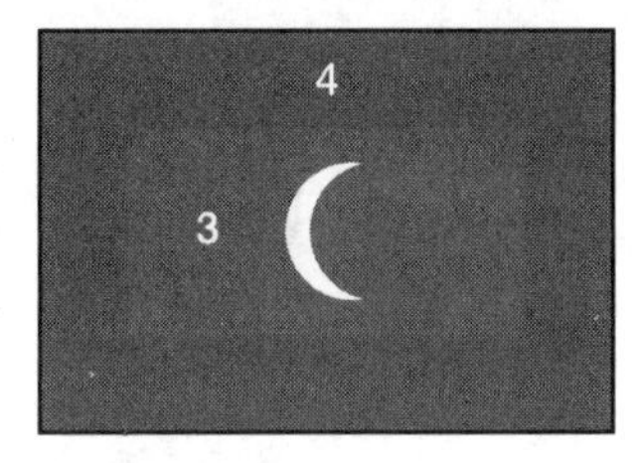

Name on driving licence: Republic of the Maldives
Capital: Malé
Population: 339,000
Dosh: Rufiyaa = 100 laari
Size: 300 km^2 (0.02 Wales)

Complete history: There are some folk – embittered types from places that get too much mist – who would contest that 1,196 coral islands do not a nation make, no matter how exuberantly populated they are. All the more reason then to overcome bitterness by acquainting yourself with the bestrewn nation's history.

Top spot: It's unlikely that the Maldives will be with us much longer given that it's the flattest nation on Earth with its highest point just over 2 metres above sea level, though a government-led plan to get every foreign tourist to donate a stilt may extend the country's existence for another 50 years with practice. **[3]**

Crowning achievement: It's Asia's smallest independent nation, and it's getting smaller every day. **[15]**

Pub fact: One of the country's islands is set aside for the detention and torture of political prisoners. **[0]**

Sad fact: Each year, one out of every hundred Maldivians gets divorced, giving the islands the highest divorce rate in the world by a country mile. This is due in large part to an adherence to Sharia Law under which a husband only has to say the words 'I divorce you' three times for the marriage to be history. Getting married four or more times has thus become the norm for the majority of Maldivians, making it akin to living in a land entirely composed of Elizabeth Taylors and Mickey Rourkes. Attractive. **[0]**

Valiantly didactic national anthem: 'We salute the colours of our Flag; Green, Red and White/Which symbolise Victory, Blessing and Success.' **[10]**

SCORE: 28 WORLD RANKING: 193

Mali

AFRICA

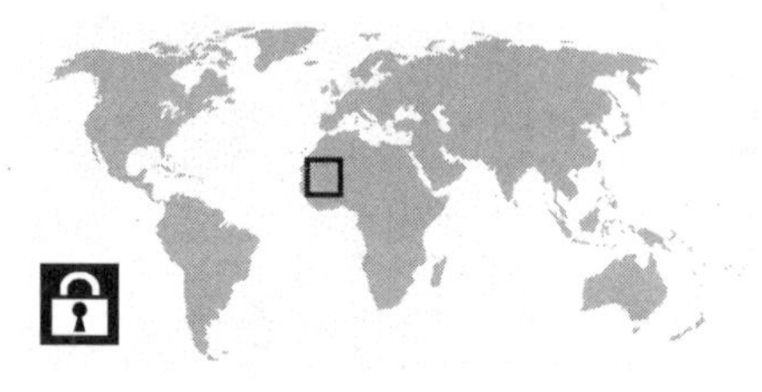

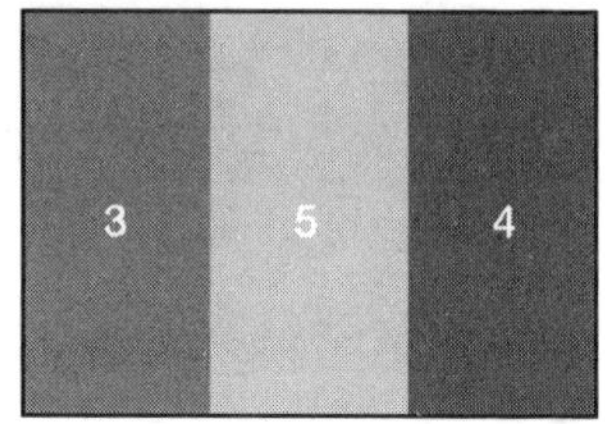

colour key: 1 = turquoise 2 = blue 3 = green 4 = red 5 = yellow 6 = orange 7 = pink 8 = purple 9 = brown

Name on driving licence: Republic of Mali
Capital: Bamako
Population: 12 million
Dosh: CFA franc = 100 centimes
Size: 1,240,000 km^2 (60 Wales)

Complete history: Maybe one day all countries will look like a cartoon dog with an oversized head scooting off over the desert sands. Until then, we'll have to content ourselves with Mali, whose comical snout juts hilariously into the Sahara Desert, no doubt in search of some uproarious capers at the expense of a hapless dromedary. Mali was first discovered by British explorer Damon Albarn, but some believe the nation's history may go back even further.

Crowning achievement: The only country in the world that is an anagram of another country's capital city. They're still pretty proud of that down Bamako way, but talk to people about it on the streets of Lima and, astonishingly, you'll be met with looks of bewildered incomprehension. **[15]**

Top spot: Timbuktu is still reassuringly remote and on the way to nowhere in particular (unless you happen to be cruising east along the Niger to the town of Gao, but you're quite clearly not). **[17]**

Pub fact: Djenné's Grand Mosque is the largest building made of mud on the planet and it has turrets and everything like a so-called 'normal' mosque would. Even more remarkable is that during the rainy season the town (boldly claimed to be the oldest city in West Africa by its measly 12,000 inhabitants) becomes an island and yet the mosque does not turn into a pile of less than holy slurry. **[14]**

Made in Mali: Sorghum, best-loved ingredient of gluten-free beer. **[12]**

National anthem lamenting the country's half-constructed football pitches: 'One people, one goal.' **[14]**

SCORE: 72 WORLD RANKING: 14

Malta

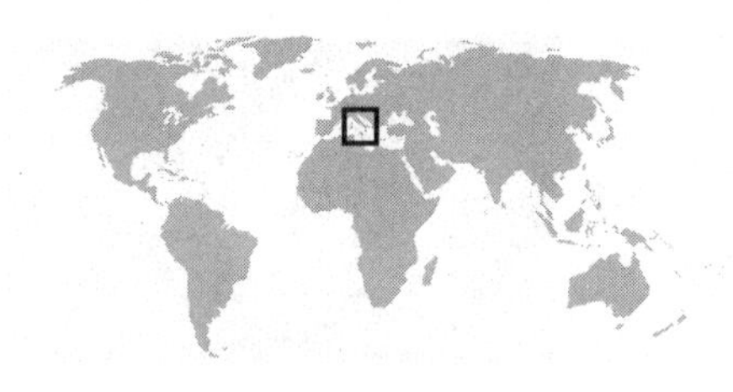

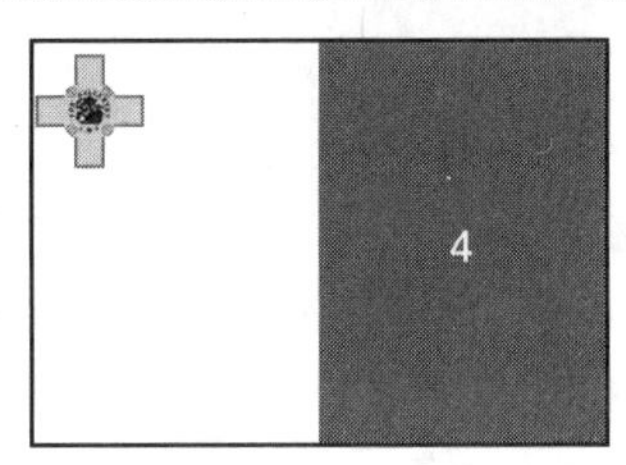

Name on driving licence: Republic of Malta
Capital: Valletta
Population: 397,000
Dosh: Maltese lira = 100 cents
Size: 320 km^2 (0.02 Wales)

Complete history: 'Tis a pity that the history of an island inhabited since the Stone Age should have been condensed to a pub-quiz question relating to the giving of the George Cross, but then posterity is a pitiless master at the best of times.

Flag fact: Malta is not only the sole country to win a George Cross, but the only one to have one on its flag, which is a bit of a coincidence. **[16]**

Crowning achievement: If you think Barbados is overpopulated – and you obviously do, given your 'Take people off Barbados, it's too crowded' T-shirt – how about Malta? 1,241 people cram themselves into each square kilometre, which gives each citizen very little room to manoeuvre. This has inevitably led to the adoption of protective clothing by all but the most foolhardy. This is worn day and night all year round but for the 'Day of Bruises' (25 February), one of the most unintentionally violent public holidays in Europe. **[10]**

Popular misconception: 'Maltesers are made of crushed up bits of Maltese people covered with chocolate.' Happily not. This practice was discontinued in the 1970s. **[12]**

Made in Malta: The Maltese dog is the result of a cruel and unnecessary cross between a Jack Russell, My Little Pony and a toothbrush. If nothing else, it acts as a stark warning of what can happen when scientists are let off the leash. **[3]**

National anthem that gave rise to the canard that all the Maltese are called Malta: 'To this sweet land, our mother/To which we owe our name.' **[14]**

SCORE: 55 **WORLD RANKING:** 110

Marshall Islands OCEANIA

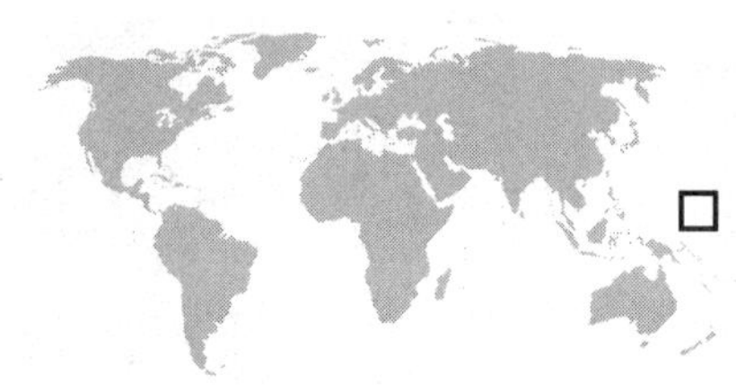

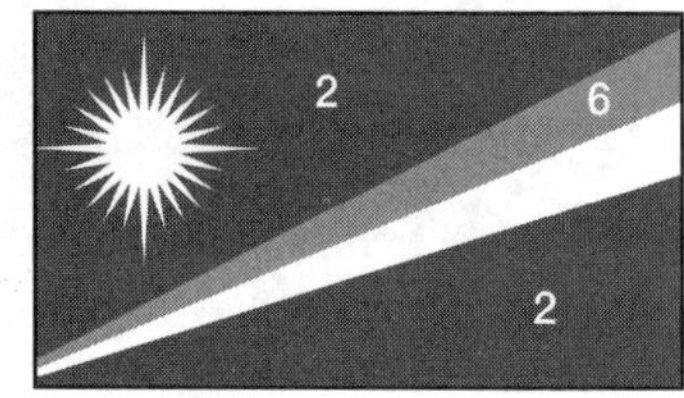

Name on driving licence: Republic of the Marshall Islands
Capital: Majuro
Population: 58,000
Dosh: US dollar = 100 cents
Size: 180 km² (0.009 Wales)

Complete history: Micronesians had been skulking around these coral atolls and islets for millennia before Alonso de Salazar spotted them in 1526. Such was the excitement generated by this event that it was only another 266 years before another European – Britain's John Marshall – had a shufti. There followed annexations by Germany, Japan (WWI) and the US (WWII). The islanders became independent in 1990 to give them something to do while awaiting World War III.

Gift to the world of Mexican fishing: Three Mexican fishermen who got lost at sea in September 2005 eventually rocked up here in August 2006. **[15]**

Made in the Marshall Islands: Breadfruit. It's a fruit. It tastes like bread. **[11]**

Sad fact: Bikini Atoll was the site of Castle Bravo, sadly not a fortress manned entirely by the opera-going classes but the detonation by the US of a thermonuclear device, 1,000 times more powerful than the Hiroshima bomb. The explosion vaporised three islands, while the fallout killed one Japanese fisherman, caused birth defects on neighbouring atolls, and has made Bikini uninhabitable ever since. Good work, chaps. **[0]**

Pub fact: The bikini is named after Bikini, since the beachwear is apparently just as explosive as the atoll's thermonuclear experiments. However, the uptake of the two-piece swimsuit as a deterrent against pre-emptive nuclear strikes has been slow, despite the obvious cost savings. **[12]**

National anthem overly influenced by 18th-century Scottish romanticism: 'My island lies o'er the ocean/Like a wreath of flowers upon the sea/ ... /Our Father's wondrous creation', O bring back my Bonny to me. **[9]**

SCORE: 47 **WORLD RANKING:** 150

AFRICA Mauritania

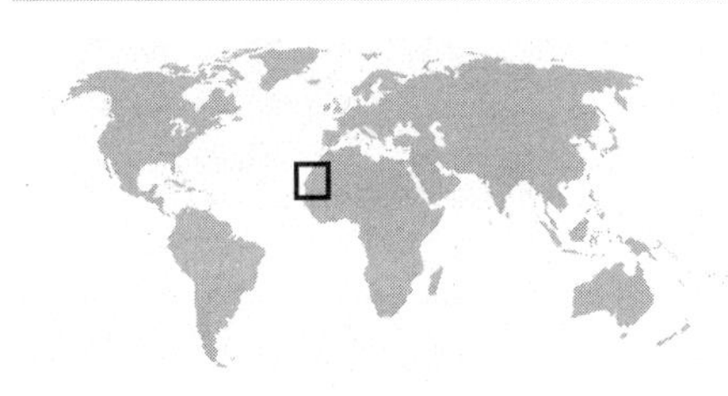

Name on driving licence: Islamic Republic of Mauritania
Capital: Nouakchott
Population: 3 million
Dosh: Ouguiya = 5 khoums
Size: 1,030,700 km² (50 Wales)

Complete history: Mauritania is a barren place, your honour, where the population gravely ponders the dunes, and 'twas ever thus.

Crowning achievement: Attempting to have a country here at all. **[10]**

Top spot: The capital, Noaukchott, where, according to travel writer Patrick Marnham: 'Sweeping the sand into heaps is about the only source of steady employment that the city can offer.' **[5]**

Sad fact: Slavery was made illegal in 1980, but no one seems to have told the slave-owners. **[0]**

Made in Mauritania: Gypsum. There's more here than anywhere else. Let joy break out. **[6]**

National anthem that immediately marks the country out as a fun place to be: 'Be a helper for God, and censure what is forbidden/And turn to the law which He wants you to follow/Consider no one useful or harmful, except for Him/And walk the path of the chosen one, and die while you are on it!/For what was sufficient for the first of us, is sufficient for the last one too/And leave those people who do evil things with respect to God/They misrepresented him by making him similar and made all kinds of excuses/They made bold claims and blackened notebooks/They let both the nomads and the sedentary people have bitter experiences/And the great sins of their [doctrinal] innovations bequeathed small/And just in case a disputant calls you to dispute with regard to their claims/Do not, then, dispute with them, except by way of an external dispute.' Possibly the strangest national anthem ever written, and as such is reproduced here in all its fascinating entirety. **[20]**

SCORE: 41 **WORLD RANKING: 174**

Mauritius

AFRICA

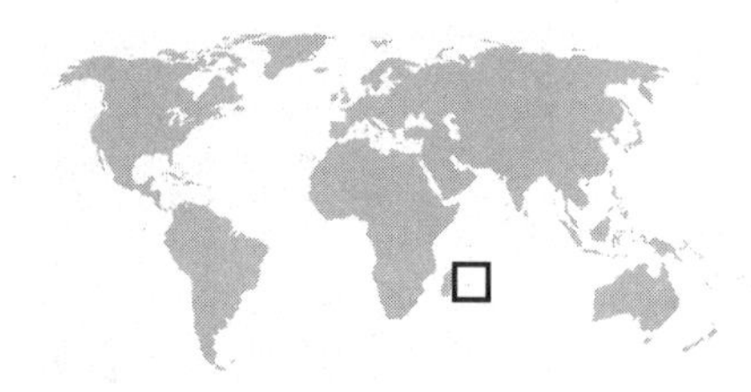

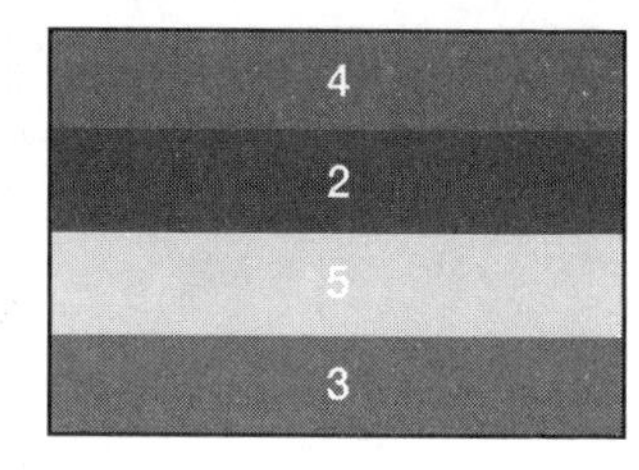

colour key: 1 = turquoise 2 = blue 3 = green 4 = red 5 = yellow 6 = orange 7 = pink 8 = purple 9 = brown

Name on driving licence: Republic of Mauritius
Capital: Port Louis
Population: 1.2 million
Dosh: Mauritian rupee = 100 cents
Size: 1,900 km² (0.1 Wales)

Complete history: Thrust a globe in front of passengers on the upper deck of the Clapham omnibus and ask them where Mauritius is and the chances are you'll end up with a lot of smudgy fingerprints around Algeria, in the mistaken belief that it's Mauritania.

Crowning achievement: The island was uninhabited when the Dutch claimed it (1598), so, for once, no one had to be slaughtered. **[13]**

Pub fact: The dodo's home until the Dutch did for it. Cruelly tagged *Didus ineptus* by Swedish botanist Linnaeus, the bird's fearlessness of humans proved to be its undoing (a lesson for us all there). The Mauritian Coat of Arms sports a 'dodo rampant', which is as grimly ironic as a symbol gets. **[3]**

Sad fact: The seeds of the tambalacoque (or dodo tree) were believed to germinate only after having passed through a dodo. However, in 1977, biologist Stanley Temple fed them to turkeys and got three seeds to grow. Two earlier biologists had apparently managed to germinate the seeds without any avian influence, but then where's the fun in that? **[10]**

Flag fact: The Mauritian flag is unique in featuring four equal bands of colour, as opposed to one colour or three (although countries known for their inexhaustible pursuit of wackiness, e.g. Switzerland, may plump for two). If you're someone who doesn't interact with flags on a strictly need to know basis: red = blood; blue = the ocean; yellow = a bright future; green = greenery. **[12]**

Opening lines of national anthem: 'Glory to thee, Motherland/O motherland of mine.' **[8]**

SCORE: 46 **WORLD RANKING:** 157

Mexico

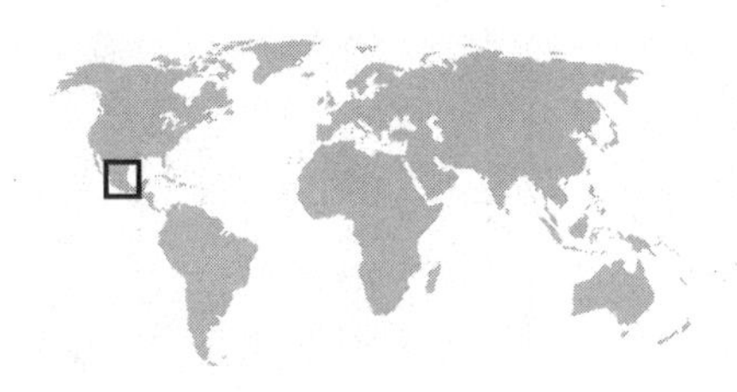

Name on driving licence: Republic of Mexico
Capital: Mexico City
Population: 105 million
Dosh: Mexican peso = 100 centavos
Size: 1,972,500 km² (95 Wales)

Complete history: All Mexicans believe their country to be a sleeping giant that will become great again some day. That day will coincide with the US returning the stolen territories of Arizona, California, Colorado, New Mexico, Nevada, Texas and Utah, so no one's holding any breath.

Top spot: Acapulco. Only a town of some merit could survive the ignominy of The Four Tops' 1989 disco smash 'Loco in Acapulco', with its rather alarming lyrics ('You can hear voices bleeding through those warm Latin nights' – what's *that* all about?). **[13]**

Customs to treasure: Every December, Oaxaca celebrates La Noche del Rábano (The Night of the Radish) in which innumerable stalls are laden with large radishes. People come from miles around to look at the radishes and be happy. **[19]**

Sad fact: More people emigrate from Mexico than anywhere else. This is not necessarily because Mexico is a horrible place – although they could probably do with fewer mariachi bands – but because it shares a 3,140-km border with the US, a country with marginally better cable television. **[5]**

Flag fact: In 2005, Sergio Witz was convicted of the criminal offence of 'insulting national symbols' after publishing a poem in which he imagined using the Mexican flag in lieu of toilet paper. **[2]**

National anthem: Alas, the good intentions of the opening lines of the anthem ('May the divine archangel crown your brow/Oh fatherland, with an olive branch of peace') have somewhat died a death by the second verse ('War, war without truce ... /The patriotic banners are saturated in waves of blood'). **[5]**

SCORE: 44 **WORLD RANKING: 167**

Micronesia OCEANIA

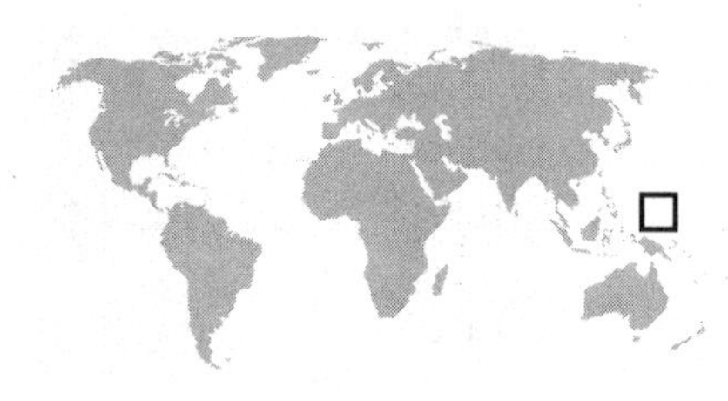

Name on driving licence: Federated States of Micronesia
Capital: Palikir
Population: 108,000
Dosh: US dollar = 100 cents
Size: 700 km^2 (0.03 Wales)

Complete history: When tackling the Federated States of Micronesia it's best to discover first what it isn't. According to the most reliable sources, it is *not* a fictional country dreamed up by Kafka (although 'When Josef K. came to he was aware only of the vast cliffs of Micronesia towering above him with the certainty of doom' does have a ring to it); or, for that matter, a medical condition in which the sufferer is apt to forget really small things. What it appears to be is a collection of 607 volcanic islands and coral atolls on which not much happened, then a lot happened in February 1944, then not much happened again.

Pub fact: Micronesian All-Around is a sort of Micronesian pentathlon comprising such varied disciplines as 'spear', 'swim', 'dive', 'husking' and perhaps most unnervingly, 'grating'. **[17]**

Sad fact: Pohnpei islanders are prone to suffer from an extreme form of colour blindness. **[8]**

Customs to treasure: As if it were not pleasure enough for the state of Yap to be so called, they also still use stone currency. The larger denominations can weigh up to five tonnes, which is why they tend to stay where they are, even when they change hands. However, don't try smuggling your own stones into Yap, since they are almost certain to be identified as counterfeits unless you really know what you're doing. **[19]**

Made in Micronesia: The seventeen Pohnpei islanders who resisted German rule in 1911 and were executed for their pains. **[20]**

Opening lines of national anthem: 'Tis here we are pledging/With heart and with hand.' **[10]**

SCORE: 74 **WORLD RANKING: 10**

EUROPE Moldova

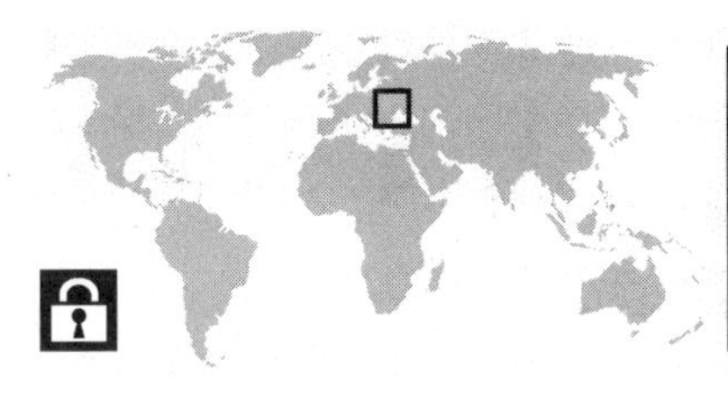

Name on driving licence: Republic of Moldova
Capital: Chisinau
Population: 4.4 million
Dosh: Moldovan leu = 100 bani
Size: 33,800 km^2 (1.6 Wales)

Complete history: Throughout the centuries, Moldova has allowed itself to be associated only with the grooviest names in the encyclopedia. Dacians, Kievan Rus, Mongols and Vlachs; Transnistria, Bessarabia, Mohyliv-Podilskyy and Prut. These are the sort of names that bring greatness upon a nation. It's just a pity it has such a poor human rights record.

Crowning achievement: Moldova has a liberal democracy so strong that the electorate has felt able to vote the Communist Party back in. If that's not a lesson in something or other, I don't know what is. **[15]**

Pub fact: The Moldovans call their language Moldovan. The Romanians call *their* language Romanian. The two languages are the same. If that's not a lesson in something or other, etc., etc. **[8]**

Made in Moldova: Zdob si Zdub, a rare example of a hardcore punk/hip-hop/Roma crossover collective to have shared a stage with the Red Hot Chili Peppers and Rage Against The Machine, only to blow it by performing a song at the 2005 Eurovision called 'Boonika Bate Toba' ('Grandmama is beating da drum-a'). **[14]**

Sad fact: In 2006, Russia banned the import of Moldovan wine, on the grounds that 60 per cent of all bottles tested contained pesticides. Presumably, given Russian taste in alcoholic beverages, this figure was not deemed high enough. **[11]**

Opening lines of national anthem: 'A treasure is our tongue that surges/From deep shadows of the past.' During the course of the anthem the people's tongues turn into burning flames, the greenest leaves, and something 'more than holy'. There's rarely a dull day in Moldova. **[15]**

SCORE: 63 **WORLD RANKING:** 61

Monaco EUROPE

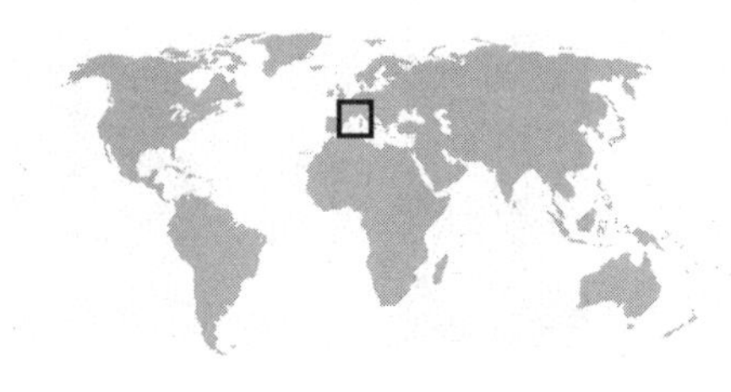

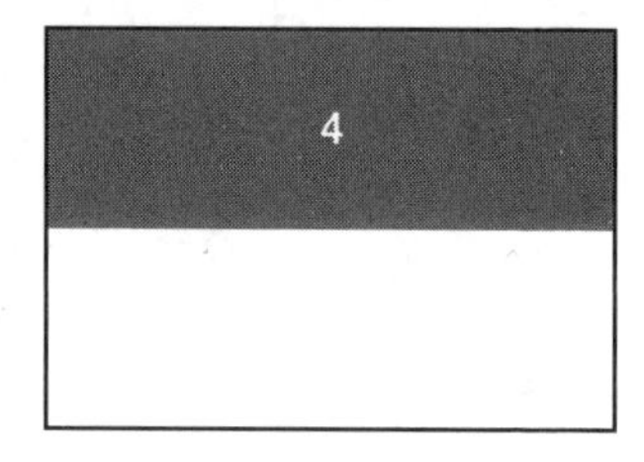

Name on driving licence: Principality of Monaco
Capital: Monaco
Population: 32,000
Dosh: Euro = 100 cents
Size: 1.95 km² (0.00009 Wales)

Complete history: Not so much a nation as a playground, Monaco is a glorious throwback to the days of city states and the sort of frenzied instability that nurtured the Renaissance. The principality has been ruled by the Genoese Grimaldi family since 8 January 1297, when François Grimaldi (the Cunning) disguised himself as a Franciscan monk to gain entry for his soldiers to the Rock of Monaco fortress.

Crowning achievement: Despite being the second-smallest nation on the planet, Monaco is still nearly five times the size of the Vatican City and has almost as many popes. **[14]**

Customs to treasure: The super-rich, for the love of large expensive watches, permatans and heart-shaped swimming pools, move to Monaco with tiresome regularity. As a result, the principality now boasts the world's greatest concentration of millionaires, so it must be frustrating for them when they discover the country doesn't possess a single golf course. **[3]**

Pub fact: Forget the moaning and groaning of Malta and Barbados, Monaco is the world's most densely populated nation. Each square kilometre has 16,410 people packed into it, giving each one a meagre 61 m² or just enough room for a luxury gold-tapped marble bathroom suite. **[6]**

Made in Monaco: The Monaco Grand Prix, more a speeded-up traffic queue than a race. **[2]**

National anthem guaranteed to get that old pulse racing: 'Forever, above our country/The same flag has flown in the wind/Forever the colours red and white/Are the symbol of our freedom/The great and the small have always respected them.' **[4]**

SCORE: 29 **WORLD RANKING: 191**

Mongolia

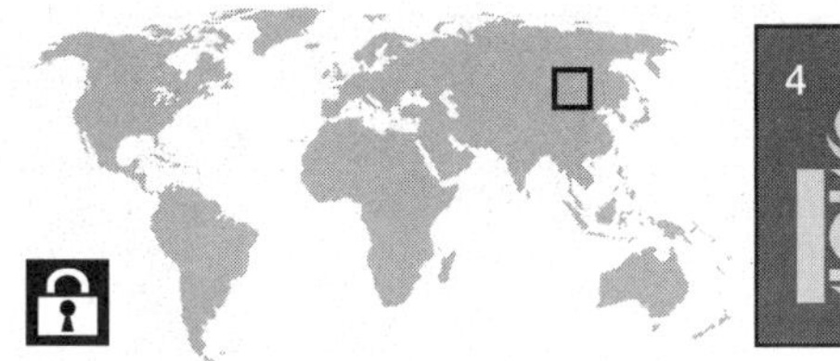

Name on driving licence: Mongolia
Capital: Ulan Bator
Population: 2.8 million
Dosh: Tugrik = 100 möngös
Size: 1,565,500 km^2 (76 Wales)

Complete history: Genghis Khan unified Mongolia in 1206 and made Karakorum the first capital of a Mongol empire which soon covered most of Asia. This was a good start, but any country squeezed between Russia and China was not going to last long. Manchurian China hoovered up both Inner and Outer Mongolia, the latter only escaping in 1911 and becoming communist, thus saving the Soviets the bother of invading.

Crowning achievement: The Turks seem to be rather proud of how Istanbul was once Constantinople and before that Byzantium, and are apt to go on about it at length over a tiny cup of foul-tasting coffee. However, Ulan Bator started life as Urga before changing to Khuree, Ikh Khuree, Kulun, Niislel Khuree and Ulaanbaatar. Even when invited to talk about the subject on chat shows, the humble Mongols maintain a dignified silence, which makes for restful television. **[15]**

Top spot: The Gobi Desert, home to the Gobi bear, the Asiatic wild ass, the Przewalski horse, the elusive snow leopard, and the wild Bactrian camel. **[18]**

Sad fact: The feared *zud*, which brings snowstorms so extreme that cattle drown in them. **[0]**

Gift to the word: *Suutei tsai*, a sort of milky tea with salt in it. **[6]**

National anthem that possibly over-eggs the pudding: 'The fortunate people of brave Mongolia have freed themselves from suffering/And enjoy happiness, the pillar of delight.' (However, the writer, Tsendiin Damdinsüren, made the four lines of the chorus and of each verse begin with the same letter (in the original), which at least shows some initiative.) **[10]**

SCORE: 49 **WORLD RANKING:** 140

Montenegro EUROPE

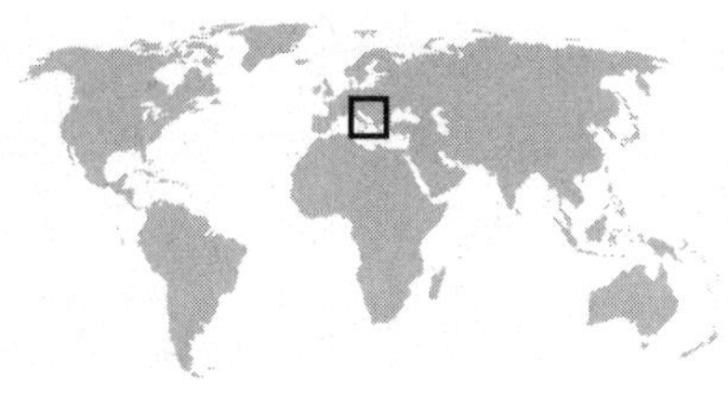

Name on driving licence: Republic of Montenegro
Capital: Podgorica
Population: 650,000
Dosh: Euro = 100 cents
Size: 13,800 km² (0.7 Wales)

Complete history: The first people to recognise the Republic of Montenegro were the Icelanders, who presumably don't have a great deal else to do but scan the world for nascent nation states and who are, no doubt, secretly thrilled about calls already being made by Serbs, Bosniaks and Albanians for their own states within Montenegro.

Crowning achievement: Kotor is the deepest fjord in southern Europe, though one feels compelled to ask what the Montenegrins are doing with one at all. As any fifteen-year-old Norwegian will tell you, fjords are created by the marine inundation of glaciated valleys, an event that is almost entirely the preserve of the northern hemisphere. This makes the Montenegrins' accomplishment all the more astonishing and accounts for Serbia's reluctance to let such a remarkable people go. **[13]**

Sad fact: The Montenegrins are the tallest people in Europe. This is apparently due to their being a somewhat remote bunch who don't get many visitors to their rather select gene pool. It's not polite to call them inbred, however. **[7]**

Customs to treasure: The Oro. A bracingly dangerous folk dance that requires participants to stand on each other's shoulders. **[15]**

Pub fact: The Serbia and Montenegro football team recorded a first at the 2006 World Cup by competing for a country that had ceased to exist. They lost all three games, including a 6-0 pasting by Argentina during which their defence also disappeared, presumably as a symbolic gesture. **[10]**

National anthem reflecting the Montenegrins' love of a good echo: 'O the bright dawn of May/O the bright dawn of May/Montenegro our mother/Montenegro our mother.' **[12]**

SCORE: 57 **WORLD RANKING: 97**

colour key: 1 = turquoise 2 = blue 3 = green 4 = red 5 = yellow 6 = orange 7 = pink 8 = purple 9 = brown

AFRICA Morocco

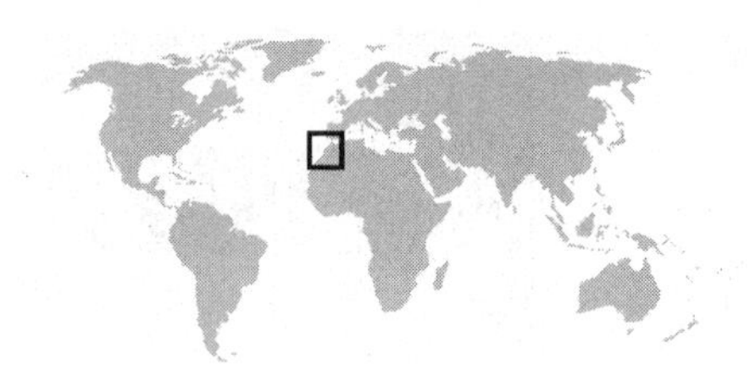

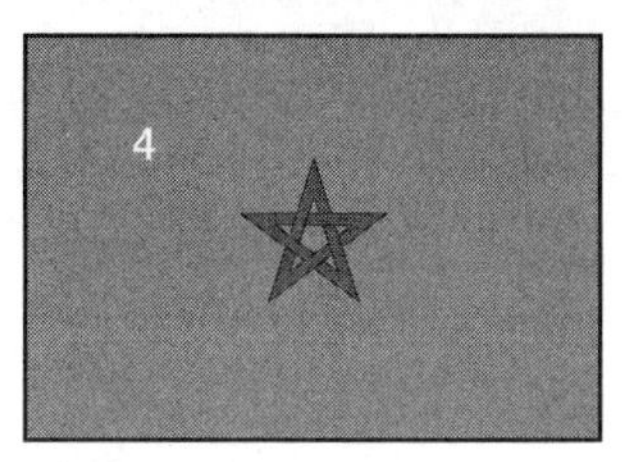

Name on driving licence: Kingdom of Morocco
Capital: Rabat
Population: 32 million
Dosh: Moroccan dirham = 100 centimes
Size: 446,300 km^2 (22 Wales)

Complete history: There was some sort of Berber culture here as long ago as 8000 BC, although there was still nothing to do on a Saturday night until the Romans arrived. Johnny-come-lately Arabs breezed in around AD 683, followed by *Juan-viene-muy-tarde* Spaniards and *Jean-est-venu-tellement-tard-qu'il-a-raté-le-fête* Frenchies.

Crowning achievement: The Al Karaouine University in Fez is the world's oldest educational institution in existence. It was founded in AD 859, so there's a good chance the world's first attempts to go a full ten weeks on just beans on toast were made within its hallowed halls. **[17]**

Top spot: Marrakesh, whether or not you know you're riding on the express that takes you there, though you might be advised to avoid the roasted sheep's heads on arrival. **[12]**

Pub fact: If they were giving out medals for pluckiness, Morocco would surely win one for granting freedom of religion to its Jewish minority – not something Arab states are renowned for – and for being the only country in Africa *not* to be a member of the African Union. **[15]**

Made in Morocco: Casablanca, though sadly only the city. The film was shot at the Warner Brothers film studio and Van Nuys Airport in California. Oh well, at least we still have Paris. **[13]**

Opening lines of national anthem: 'Fountain of Freedom, Source of Light/Where sovereignty and safety meet.' The anthem continues really rather prettily, free from the usual entreaties to put enemies to the sword or die gloriously in the attempt (preferably in pools of one's own blood). **[14]**

SCORE: 71 **WORLD RANKING: 16**

Mozambique AFRICA

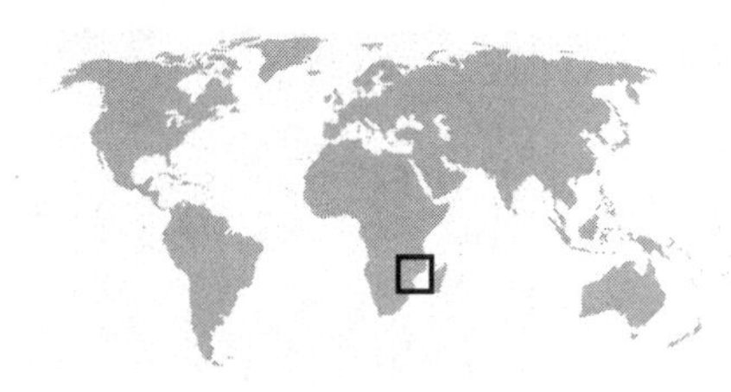

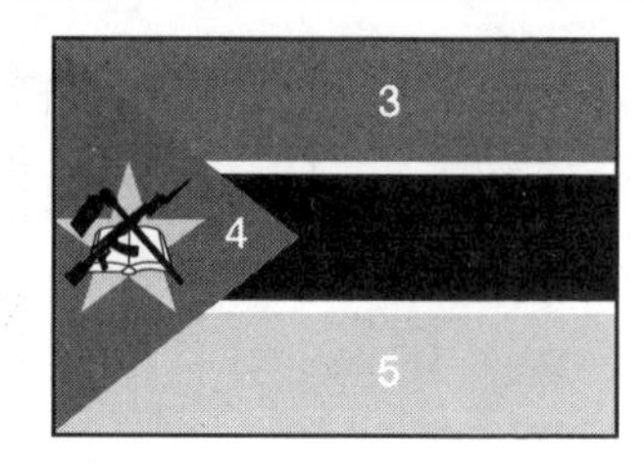

Name on driving licence: Republic of Mozambique
Capital: Maputo
Population: 19 million
Dosh: Metical = 100 centavos
Size: 801,600 km^2 (39 Wales)

Complete history: The key to Mozambican history is to sort your FRELIMOs from your RENAMOs. FRELIMO, the Front for the Liberation of Mozambique, waged an ultimately successful campaign against the Portuguese colonists and took over in 1975. RENAMO, the Mozambique National Resistance Movement, was established by Rhodesia and (white) South Africa to wage an ultimately unsuccessful campaign against FRELIMO. About a million people were killed.

Top spot: The Chapel of Nossa Senhora de Baluarte (1552), built by the Portuguese on Mozambique Island, is purportedly the oldest European building in the southern hemisphere. However, given that there is an awful lot of the southern hemisphere, and that the early Europeans pushed their noses into a great deal of it, this claim is always going to be easier to make than to prove. Still, does no harm to the tourist trade which is not exactly booming in Mozambique. Perhaps it's the landmines. **[13]**

Pub fact: Maputo is Africa's second-largest harbour, which makes one wonder whether they aren't tempted to add a few more staithes just to inch it past Durban. **[11]**

Sad fact: 'Mozambique', from Bob Dylan's 1976 album Desire, was allegedly written as a game to see how many *-ique* rhymes existed. (The answer apparently being a measly three – cheek, speak and peek.) **[9]**

Made in Mozambique: Prawns, cotton, aluminium. It's the healthy afternoon snack that won't scupper your appetite. **[8]**

National anthem caught in the twilight zone between pretentiousness and the use of hallucinogens: 'The dream grows, waving in the flag/It goes on cultivating in the certainty of tomorrow.' **[14]**

SCORE: 55 **WORLD RANKING:** 110

AFRICA Namibia

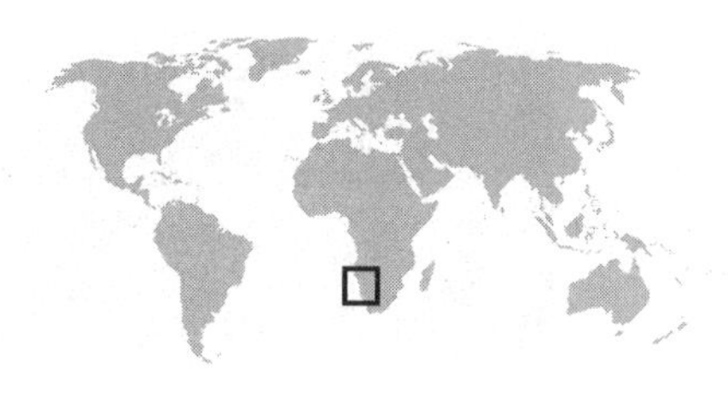

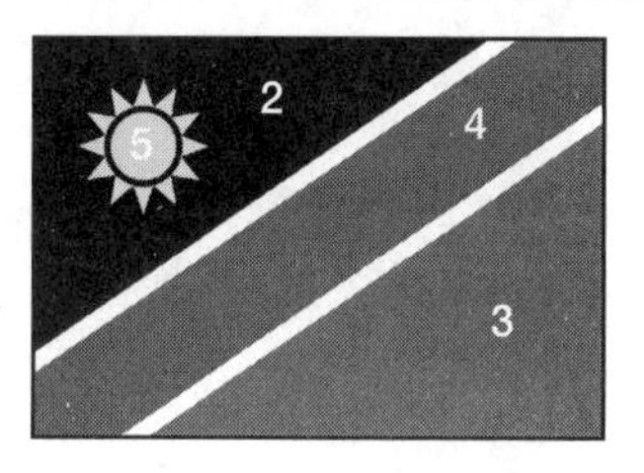

Name on driving licence: Republic of Namibia
Capital: Windhoek
Population: 2 million
Dosh: Namibian dollar = 100 cents
Size: 825,400 km² (40 Wales)

Complete history: Germans! *Sie sind* mad people. They didn't get around to annexing 'German South West Africa' (such lively imaginations) until 1884, so one must assume that their famed towels-first-on-beach reputation is one they've developed only recently. They then embarked on some ethnic cleansing of the Damara and Herero people, killing 65,000 of the latter. The invasion by South African troops in 1915 was thus something a relief, albeit that they hung around until 1990.

Crowning achievement: According to the magnificently arcane Duckworth-Lewis method, Namibia were ahead of England for twelve overs during the 2003 Cricket World Cup, but even this effort is outshone by the fact that the Namibian Constitution was the first to incorporate green values. **[15]**

Pub fact: The Namib Desert is the oldest desert on the planet. This begs a question, viz.: 'How do we know?' The answer is simple. We know because we once read it in a book. **[14]**

Top spot: Damaraland, home to a 200-million-year-old petrified forest, a 35-metre-high natural obelisk in the shape of a crooked finger, and one of the most substantial meteorites that ever clunked into Earth without killing off the dinosaurs. **[18]**

Sad fact: The famous Skeleton Coast – the desolate nothingness between the Atlantic and the Namib Desert – is so named because survivors of shipwrecks on the reef had no way of returning to sea and no chance of making it through the Namib to civilisation, which is a grim thought. **[6]**

National anthem: 'Namibia our country/Namibia motherland/We love thee.' All right, chaps, you've made your point. **[7]**

SCORE: 60 **WORLD RANKING:** 76

Nauru

OCEANIA

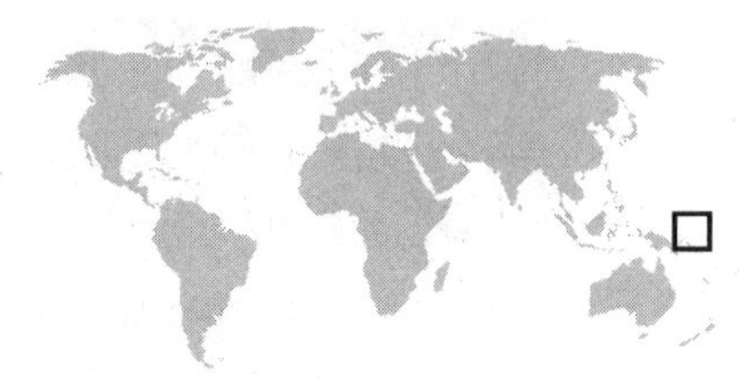

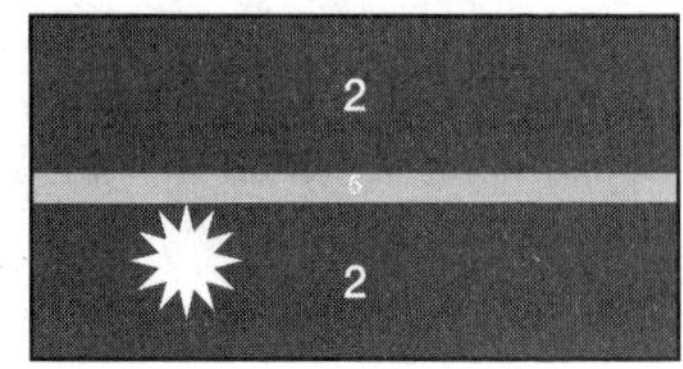

Name on driving licence: Nauru
Capital (unofficial): Yaren
Population: 13,000
Dosh: Australian dollar = 100 cents
Size: 21 km^2 (0.001 Wales)

Complete history: Pity Nauru, the living parable. Following independence in 1968, Nauruans grew rich on the proceeds of the island's phosphate mines. They proceeded to blow their wealth on imported luxury items. However, the mines have made four-fifths of the country uninhabitable, deforestation has brought drought and now the phosphate is almost exhausted.

Crowning achievement: Nauru is not the world's smallest nation, but it is the world's smallest republic, the world's smallest island nation and the only country in the world without an official capital. These are just the sort of facts that Guinness Book of Records' compilers turn to for solace after spending fifteen days officiating by the side of a swimming pool watching some lost soul play 47,000 games of underwater solitaire while humming his country's national anthem backwards. **[19]**

Made in Nauru: Pandanus trees, the fibrous fruit of which can be eaten or used as dental floss as required. **[10]**

Customs to treasure: On Angam Day (26 October) Nauruans celebrate the anniversary of the occasion in 1932 when the population swelled to 1,500 (believed to be the minimum number of people a group needs to survive) after a disastrous influenza epidemic in the 1920s. The Nauruans dipped below the bar again during World War II but re-angammed in March 1949. **[15]**

Sad fact: In return for aid, Nauru acts as a detention centre for refugees seeking asylum in Australia, a country evidently too small to cater for such people. **[0]**

Opening lines of national anthem: 'Nauru our homeland, the land we dearly love/We all pray for you and we also praise your name.' **[10]**

SCORE: 44 **WORLD RANKING:** 167

Nepal

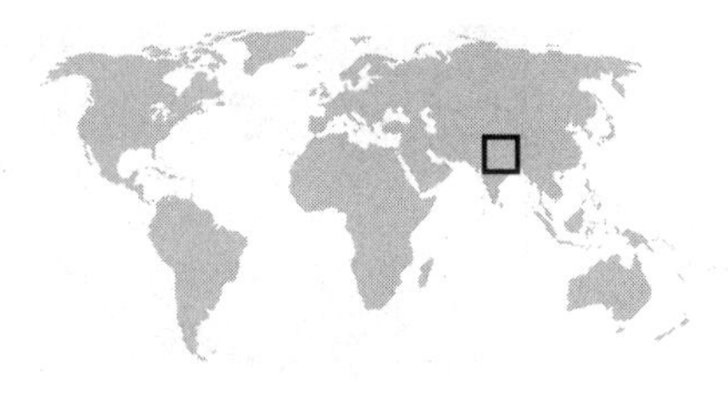

Name on driving licence: Kingdom of Nepal
Capital: Kathmandu
Population: 27 million
Dosh: Nepalese rupee = 100 paisa
Size: 140,800 km² (7.1 Wales)

Complete history: 'Schism' rather than 'Unity' has been the watchword of Nepal over the 9,000 years of human existence there. Rather breaking with this tradition, the decade-long civil war between monarchist forces and the (Maoist) Communist Party of Nepal came to an end on 21 November 2006. Nobody breathe.

Flag Fact: Created in the 19th century by wedging two triangular pennants together. No one, it seems, thought to form a rectangle by turning one of the pennants over and sticking the two hypotenuses together, for which let us give thanks or we would be left without a single flag on the planet that resembled the profile of a rosy-cheeked face. **[19]**

Top spot: Everest, one would imagine. **[20]**

Religious affiliations: The Buddha, aka Prince Siddartha Gautama, was born in Lumbini, in what is now southern Nepal, around 563 BC and some 35 years later attained enlightenment. This is a state reached by a very few so it's really no wonder that a religion sprung up around him. **[13]**

Pub fact: Nepal is an anagram of 'Alpen', which is so nearly apt that its closeness might become a frustration if you weren't careful, so it's just as well we all are. **[7]**

Opening lines of national anthem: 'May glory crown you, courageous Sovereign/You, the gallant Nepalese.' Due to various revisions, the words to this anthem don't scan at all well, making the singing of it a somewhat embarrassing event. Thankfully, since the king has been somewhat out of favour of late, the Nepalese have rather abandoned the anthem, thus neatly sidestepping the difficulty. **[8]**

SCORE: 67 WORLD RANKING: 37

Netherlands

EUROPE

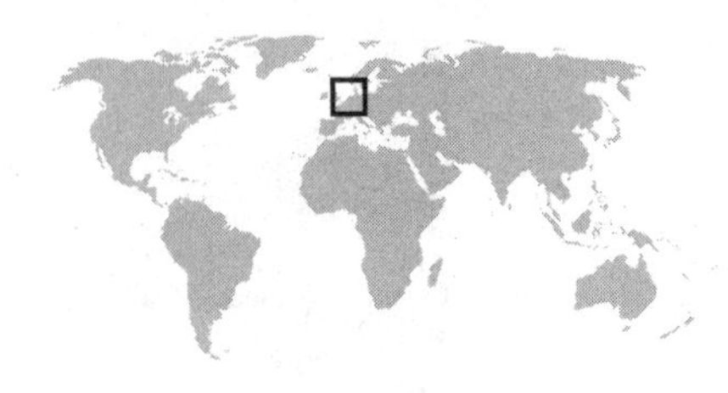

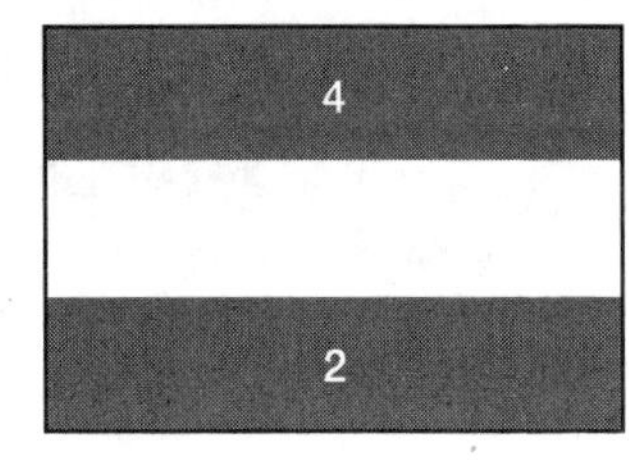

colour key: 1 = turquoise 2 = blue 3 = green 4 = red 5 = yellow 6 = orange 7 = pink 8 = purple 9 = brown

Name on driving licence: Kingdom of the Netherlands
Capital: Amsterdam
Population: 16 million
Dosh: Euro = 100 cents
Size: 41,500 km^2 (2 Wales)

Complete history: The Netherlands has spent an unhealthy percentage of its past intermingled with Belgium. However, reports that this triggered the decision in 2002 to become the first country in the world to legalise euthanasia are probably exaggerated.

Crowning achievement: No one in the world loves a polder like a Netherlander. Not only do they furnish the country with 4 per cent more land by biting into the Zuider Zee, but they have also inspired the 'polder model', a uniquely Dutch form of consensus decision-making. **[17]**

Made in Holland: Sunflowers, starry starry nights, potato eaters, ears. **[18]**

Pub fact: At the 2006 Dutch general elections the Partij voor der Dieren (Party for the Animals) won two seats, making it the first animal-rights party in the world to be represented in parliament. **[20]**

Gifts to the world: The first navigable submarine (1620 – *dank u*, Cornelius Drebbel) and the electrocardiogram (1903 – ditto, Willem Einthoven). Scandalously, to date no one has put the two together so that you can get your heart checked out while stalking a fishing fleet off the Japanese coast. **[14]**

National anthem: The Dutch have one of the oldest anthems known to the human ear. It has fifteen verses, each one beginning with the first letter of the name Willem van Nassov (William of Orange). Sweetly, the English translation retains this acrostic, using the anglicised William of Nassau. Those who make it alive to the end of the fifteenth verse are treated to the inspiring closing lines: 'Obedience first and latest/For Justice wills it so.' **[16]**

SCORE: 85 WORLD RANKING: 1

OCEANIA New Zealand/Aotearoa

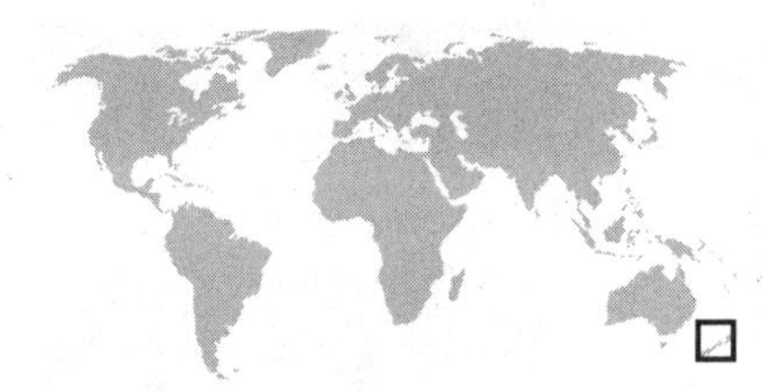

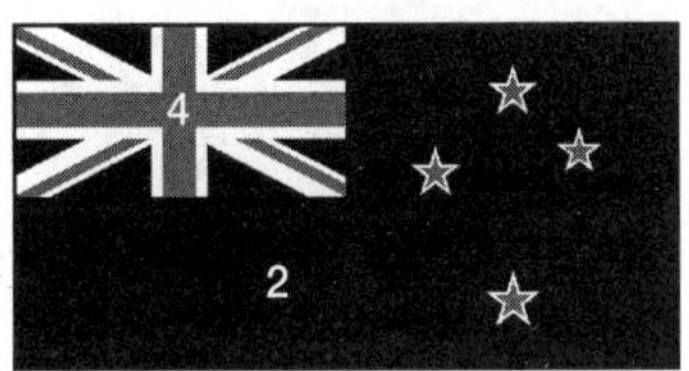

Name on driving licence: New Zealand
Capital: Wellington
Population: 4 million
Dosh: New Zealand dollar = 100 cents
Size: 268,700 km² (13 Wales)

Complete history: Aotearoa (Maori for 'Place Where They Filmed *The Lord of the Rings*') was visited in 1642 by Dutchman Abel Tasman, but it was 127 years before the arrival of the next European, Captain Cook, though since he was only born in 1728, it seems a bit ripe to blame him for the delay. His countrymen followed and, ere long, New Zealand was engulfed in the loving embrace of the British Empire.

Top spot: Taumatawhakatangihangakoauauotamateaturipukakapiimaungahoronukupokaiwh enuakitanaahu. [13]

Made in New Zealand: Bagpipe bands. Apparently New Zealand has more than Scotland, though nobody seems too keen to find out for sure. [9]

Customs to treasure: The All Blacks rugby team perform the *haka* before ritually slaughtering their adversaries. They get away with this because in almost every other arena of life, New Zealand is patronised as a land of sheep and boredom. For instance, one can't imagine it going down well if the *French* rugby team performed a dance in which, say, they mimed guillotining their opponents with a sharpened copy of *La Nausée*. [12]

Pub fact: New Zealand Sign Language is one of the country's three official languages. Disappointingly, however, it's not represented on the country's road signs. [14]

National anthems: New Zealanders have two national anthems, the lucky things. There's 'God Save the Queen' and 'God Defend New Zealand'. The latter begins: 'God of Nations at Thy feet/In the bonds of love we meet', which is nice. The Maori 'translation' just about reins itself in from becoming 'God Defend Us From New Zealanders'. [10]

SCORE: 58 **WORLD RANKING:** 91

Nicaragua

AMERICAS

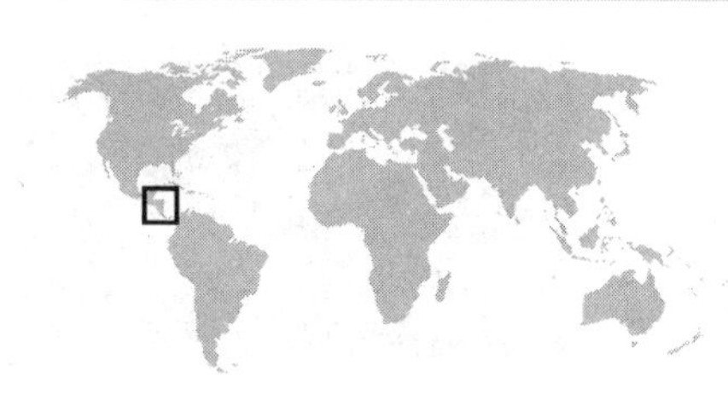

colour key: 1 = turquoise 2 = blue 3 = green 4 = red 5 = yellow 6 = orange 7 = pink 8 = purple 9 = brown

Name on driving licence: Republic of Nicaragua
Capital: Managua
Population: 5.4 million
Dosh: Córdoba oro = 100 centavos
Size: 129,500 km^2 (6.3 Wales)

Complete history: There's something about Nicaragua that encourages ravaging. The Spaniards ravaged first, of course, then came the Brits and Americans in the 19th century. Come the 1980s and the Americans were ravaging by proxy with that nice Oliver North selling arms to the Iranians and channelling the profits to the Contras. Toppled president Daniel Ortega appears to have laughed last, however, getting himself elected again in 2006.

Crowning achievement: The Sandinistas. All right, so it did go belly up, but their hearts were in the right place, and red-and-black does look very striking on one's bedroom wall. **[15]**

Top spot: Lake Nicaragua contains sea creatures including sharks, a feat no other freshwater lake in the world has even attempted. **[14]**

Pub fact: Nicaragua is an anagram of 'iguana car', which is ironic since, under Nicaraguan law, lizards are not allowed to drive. **[10]**

Customs to treasure: Since the catastrophic 1972 earthquake, addresses in Managua have been a mite haphazard. For example, letters to Comercial Gilkar stationers should be addressed: '*de donde fue Cine Salinas 4c Abajo 1½c al Lago, Managua*' ('from where the Salinas Cinema used to be, then four blocks down and 1½ blocks towards the lake, Managua'). Other addresses begin: 'from the little tree', 'from where the yellow van was' and, more alarmingly, 'from where the shoot-out occurred'. **[18]**

Opening lines of national anthem: Refreshingly, the government-run national anthem contest stipulated that entries had to be short and about peace. Look and learn, Latin America: 'Hail to you, Nicaragua. On your soil/The cannon's voice no longer roars.' **[13]**

SCORE: 70 WORLD RANKING: 22

Niger

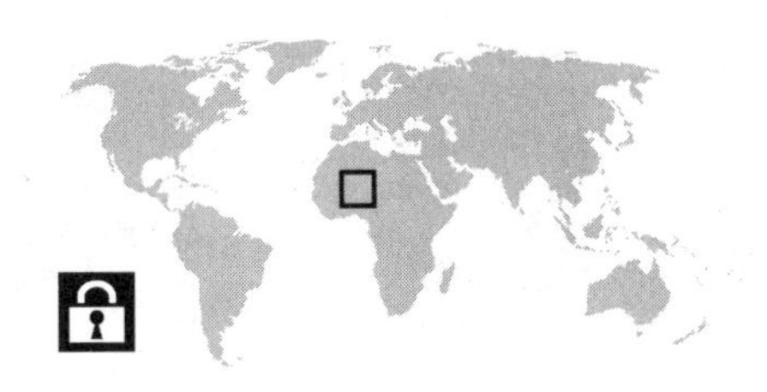

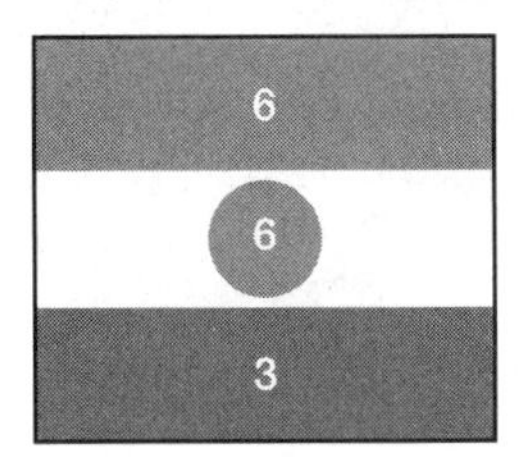

Name on driving licence: Republic of Niger
Capital: Niamey
Population: 11 million
Dosh: CFA franc = 100 centimes
Size: 1,267,000 km^2 (61 Wales)

Complete history: Try to pick the odd group out of this Niger-related list: Tuareg, Hausa, Songhai, Kanem, Bornu, Fulani, FRENCH. Yes, that's right, it's the French. None of the others drink hot chocolate from a bowl.

Crowning achievement: The rock art of Aïr. The enormous 9,000-year-old engraving at Dabous Rock is particularly highly regarded. First believed to be a self-portrait, it is now generally recognised that the two giraffe were carved by Tuareg artists. **[13]**

Made in Niger: Almost extinct in the wild, about 130 addax can be found scattered around the Sahara in the north-east of the country. Apparently these desert antelopes never have to drink, which is a handy trick but which, on reflection, doesn't seem to have helped them all that much. **[12]**

Popular misconception: 'Niger is so called because French colonists noticed that the people there were (ahem) negro.' *Pas du tout, mon vieux.* It's derived from the word *n'eghirren* ('flowing water'), the Tuareg name given to the River Niger. **[10]**

Pub fact: For decades, the pubs, bars and discotheques of the world have resounded to the debate, often quite heated, surrounding the adjectival form of Niger. Go for 'Nigerian' and everyone thinks you're referring to Nigeria; opt for 'Nigerois' and no one knows what you're talking about; while plumping for plain old 'Niger' just doesn't sound right. Apparently, the answer is to aim at a noise approximating to 'Nigérien'. Good luck. **[12]**

At last, a national anthem with a bit of common sense: 'Let us avoid vain quarrels/In order to spare ourselves bloodshed.' **[17]**

SCORE: 64 **WORLD RANKING: 54**

Nigeria

AFRICA

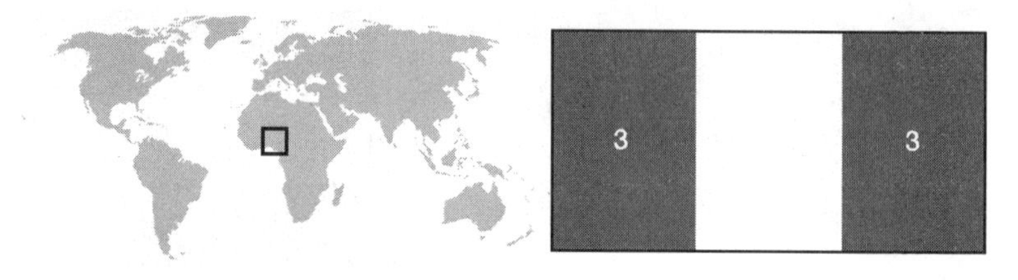

Name on driving licence: Federal Republic of Nigeria
Capital: Abuja
Population: 130 million
Dosh: Naira = 100 kobo
Size: 923,800 km^2 (45 Wales)

Complete history: Originally, Nigeria specialised in civilisations with brief, clipped names – like Nok and Ife – that didn't waste people's time. Things only started to go awry with the arrival of the British and their excessively lettered nomenclature.

Popular misconception: 'Nigeria is the world's most corrupt country.' For goodness' sake, bring your prejudices up to date, please. Nigeria has got its act together and now officially ranks above Bangladesh and Chad (however, it is possible that both these countries have been bribed by Nigeria to appear more corrupt). **[2]**

Sad fact: According to the Nigerian government commission set up to counter so called '419 scams' in which emails ask for help in transferring MILLIONS OF US DOLLARS out of some troubled African country or other: 'There are only between 50,000 and 100,000 people involved in this thing.' Which is a relief. **[1]**

Made in Nigeria: Author Chinua Achebe, who declared of his own country: 'It is one of the most corrupt, insensitive, inefficient places under the sun ... It is dirty, callous, noisy, ostentatious, dishonest and vulgar. In short, it is among the most unpleasant places on earth.' Not to be outdone, Nobel laureate and fellow Nigerian, Wole Soyinka, went for: 'Only a masochist with an exuberant taste for self-violence will pick Nigeria for a holiday.' **[4]**

Pub fact: Out of every seven Africans, one is Nigerian. **[10]**

National anthem: Cobbled together from no fewer than five 'winning' entries in a 'write our national anthem' contest, so in the circumstances, the opening lines got away quite lightly: 'Arise, O compatriots/Nigeria's call obey.' **[4]**

SCORE: 21 WORLD RANKING: 199

colour key: 1 = turquoise 2 = blue 3 = green 4 = red 5 = yellow 6 = orange 7 = pink 8 = purple 9 = brown

EUROPE (*see also* United Kingdom) Northern Ireland

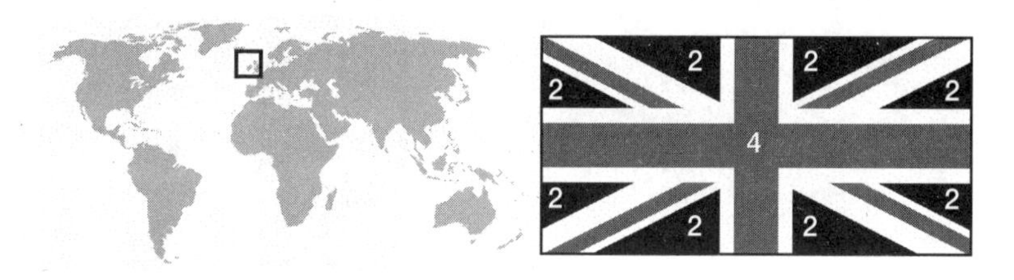

Name on driving licence: Northern Ireland
Capital: Belfast (*Béal Feirste*)
Population: 1.7 million
Dosh: Pound sterling = 100 pence
Size: 13,800 km^2 (0.7 Wales)

Complete history: Northern Ireland is a construct of the dear old English who so loved Ireland that they began invading it as early as 1169, although to be fair, that sortie was led by a naturalised Welsh Norman, Richard de Clare. However, the Irish sensed that the English were probably to blame anyway and kept up one heck of a kerfuffle until in 1921 the invaders had to give them most of their island back. However, many Irish people have since declared that it would be nice if they could have the whole thing, even though part of it is now packed with curious types who'd rather burn their own grandmothers alive than come under Dublin rule.

Popular misconception: 'Belfast is still very scary.' Not so. Less than 20 per cent of Belfast is now officially rated 'very scary' and visitors can look at those bits from behind a sofa if needs be. **[6]**

Top spot: The Mountains of Mourne, if all goes well, are soon to be Northern Ireland's first national park. **[13]**

Made in Northern Ireland: The Undertones. Bless. **[20]**

Pub fact: Kenneth Branagh was born in Belfast, which may explain why he turned down an OBE. **[15]**

Anthem: Singers of 'A Londonderry Air' (aka 'A London *Derrière*') wish themselves to be 'the tender apple blossom', 'a little burnish'd apple' and 'a happy daisy' all ardently essaying to touch a queen as she walks through a garden. Indeed, the daisy is quite happy to be squashed to death beneath her silver foot. Strangely moving. **[19]**

SCORE: 73 **WORLD RANKING: 11**

Norway

EUROPE

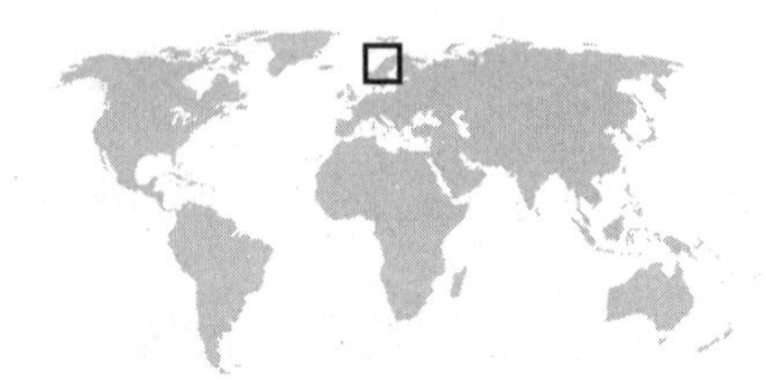

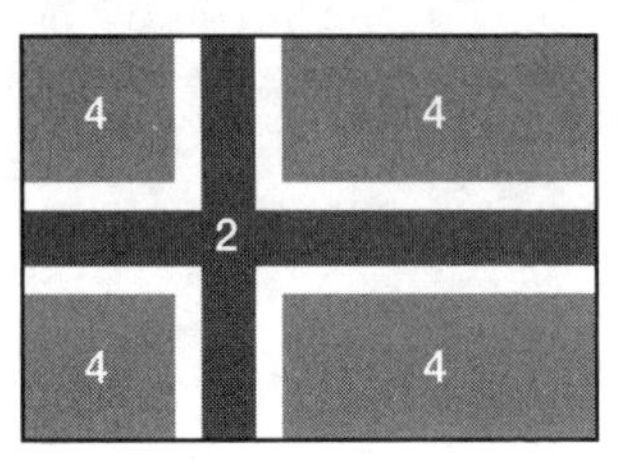

Name on driving licence: Kingdom of Norway
Capital: Oslo
Population: 4.6 million
Dosh: Norwegian krone = 100 øre
Size: 324,200 km^2 (16 Wales)

Complete history: Vikings (all happy and slaughterous); union with Denmark/Sweden (happy-ish); Danish rule (not at all happy); Swedish rule (livid); free (happy happy); brief visit by Germans (briefly livid again).

Top spot: Bergen. At the top of the Mount Floyen funicular there's a ginger cat who will keep you company as you gaze down on the city and reflect on the opportunities you've missed and what might have become of your life if you'd only taken hold of them when you had the chance. **[17]**

Customs to treasure: The *yoik*, a singing style practised by the itinerant Sami. To the untrained ear, it can sound a bit like wolves in some sort of distress – their paws caught underneath an elk, perhaps. Try starting with a *yoik* crossover artist like Mari Boine and work your way up. **[13]**

Sad fact: The country is nothing but fjords and yet Norway has only the world's second-deepest (Sognefjord). **[8]**

Made in Norway: The globally-warmed-and-troubled-about-it polar bear. Many of them loiter with intent on Spitsbergen where, coincidentally, the Norwegian government is setting up a 'doomsday seed bank' for when we bring about ecological/nuclear disaster upon ourselves. **[12]**

National anthems: The national anthem begins as a rebuttal to that oft-asked question: 'Do you really love Norway, even as it rises?' The royal anthem, meanwhile, is more monarch-based, as might be expected, and is apparently a popular drinking song. National anthem: 'Yes, we love this country/As it rises.' Royal anthem: God bless our good King/Bless him with strength and courage.' **[14]**

SCORE: 64 **WORLD RANKING:** 54

colour key: 1 = turquoise 2 = blue 3 = green 4 = red 5 = yellow 6 = orange 7 = pink 8 = purple 9 = brown

Oman

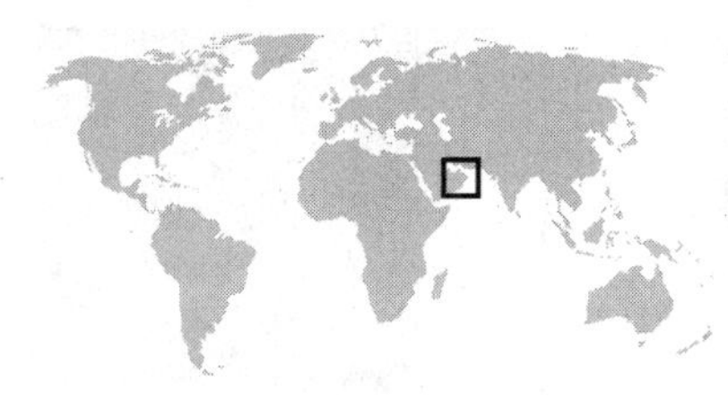

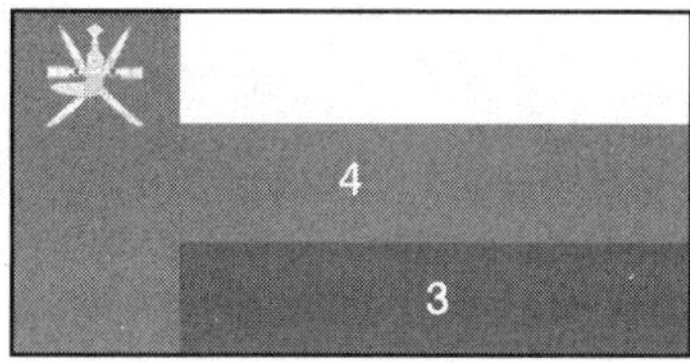

Name on driving licence: Sultanate of Oman
Capital: Muscat
Population: 2.9 million
Dosh: Omani rial = 1,000 baizas
Size: 212,500 km^2 (15 Wales)

Complete history: The Persian empire brought rule by *satraps* and a change of name from Magan, which Omanis had always felt made their country sound like a low-quality Slovakian dessert wine. There followed periodic interventions by Sassanids (really just Persians again), Arabs, Portuguese (eventually sent packing), Ottomans (ditto), Persians *again*, and the British, who banned the Omanis from trading slaves, so good for them. The nation briefly became Muscat and Oman, but this only made it sound like a low quality Slovakian cava, so they wisely reverted to Oman.

Crowning achievement: Oman is mostly made of gravel, so it's done very well to make itself into a country at all. Yoke this with the fact that it's the only nation in the world that starts with an O and it's no wonder Omanis have been accused of 'looking permanently smug for no apparent reason'. **[11]**

Sad fact: Until recently, Oman was closed to everyone but travellers on official business or business business – a strategy that used to be known as 'doing a Japan'. It remains almost impossible for a foreigner to become an Omani national. **[4]**

Pub fact: Oman is the easternmost Arabian country. Make as much of this as you can. **[6]**

Popular misconception: 'Said bin Taimur – sultan from 1932 to 1970 until he was deposed by his son – is buried somewhere in Oman.' Dead wrong. He's interred at Brookwood Cemetery, Woking, Surrey. **[15]**

Rare example of a state whose national anthem includes a naked threat aimed at its own people: 'Be happy! Qaboos has come.' **[16]**

SCORE: 52 **WORLD RANKING:** 125

Pakistan

ASIA

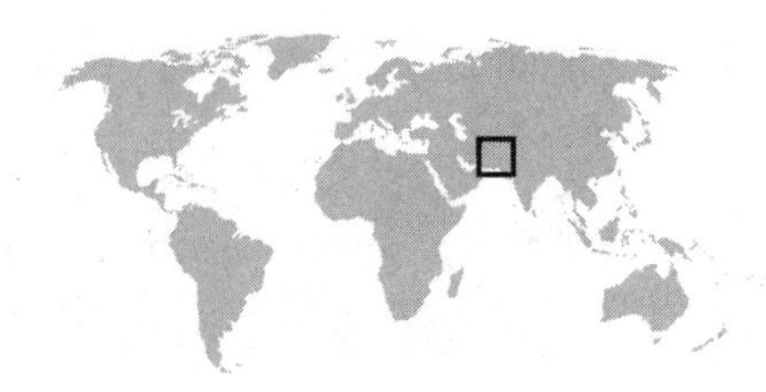

Name on driving licence: Islamic Republic of Pakistan
Capital: Islamabad
Population: 165 million
Dosh: Pakistani rupee = 100 paise
Size: 803,900 km² (38 Wales)

Complete history: Until 14 August 1947, Pakistan was merely part of India. After that, there was all manner of kerfuffle, general rioting and, finally, warfare between the two countries, mainly over Kashmir. Nowadays, of course, they love each other.

Crowning achievement: Pakistan has a population greater than that of the world's largest country, Russia. To keep this fact hidden from the rest of the world, and thus enable them to go into sporting events as the plucky underdog, Pakistani citizens are encouraged to hide during the day, only coming out when it's too dark for spy satellites to make them out properly. **[14]**

Customs to treasure: Bollywood films have in no sense been banned since 1965 in order to boost the home-grown film industry (based at Lahore (Lollywood) and Peshawar (Pollywood) – perleease). Nor do Pakistanis hold grudges. Oh no. **[5]**

Pub fact: The name 'Pakistan' was coined in 1933 and is officially the cleverest country name in existence. Not only does it mean 'land of the pure' in various local languages, but it also stands for the five provinces of India – Punjab, Afgania, KashmIr, Sind and BlouchisTAN – that formed the country. **[15]**

Flag fact: Never let it be said that non-Muslims are under-represented on the flag of Pakistan. Only the colour green, the crescent moon and the star are related to Islam: the other religions get a sliver of white down the left-hand side all to themselves. **[10]**

National anthem in serious danger of overreaching itself: 'Interpreter of our past/Glory of our present/Inspiration of our future.' **[12]**

SCORE: 56 **WORLD RANKING: 101**

OCEANIA

Palau

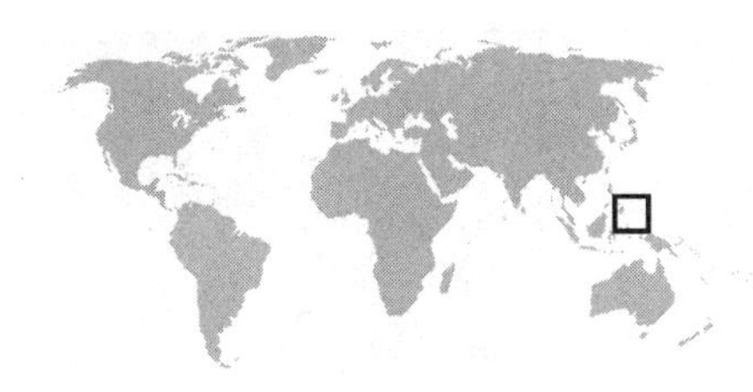

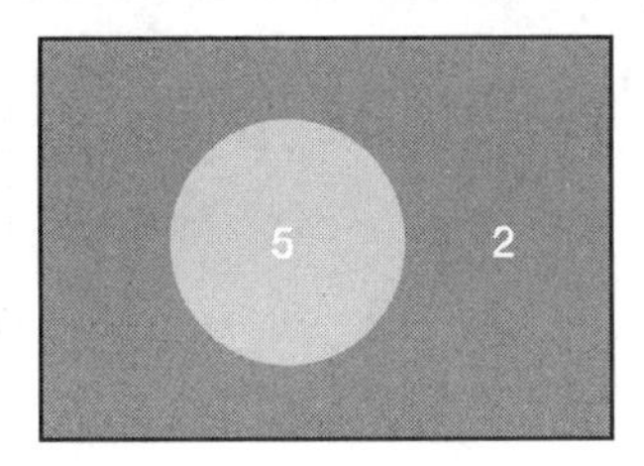

Name on driving licence: Republic of Palau
Capital: Koror
Population: 20,000
Dosh: US dollar = 100 cents
Size: 450 km² (0.02 Wales)

Complete history: The Spaniard Ruy López de Villalobos may or may not have spotted one of Palau's 300-odd islands in 1543. The islanders ducked when they saw his ship so he couldn't be sure. They repeated the operation for Sir Francis Drake and 'Pelew' was only inked in on maps of the world. Captain Henry Wilson rocked up, quite literally, in 1783. The shipwrecked Englishmen and the Pelewans got on like a house on fire, and both were sad when the former eventually sailed off in their repaired ship. After that, it all rather went to pot, with the islands changing hands between the Spaniards, the Germans, the Japanese and the Americans before independence in 1994.

Crowning achievement: Nobody has ever heard of Palau. This is because Palauans have taken the precaution of ensuring that nothing newsworthy ever happens there. If there is any danger of an event occurring on the islands, it is quickly moved to the nearest Indonesian islet under cover of darkness. The Indonesian army, complicit in this cover-up, destroy the event in a controlled explosion, and the danger passes. **[16]**

Actual motto: The rigorously enigmatic 'Rainbow's End'. **[15]**

Historical high: Palau declared itself a nuclear-free zone in 1979, which rather rankled the Americans, who enjoy blowing up islands thousands of miles away from home for the good of mankind. **[20]**

Sad fact: Palauans are at risk from earthquakes, volcanoes, tropical storms and global warming-related flooding. **[3]**

National anthem evidently designed by committee: 'Let's build our economy's protecting fence/With courage, faithfulness and diligence.' **[4]**

SCORE: **58** WORLD RANKING: **91**

Panama AMERICAS

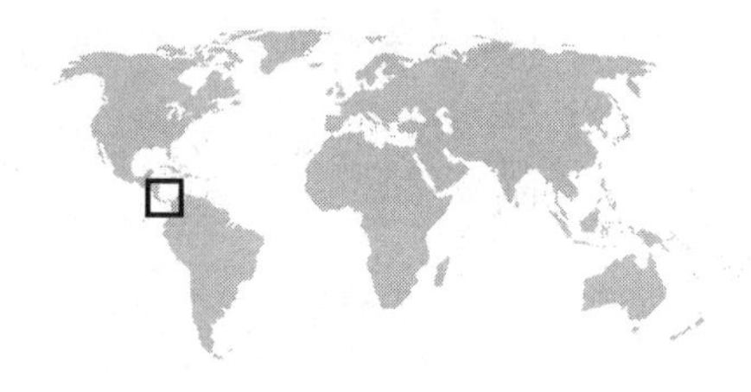

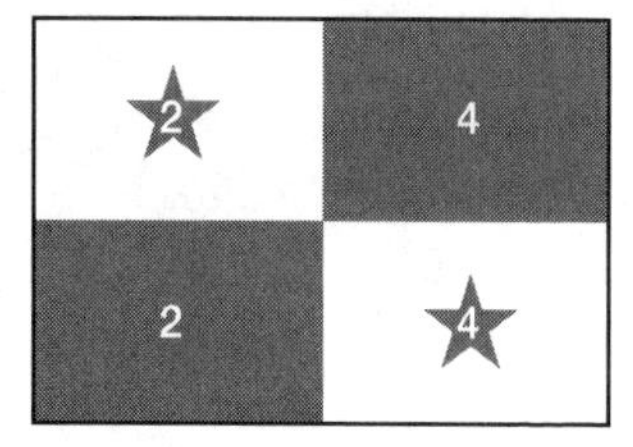

Name on driving licence: Republic of Panama
Capital: Panama City
Population: 3 million
Dosh: US dollar and Balboa = 100 centésimos
Size: 78,200 km^2 (3.7 Wales)

Complete history: There was loads of history before Christopher Columbus pressed his ugly thumbprint into the side of Panama in 1502. For instance, there was the coming of the first European, Rodrigo de Bastidas, the year before. Between them, the Spaniards wiped out the Cuevan, Chibchan and Chocoan populations and so it is fitting, if somewhat sad, that Colón – the city named in honour of the adventurer – has become a violent, poverty-stricken dump.

Pub fact: Depending on whom you ask, the Panama Canal takes about 11,000 km off the sea voyage between Liverpool and Tokyo, which is a not insubstantial saving if you find yourself having to row. **[15]**

Crowning achievement: It's a toss up between the actual building of the Panama Canal, and the succinct history of the project, viz.: 'A man, a plan, a canal – Panama!', which also doubles as one of the most sublime palindromes in the English language. **[18]**

Gift to the world: The panama, a hat worn by the sort of gent who can wear horn-rimmed glasses and safari suits without recourse to irony. **[10]**

Sad fact: The Colón Free Trade Zone – a walled-off area in which companies can set up sweat shops and the like free of nasty things like taxes and import/export duties – is the second largest in the world after Hong Kong. **[1]**

National anthem revealing disturbing delusional tendencies centring on an imagined deification of the country: 'It is fitting to cover with a veil/The Calvary and the cross of the past.' **[3]**

SCORE: 47 **WORLD RANKING: 150**

OCEANIA Papua New Guinea

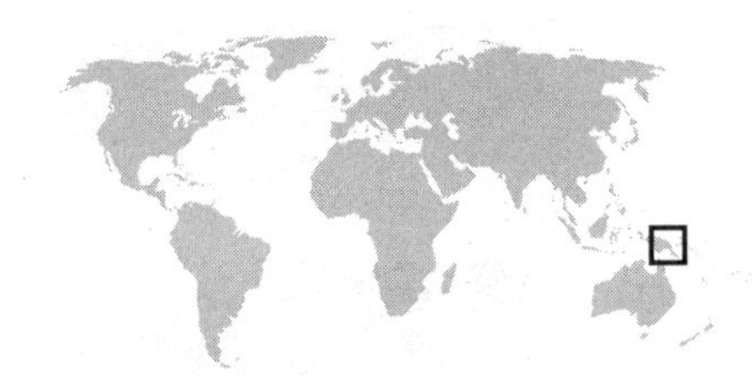

Name on driving licence: Independent State of Papua New Guinea
Capital: Port Moresby
Population: 5.9 million
Dosh: Kina = 100 toea
Size: 462,800 km^2 (22 Wales)

Complete history: Papua New Guinea, just like Haiti/Dominican Republic and Harris/Lewis, has to share an island with a neighbour (Indonesia). Unlike Harris or Lewis, however, Papua New Guinea broke free from Australian rule in 1975 and is now about as close to social and economic ruin as you'd ever wish.

Where to avoid: Port Moresby, the admirably consistent winner of every 'Worst Capital City on Earth' survey going. It's usually the extremely high levels of murder, rape and robbery that attract the wrath of the survey compilers, although the estimated 60–90 per cent unemployment rate sometimes raises an eyebrow too. The city is twinned with a place in Australia called Townsville, which probably tells you all you need to know about its cultural life. **[2]**

Gift to the world: PNG Pidgin English, the archly descriptive language in which the term for 'a weed' is rendered '*tekimaut gras nogut*', while 'rich' is '*i gat planti samting*', and 'railway platform' translates as '*ples bilong wetim tren*'. **[16]**

Made in Papua New Guinea: The hooded pitohui, the world's only poisonous bird. More curious still, the poison it emits from its wings, homobatrachotoxin, is exactly the same as that manufactured by the poison arrow frog. **[18]**

Pub fact: With over 700 indigenous languages, PNG is the most linguistically diverse nation on the planet. **[15]**

National anthem in danger of overvaluing the benefits of bellowing the name of one's country to the four winds: 'Shout our name from the mountains to seas/Papua New Guinea/Let us raise our voices and proclaim/Papua New Guinea.' **[10]**

SCORE: 61 **WORLD RANKING: 70**

Paraguay

AMERICAS

colour key: 1 = turquoise 2 = blue 3 = green 4 = red 5 = yellow 6 = orange 7 = pink 8 = purple 9 = brown

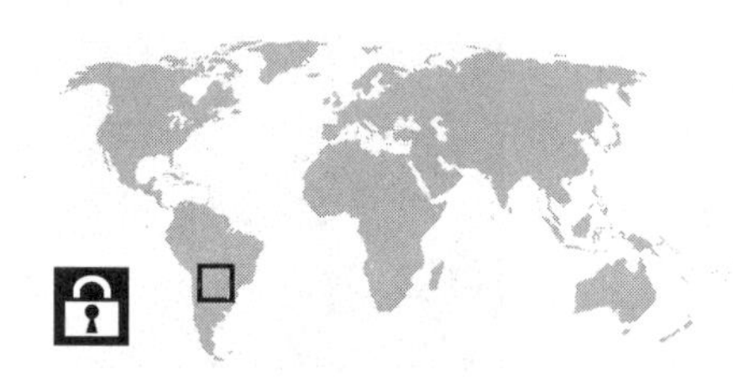

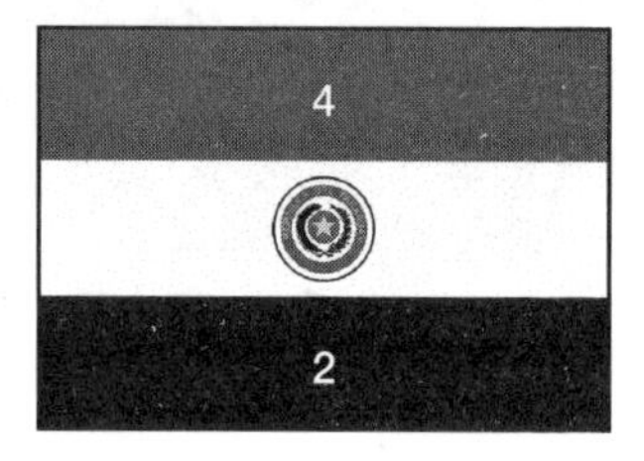

Name on driving licence: Republic of Paraguay
Capital: Asunción
Population: 6.2 million
Dosh: Guaraní = 100 céntimos
Size: 406,800 km^2 (20 Wales)

Complete history: The Guaraní got here first; the Spaniards came next; the country has been in an almost permanent state of war ever since.

Crowning achievement: Prosecuting the most fantastically suicidal war of all time. Taking on Brazil or Argentina or even Uruguay might be ill-advised for a relatively small country like Paraguay. Attempting to invade all three *at the same time* might best be described as foolhardy. It is estimated that the War of the Triple Alliance (1864–70) brought about the death of 90 per cent of the male population. **[1]**

Top spot: The Chaco, the world's second-largest forest. **[13]**

Pub fact: To make up for the Chaco disappointment, Paraguay made sure that the Itaipú Dam was the world's biggest hydroelectric complex. As a bonus, Philip Glass wrote some cantata about it. **[10]**

Sad fact: Alfredo Stroessner, dictator from 1954 to 1989. A man happy to harbour Nazis (some 200,000 Germans settled after World War II, not all of whom can be said to have been 'good Germans') and occasionally genocidal (ask the Aché people), with a fondness for torturing and 'disappearing' his opponents (real and imagined), and lining his own pocket. His entire life appears to have been dedicated to making General Pinochet look cuddly. **[0]**

Opening lines of national anthem: 'The unfortunate peoples of America/Were under an oppressive reign for three centuries', whereas the oppressive reign of the anthem lasts for a mere seven verses and a chorus that has the singers proclaiming they would choose death rather than lose their status as a republic. Just stop it now, please. **[1]**

SCORE: 25 WORLD RANKING: 198

Peru

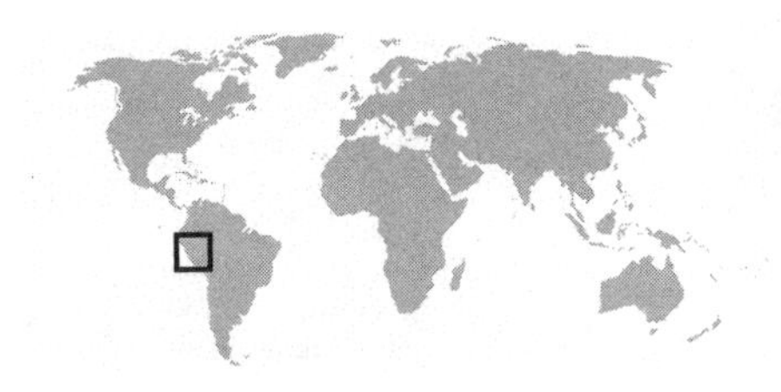

Name on driving licence: Republic of Peru
Capital: Lima
Population: 28 million
Dosh: Nuevo sol = 100 céntimos
Size: 1,285,200 km² (62 Wales)

Complete history: Peruvians suffer from a form of collective masochism that manifests itself every few years in a desire to vote tyrannical (Alberto Fujimori) or just plain useless (Alan García) presidents back into office. Democracy is wasted on some countries, you know.

Crowning achievement: The Nazca Lines. Good old ancient people – not only did they not jib at an average life expectancy of around 20, they also etched huge pictogram things in the pampas to fool future loons such as Erich von Däniken that they had been communing with extraterrestrial aliens. Spot on. **[19]**

Top spot: Machu Picchu, former stronghold of the Incas and one of the most astonishing places on Earth, even when invaded by Gap Year Garys drunk on pisco sour and the novelty of, like, EVERYTHING. **[20]**

Flag fact: According to legend, the red and white of the nation's flag was inspired by a flock of flamingos that flew over Argentine General José de San Martín as he was on his way to liberate Peru in 1820. This is fine up to a point, and that point is the fact that flamingos are pink. **[8]**

Customs to treasure: Fronton, a sort of squash played on a court wherein a flock of red flamingos has destroyed three of the walls. **[13]**

Opening line of national anthem: 'We are free, may we be so forever, may we be so forever.' Having begun in this rather desperate manner, the anthem takes six verses recovering its composure without ever quite doing so. **[8]**

SCORE: 68 **WORLD RANKING:** 31

Philippines

ASIA

colour key: 1 = turquoise 2 = blue 3 = green 4 = red 5 = yellow 6 = orange 7 = pink 8 = purple 9 = brown

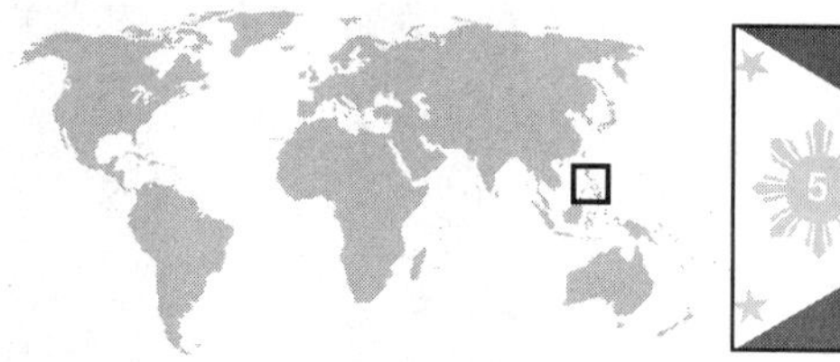

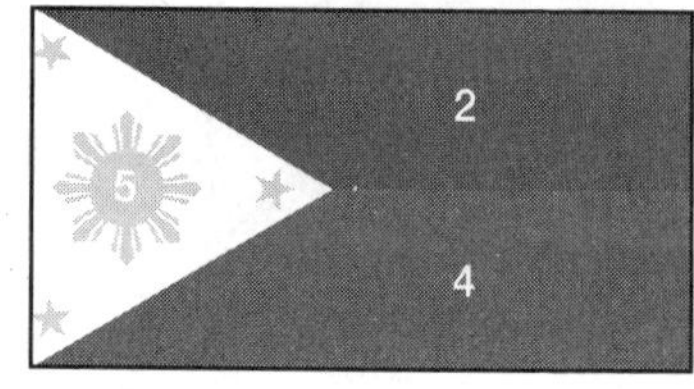

Name on driving licence: Republic of the Philippines
Capital: Manila
Population: 86 million
Dosh: Philippine peso = 100 centavos
Size: 300,000 km^2 (15 Wales)

Complete history: There's more to the Philippines than Imelda Marcos' shoe collection. There's also the fact that she once bought a bulletproof bra. Previous to that, there were lots of shoeless Austronesians to whom it never occurred that someone might shoot at their chests, until the Spanish came along. Happily, they'd got accustomed to the idea by the time the Americans and Japanese arrived.

Made in the Philippines: Rattan. No one knows what it is, but you can make chairs out of it. **[15]**

Top spot: Anywhere some distance from the 21 active volcanoes that dot the islands like smouldering teepees of death. **[7]**

Pub fact: In one of those amusing quirks of the English language that have kept men with elbow patches and flaxen moustaches quietly chuckling by their firesides since time immemorial, Philippines is spelt with a Ph, but Filipino comes with an *F*. **[11]**

Sad fact: Filipino mobile phone owners send more text messages than any other nationality, with some networks reporting an average of 12,000 texts per person per year. This probably says something about modern Filipino society that it unfortunately won't be able to hear over the frenzied clicking of 80 million thumbs. **[3]**

National anthem: Starts off promisingly with 'Land of the morning/Child of the sun returning', but ends with a lot of guff about how glorious it would be to die for the country if it were ever wronged. Perhaps all those who write anthems along such lines should agree to perish for their country within the year, just to check that they really mean it. **[3]**

SCORE: 39 WORLD RANKING: 178

EUROPE Poland

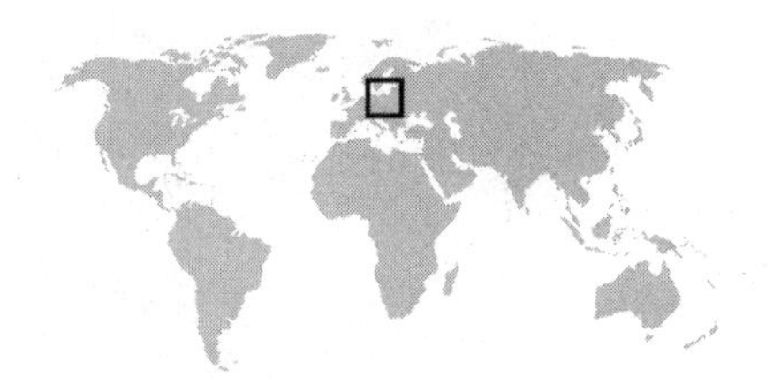

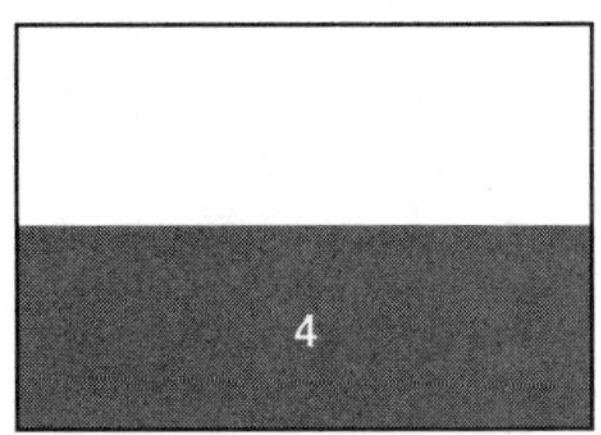

Name on driving licence: Republic of Poland
Capital: Warsaw
Population: 39 million
Dosh: Zloty = 100 groszy
Size: 312,700 km^2 (16 Wales)

Complete history: Poor Poland, since its birth in 966 it keeps splitting up, getting itself together again, only to fragment once more or find itself a victim of its neighbours' need for *Lebensraum* (the handy German term for the amount of room a nation needs to be able to get really comfortable).

Crowning achievement: Young Jadwiga of Poland can claim two crowning achievements: the second was achieving sainthood (1997), the first was being crowned (1384). More impressive still, she was crowned King of Poland, apparently to prevent folk from thinking she was a bit of a soft touch. She died, fifteen years and several miracles later, at the age of 25. **[18]**

Sad fact: The Poles have been the butt of sad puns ever since the British came up with a word for a long straight piece of wood or metal usually narrow enough to be held in the hands. This is why they can often be seen in dark corners of bars gently weeping on the shoulders of understanding Swedes. **[4]**

Made in Poland: Aleksander Wolszczan, the first person to discover planets outside our solar system. **[15]**

Pub fact: Poland was reunited in 1320 by someone called Władysław the Elbow-high, a not inconsiderable feat for a man with two apparently diseased łs in his name, and one in the eye for all those so-called tall people who think the only short leaders of any note were Hitler, Mussolini and Napoleon Bonaparte. **[17]**

Opening line of national anthem that could be construed as somewhat pessimistic: 'Poland has not yet succumbed.' **[6]**

SCORE: 60 **WORLD RANKING:** 76

Portugal

EUROPE

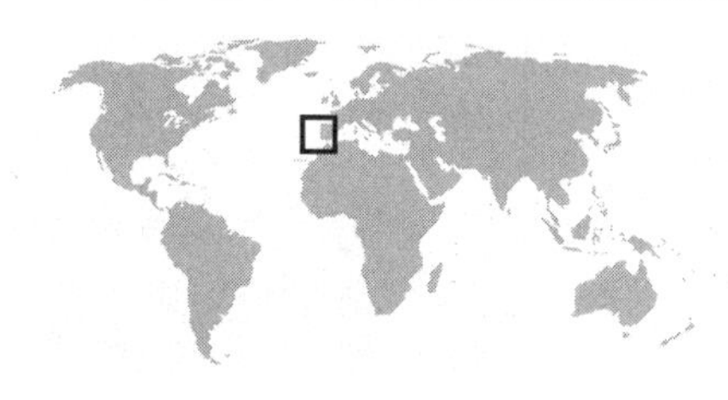

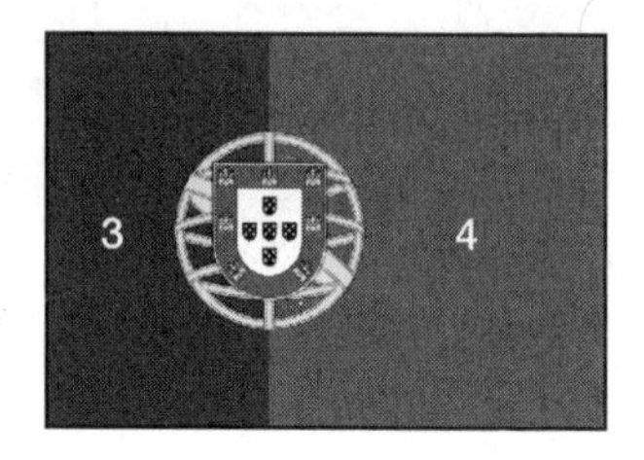

Name on driving licence: Republic of Portugal
Capital: Lisbon
Population: 11 million
Dosh: Euro = 100 cents
Size: 92,400 km^2 (4.3 Wales)

Complete history: The Romans, on taking over, named the country Lusitania, presumably because it reminded them of an ocean liner that the Germans would sink, thus precipitating US involvement in World War I. Seeking revenge, the Portuguese invaded everywhere that hadn't already been bagsied by the Spanish, British, French or Dutch, and ruined it.

Crowning achievement: Managing to be England's oldest ally (a formal arrangement since the Treaty of Windsor in 1386) without being universally hated. Even the Scots don't seem to have noticed. **[11]**

Sad fact: Crown Prince Luís Filipe was de facto King of Portugal for less than half an hour in 1908. His somewhat truncated stay at the top came about when he and King Carlos I were shot by two Republican Party assassins in Lisbon. The king died immediately, while Luís Filipe, his heir, managed to hang on for about twenty minutes. This makes his the shortest reign in the world, which must have been some consolation as he drew his final agonised breath. **[16]**

Gift to the world: *Fado*, a maudlin musical style that makes ordinary blues singers seem like Pollyanna on uppers. **[12]**

Pub fact: There is more cork made in Portugal than anywhere else, which seems a mite tough on Cork, which doesn't produce a drop of port. It's all horribly skewed. **[6]**

Opening lines of national anthem bristling with braggadocio followed by a chorus of pathetic sabre rattling: 'Heroes of the sea, noble race/Valiant and immortal nation/ … /To arms, to arms/On land and sea!/To arms, to arms/To fight for our Homeland!/To march against the enemy guns!' **[0]**

SCORE: 45 **WORLD RANKING: 163**

Qatar

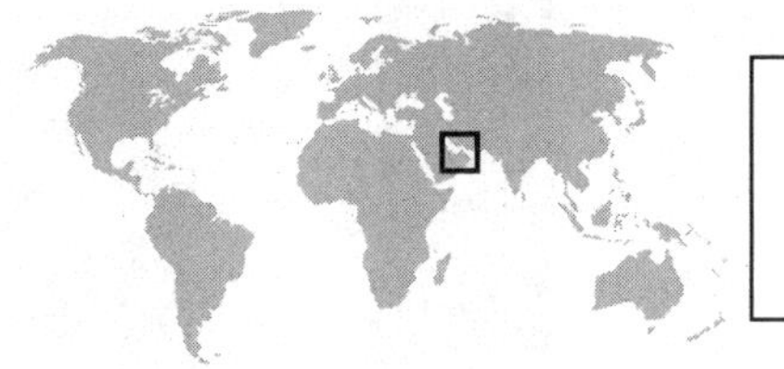

Name on driving licence: State of Qatar
Capital: Doha
Population: 522,000
Dosh: Qatari riyal = 100 dirham
Size: 11,400 km² (0.6 Wales)

Complete history: The crumpled fag end that is Qatar lounges insouciantly in the ashtray of the Persian Gulf and has done for ages. The lounging is about all that's happened here if you don't count the fact that the serrated bit of the flag was put in at the request of the British, who'd never seen such a thing done before and wondered what it would look like.

Crowning achievement: Qatar is the world's only Q (until the Quebecois get their way), and one of a select band with no income tax. The reason for this is that in 1940, deep under the sands, black gold was found. This soon ran out, but they've made do with oil reserves ever since. **[9]**

Sad fact: Due to the many expatriate petrochemical employees in Qatar, the country suffers from the greatest imbalance between the sexes anywhere in the world – 1.88 men to 1 woman. So, if you're a woman and wish to marry 1.88 oil workers, you know where to go. **[7]**

Customs to treasure: Camel racing. At the Shahaniya course, the camels whizz around an 18-km circuit while spectators whizz around behind them in cars. It's great. **[2]**

Pub fact: Qatar will soon become the world's leading supplier of liquefied natural gas, which is ace because it's 600-times denser than ordinary gas, thus giving you more room in your bedroom for your hi-fi and stuff. **[8]**

National anthem that kicks off with an unusual amount of swearing: 'Swearing by God who erected the sky/Swearing by God who spread the light.' **[3]**

SCORE: 29 **WORLD RANKING:** 191

Romania EUROPE

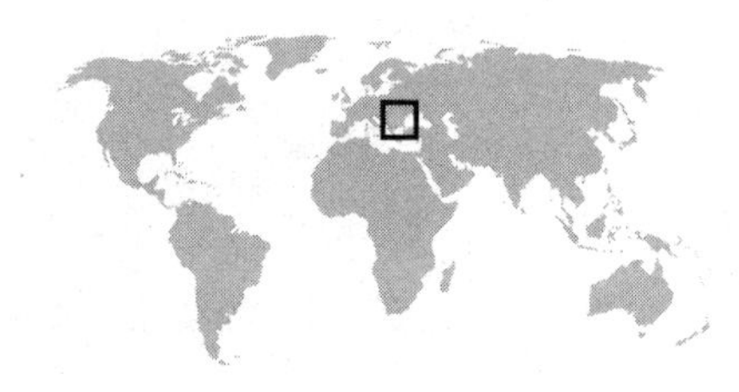

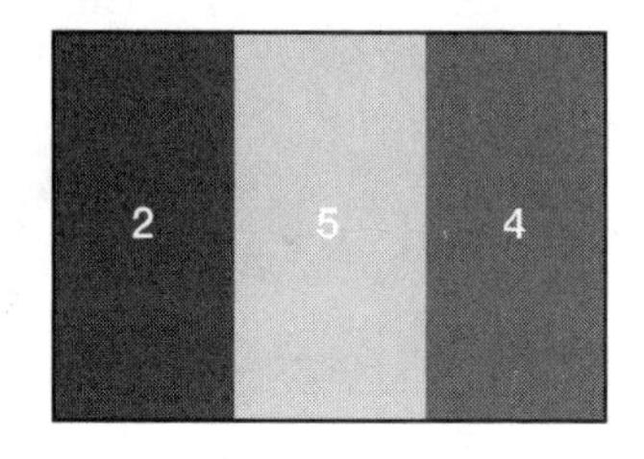

Name on driving licence: Romania
Capital: Bucharest
Population: 22 million
Dosh: Leu = 100 bani
Size: 237,500 km^2 (12 Wales)

Complete history: The Dacians so loved their Roman conquerors, they renamed their country Romania. They didn't much care for the Ottoman-Turks though.

Top spot: Transylvania, birthplace of Vlad III the Impaler whose surname, *Draculea*, was half-inched by Bram Stoker. Though keen on impaling his enemies, Vlad is not known to have been a light-fearing, blood-sucking, reanimated corpse. **[12]**

Crowning achievement: A wholly refreshing attitude to nomenclature. The dogged former Soviet satellite is not the 'Republic of Romania', the 'Kingdom of Romania', the 'Independent State of Romania', or even the 'Third Coming of the Holy Romanic Empire', but just Romania. It kind of makes you like them more already. **[18]**

Made in Romania: Tennis player and sometime novelist Ilie Nastase, winner of the US and French Open events but still best known for donning a rugby shirt at Wimbledon. For the record, he wore a red one for Romania, while his doubles partner Jimmy Connors wore green for Ireland, which was a bit of a shock for the Irish who had, up until that point, assumed he was American. **[13]**

Sad fact: After decades of living in an officially atheist country, Romanians naturally forgot how to celebrate Christmas. The worst example of failed Christmas Day revelries occurred in 1989 when they attempted to cheer themselves up by executing Nicolae and Elena Ceauşescu with a sub-machine gun. **[0]**

Opening lines of national anthem: 'Awaken thee, Romanian, shake off the deadly slumber/The scourge of inauspicious barbarian tyrannies.' The anthem continues in like manner for eleven verses, by which time the deadly slumber has well and truly returned. **[4]**

SCORE: 47 **WORLD RANKING:** 150

EUROPE/ASIA Russia

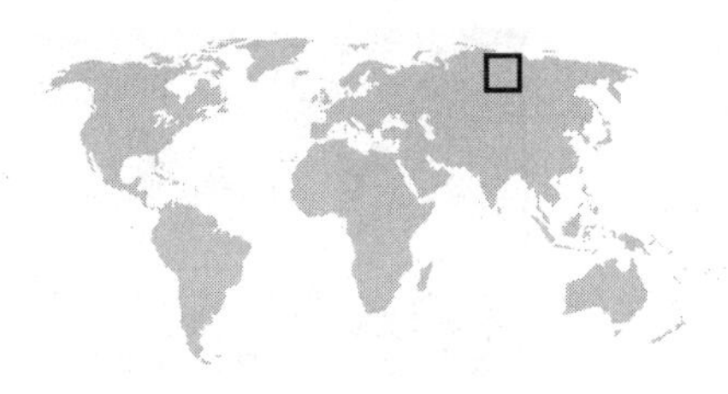

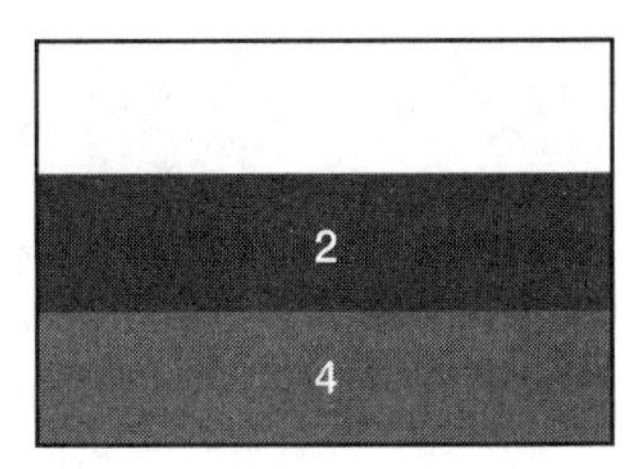

Name on driving licence: Russian Federation
Capital: Moscow
Population: 144 million
Dosh: Russian rouble = 100 kopeks
Size: 17,075,200 km² (824 Wales)

Complete history: Mother Russia started really late, as Kievan Rus, in about 880. However, she soon made up for her dilatoriness by giving birth to Ivan the Terrible and, much later, Joseph Stalin. She also owns Kaliningrad, an exclave that all the other Baltic coast countries pretend isn't really there.

Crowning achievement: Russia is the world's largest country, sharing borders with Norway and North Korea. Indeed, before it sold Alaska to the US in 1867 (for US$7.2 million which, even at today's equivalent of US$1.7 billion is a bit of a bargain if you like snow), Russia was in the unique position of forming part of three continents. **[13]**

Top spot: Siberia, possessor of 53,000 rivers and more than a million lakes and, even more cheerily, no longer home to the Soviet Gulag. **[14]**

Pub fact: The 9,288-km-long Trans-Siberian railway is the longest in the world. It passes through seven time zones, although this is not the only reason it seems to take forever to get to Vladivostok. **[16]**

Gifts to the world: The backpack parachute (Gleb Kotelnikov, 1911), synthetic rubber (Sergei Lebedev, 1928), t.A.T.u. (Elena Katina and Yulia Volkova, 1999). **[15]**

Opening lines of national anthem: 'O Russia, you are forever a strong sacred country!/O Russia, forever the land that we love!' I don't know about you, but these seem disappointingly bland lyrics for a former superpower. Better, surely, to have settled for a couple of defiant choruses of 'They're not gonna get us' (see Gifts to the world, above) which would, if nothing else, liven up international sporting events. **[5]**

SCORE: 63 **WORLD RANKING: 61**

Rwanda

AFRICA

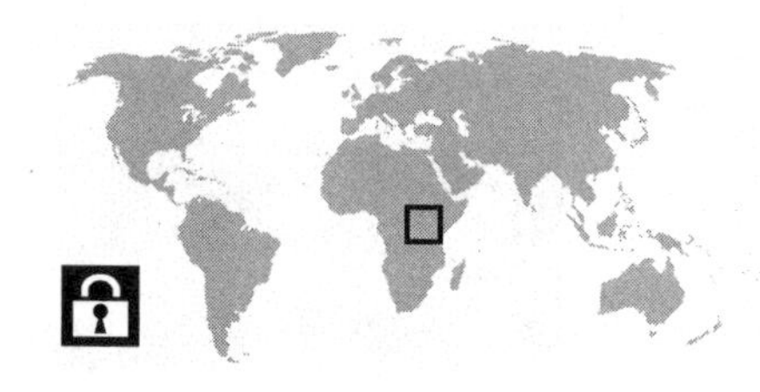

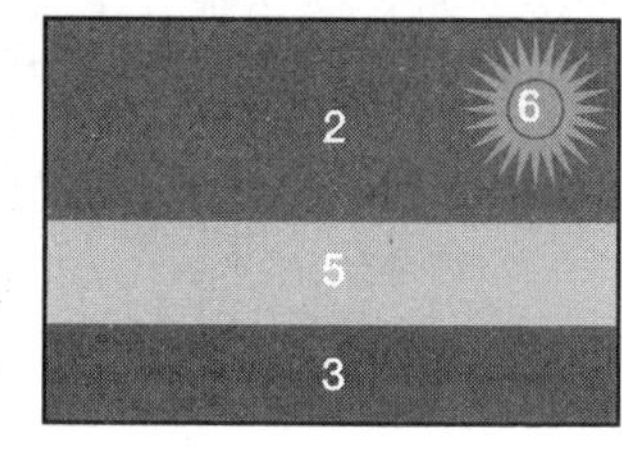

Name on driving licence: Republic of Rwanda
Capital: Kigali
Population: 8 million
Dosh: Rwandan franc
Size: 26,300 km^2 (1.3 Wales)

Complete history: The 'Land of a Thousand Hills', like Burundi, has a Hutu/Tutsi thing going on, only more disastrously so. It wasn't always thus. There were the poor put-upon Twa pygmies first, and then the rather less put-upon Germans (1890), followed by the not-put-upon-but-the-butt-of-occasional-jokes Belgians (1916). A Hutu/Tutsi civil war (won by the Hutus) claimed 150,000 lives as far back as 1959, so it's a wonder that, come 1994, there were still 800,000 Tutsis left in the country to be slaughtered.

Top spot: Mulindi, as featured in the internet video phenomenon (wherethehellismatt.com) wherein a rather personable American called Matt dances in front of a vast array of places in the world he has visited. It doesn't sound compelling but it is. **[15]**

Pub fact: The Nyungwe Forest resounds to the tweet of the strange weaver (*Ploceus alienus*), a bird that cannot weave, as far as can be established, and whose strangest feature is the fact that its scientific name used to be *Hyphanturgus alienus*, which also means strange weaver. **[6]**

Sad fact: Between 2004 and 2006, at least seven feature films about the 100-day genocide were released, including one (*Back Home*) made by a survivor. **[14]**

Made in Rwanda: The mountain gorilla, which would almost certainly have died out altogether in Rwanda were it not for primatologist Dian Fossey, of *Gorillas in the Mist* fame, whose 1985 murder, probably at the hands of those who wanted to exploit the gorillas for tourism purposes, remains unsolved. **[12]**

Opening lines of national anthem: 'Rwanda, our beautiful and dear country/Adorned by hills, lakes and volcanoes.' **[8]**

SCORE: 55 WORLD RANKING: 110

colour key: 1 = turquoise 2 = blue 3 = green 4 = red 5 = yellow 6 = orange 7 = pink 8 = purple 9 = brown

St Kitts & Nevis

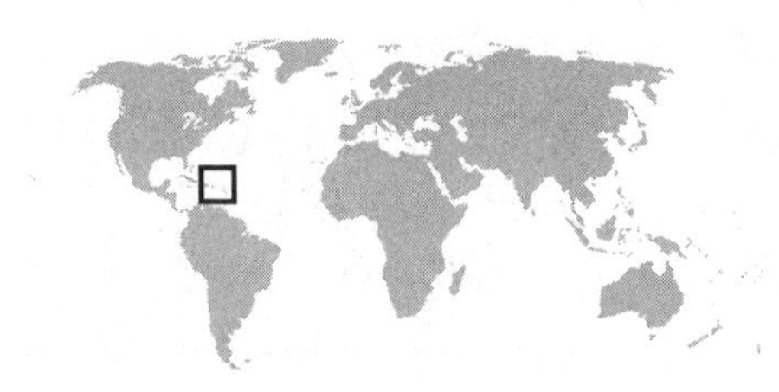

Name on driving licence: Federation of Saint Christopher and Nevis
Capital: Basseterre
Population: 39,000
Dosh: East Caribbean dollar = 100 cents
Size: 260 km² (0.01 Wales)

Complete history: While St Kitts is merely the 'Gibraltar of the West Indies', St Nevis is the 'Queen of the Caribbean', which is possibly why the majority of the 12,106 Nevisians are keen to secede.

Crowning achievement: St Kitts was the first Caribbean island that French and British settlers felt able to share without ripping each others' throats out (1627–1713). The fact that they used the island to continue their evil colonial deeds can hardly be said to be the fault of the island itself. **[10]**

Top spot: Brimstone Hill Fortress (built from around 1690 to 1800), a fine example of the then revolutionary 'polygonal system', was forever being attacked by the lunatic French. Built on and mainly *of* the horribly precipitous Brimstone Hill, the castle stands as tribute to British engineering and architecture, though it's unlikely the African slaves who died constructing it ever quite saw it in that light. **[8]**

Made in St Kitts and Nevis: The mongoose, the islands' most prevalent wild animal, was introduced to keep the rats from the sugar. Foolishly, as it turned out, because mongooses are diurnal, while rats are nocturnal, so the twain only ever met at dusk and dawn. **[14]**

Pub fact: St Kitts and Nevis was the last country in the Eastern Caribbean to base its economy entirely on sugar, calling a halt to proceedings only in 2005. Now the islands cultivate tourists, though without necessarily harvesting them by setting fire to them and hacking them down with machetes. **[7]**

Anthem proud of the nation's free-standing children: 'O Land of Beauty!/ ... /Thy children stand free.' **[12]**

SCORE: 51 **WORLD RANKING: 130**

St Lucia

AMERICAS

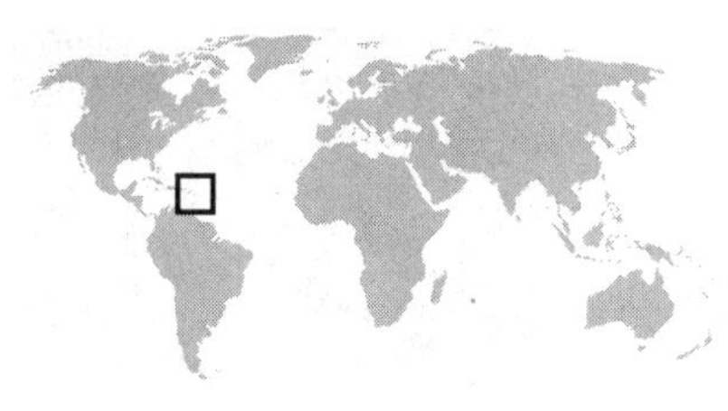

Name on driving licence: Saint Lucia
Capital: Castries
Population: 162,000
Dosh: East Caribbean dollar = 100 cents
Size: 620 km^2 (0.03 Wales)

Complete history: After the usual Arawak/Carib business, the British and French fought fourteen separate wars for St Lucia before the latter took it 'for good' in 1814. 'For good' lasted until 1979, when the St Lucians got to have a go.

Crowning achievement: St Lucia boasts the greatest concentration of Nobel Prize winners per head of population. Step forward Derek Walcott (Literature, 1992) and Sir William Lewis (Economics, 1979). **[17]**

Top spot: Qualibou Soufrière, which the local tourist board delights in calling a 'drive-in volcano', as if that were a good thing. It's actually a collection of steaming jets and boiling pools in a caldera that last erupted around 40,000 years ago. **[8]**

Not made in St Lucia: St Lucia. She was martyred at the age of 21 at Syracuse, Sicily, in AD 304, probably meeting her end by having a sword thrust through her throat. Despite this, she is not the patron saint of scarf-makers but of blindness, and portraits often show her carrying a tray with two eyes on it, which *is* a good thing and not at all weird. **[16]**

Flag fact: Although St Lucians protest that their flag represents the island's Pitons Mountains surrounded by the blue of the Caribbean, the design has quite clearly been lifted from the insignia worn by the infinitive-splitting James T. Kirk and his doughty crew. **[15]**

Opening lines of national anthem masquerading as a nursery rhyme: 'Sons and daughters of Saint Lucia love the land that gave us birth/Land of beaches, hills and valleys, fairest isle of all the earth.' **[12]**

SCORE: 68 WORLD RANKING: 31

St Vincent & the Grenadines

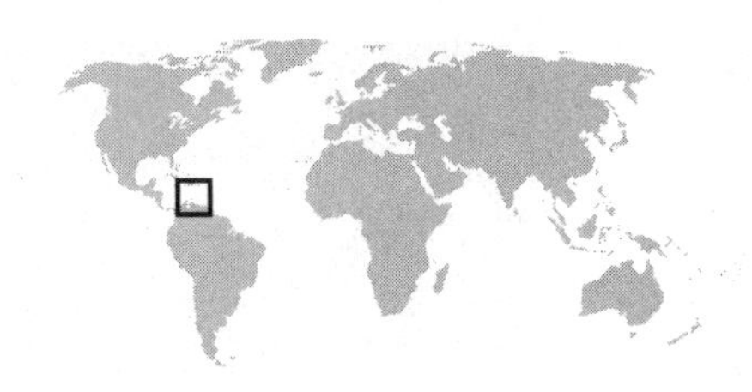

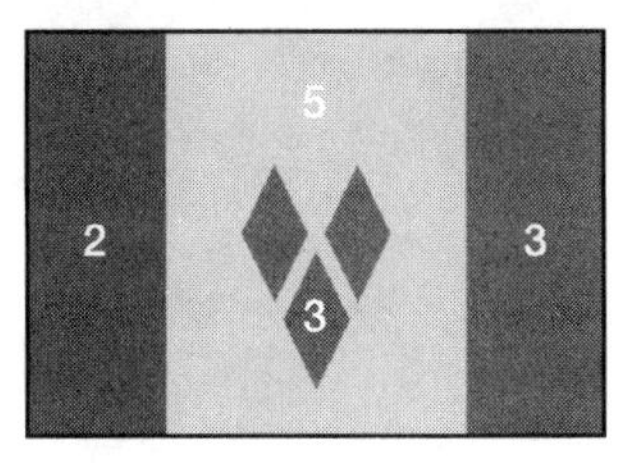

Name on driving licence: St Vincent and the Grenadines
Capital: Kingstown
Population: 117,000
Dosh: East Caribbean dollar = 100 cents
Size: 390 km^2 (0.02 Wales)

Complete history: St Vincent was named Hairoun ('land of the blessed') by the marauding Carib invaders who, along with escaped African slaves, fought off all-comers until the early 18th century. Good effort, chaps.

Top spot: One of the Grenadines is Mustique, the privately owned island frequented by Princess Margaret, Mick Jagger, Kate Moss et al. (and, err, Bryan Adams). **[11]**

Pub fact: The 'St Vincent' in question was St Vincent of Saragossa, whose feast day coincided with Columbus' arrival in 1498 and who, coincidentally, was martyred in the same year as St Lucia (see previous page). The choice of Vincent of Saragossa is a persistent source of upset and consternation to followers of the 28 other St Vincents, who include such notables as Vincent Ferrer (patron saint of brick-makers and pavement workers) and Vincent de Paul (horses, hospitals and lepers), star of his own Jean Anouilh film. **[10]**

Sad fact: The Soufrière Volcano erupted in 1902, killing 1,700 people. Like Elvis, Soufrière enjoyed a 70s revival, erupting in both 1971 and 1979. **[6]**

Pub fact: St Vincent is a volcanic island, whereas the Grenadines are flat coral reefs that look just like grenadines if you've never actually seen a grenadine and believe it to be a sort of flat coral reef of a fruit. **[11]**

Opening line of national anthem: 'Saint Vincent! Land so beautiful.' Grenadines fans have to wait until the last verse for a mention and then it's only the rather patronising: 'Our little sister islands are/Those gems, the lovely Grenadines.' **[7]**

SCORE: 45 **WORLD RANKING: 163**

Samoa

OCEANIA

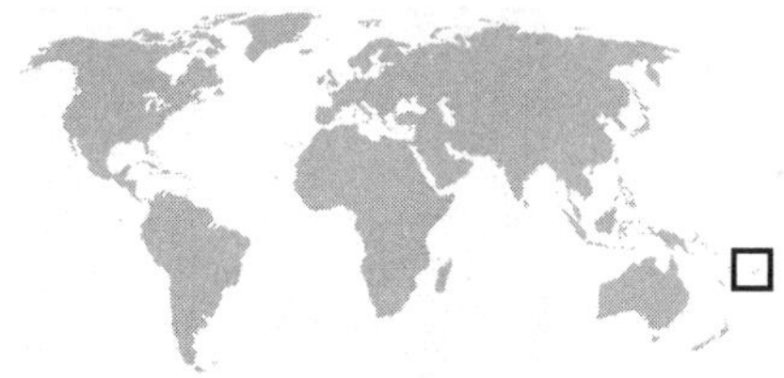

Name on driving licence: Independent State of Samoa
Capital: Apia
Population: 178,000
Dosh: Tala = 100 sene
Size: 2,850 km^2 (0.1 Wales)

Complete history: Samoa (lit. 'sacred chickens') is home to a bellicose people who fought other Polynesian islanders, the British, the French, the Americans and each other. The Germans took over the western isles, calling them German Samoa (lit. 'German sacred chickens'), while the eastern isles became American Samoa (lit. 'American sacred chickens'). The Kiwis kicked the Germans out in 1914, made a pig's ear of running the country, and were themselves asked to leave by Mau (lit. 'strongly held opinion'), a non-violent popular movement. Since independence (1962), Samoans have devoted themselves to beating much larger countries at rugby.

Top spot: The enormous Alofa'aga Blowholes. These act in a very similar way to the ones on whales' backs, except that our cetacean friends are not so dependent on a really big tide in order to look impressive. **[10]**

Made in Samoa: An adult *Patu* measures 0.43 mm across its mighty frame, making it the world's smallest spider. **[13]**

Customs to treasure: *Kirikiti*, a Samoan form of cricket now popular throughout Polynesia. The bat used is a three-sided affair, making the trajectory of the thwacked rubber ball pleasingly unpredictable. Teams can be of any number and most rules are flexible, the only constant being that the home side forfeit the game if they cannot adequately feed their opponents. **[16]**

Popular misconception: 'It's the home of the samosa.' **[12]**

Opening lines of national anthem that appear not to blush at their own control freakery: 'Samoa, arise and raise your banner that is your crown!/Oh, see and behold the stars on the waving banner!' **[5]**

SCORE: 56 **WORLD RANKING:** 101

EUROPE San Marino

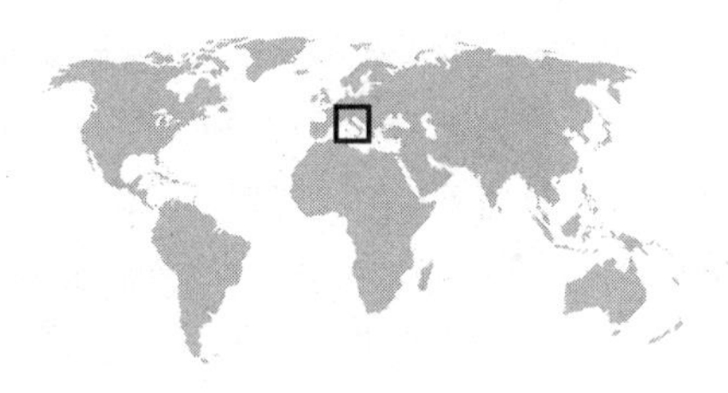

Name on driving licence: The Most Serene Republic of San Marino
Capital: San Marino
Population: 29,000
Dosh: Euro = 100 cents
Size: 61 km² (0.003 Wales)

Complete history: Saint Marinus, the Dalmatian stonemason, is credited with founding a community on Monte Titano on 3 September 301, the precision of the date being a feature of which the Sammarinese are justly proud. They cleverly escaped being amalgamated into Garibaldi's new-fangled 'Italy' (which now surrounds them) by giving him refuge from the dastardly Austrians in 1849 and thus claiming later on that he 'owed them one'.

Crowning achievement: Forging the world's longest-lived republic out of what is, ostensibly, a large hill. **[18]**

Made in San Marino: By a convoluted system of pulleys, mirrors and trap-doors, the locals have managed to avoid producing a single famous son or daughter since the military architect Giambattista Belluzzi (1506–54) burst onto the scene with his designs for useful wartime buildings. **[16]**

Pub fact: An estimated 10 per cent of San Marino's gross national product comes from the sales of postage stamps and coins. **[10]**

Sad fact: San Marino has no home-grown *carabinieri*, so the country has to depend on the Italian police force, an institution incapable of appearing in a sentence that does not also feature the word 'corrupt'. See? It's done it again. **[5]**

Opening lines of national anthem: None. Indeed, there are no closing lines either, for the Sammarinese have wisely chosen a wordless 10th-century chorale for their anthem. A further advantage is that this leaves them with the option to hum, la-la-la, or whistle along to the tune at monthly state occasions such as the ceremonial 'first licking of the new postage stamp'. **[18]**

SCORE: 42 **WORLD RANKING: 171**

São Tomé and Príncipe AFRICA

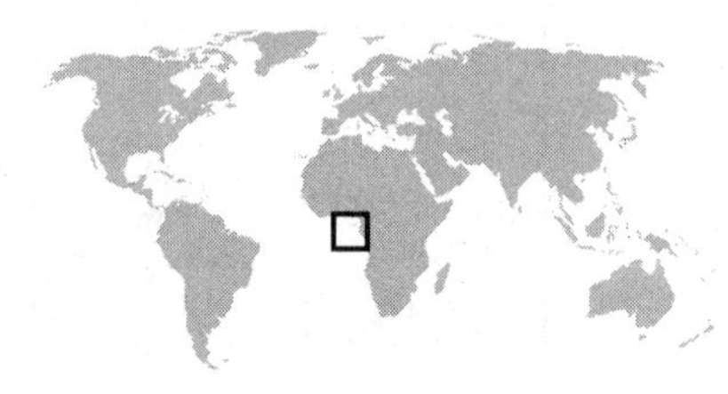

Name on driving licence: Democratic Republic of São Tomé and Príncipe
Capital: São Tomé
Population: 162,000
Dosh: Dobra = 100 céntimos
Size: 1,000 km^2 (0.05 Wales)

Complete history: You have to raise some questions about islands that were uninhabited until the late 15th century. For one thing, they have never provided a very prosperous home for anyone but the landowning elite. They've also suffered the usual litany of European exploitation (by the Portuguese mainly, but the Dutch and French also chipped in), slave revolts, massacres and cocoa. By the late 19th century São Tomé had became the world's second-largest cocoa exporter, but things fell apart in 1909 when the chocolate company Cadbury boycotted Santomean produce in protest at slave-like working conditions.

Top spot: Obô, classified in 1988 as Africa's second-most important forest in terms of biodiversity. **[12]**

Customs to treasure: Auto de Floripes, which takes place twice every August, involves the greater part of the population of Príncipe in a re-enactment of past tussles between Moors and Christians. Bring plasters. **[10]**

Sad fact: Cocoa accounts for 90 per cent of export income, which is fine up to but not including the point where: a) the crop fails, or b) it is discovered that chocolate causes cancer. **[5]**

Pub fact: São Tomé is unique among nations in knowing what day of the year it was discovered but not which year. The name indicates that the island was first seen by Portuguese explorers on the feast day of St Thomas (21 December), but no one knows for sure if it was in 1470 or 1471. **[13]**

Opening lines of national anthem: 'Total independence/Glorious song of the people.' As sung since 1975. **[11]**

SCORE: 51 **WORLD RANKING:** 130

MIDDLE EAST Saudi Arabia

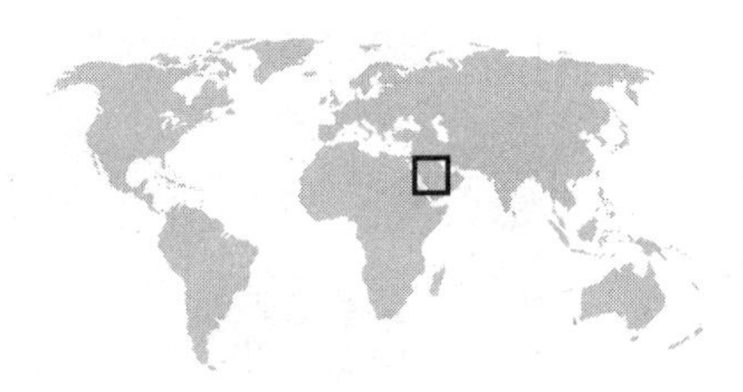

Name on driving licence: Kingdom of Saudi Arabia
Capital: Riyadh
Population: 26 million
Dosh: Saudi riyal = 100 halalas
Size: 1,960,600 km^2 (104 Wales)

Complete history: Saudi Arabia as we know it didn't really get going until 1744. Nowadays, King Abdullah runs the country along the austere lines of the Hanbali interpretation of sharia law, which, conveniently enough, sidesteps the fact that sharia law forbids the existence of monarchies.

Top spot: The city of Mecca, a trip to which is one of the five pillars of Islam, though don't bother if you're not a Muslim because you won't be allowed in, which seems a mite harsh. After all, most cinemas will let you in to see a Star Wars film even if you didn't declare your religion as 'Jedi' in the last census. **[3]**

Sad fact: Among other restrictions on their lives on account of their sex, women are not allowed to drive or travel without written permission from a male guardian. **[0]**

Pub fact: Saudi Arabia has more oil reserves than anywhere else, which is why the developed world doesn't get too hot under the collar about the rights of Saudi women, or indeed the kingdom's complete, total and utter lack of anything even smelling of democracy. **[2]**

Made in Saudi Arabia: Ar Rub' al-Khali ('Empty Quarter'), the largest uninterrupted expanse of sand anywhere in the world. Yep, including the Sahara. **[16]**

Opening lines of national anthem: 'Hasten to glory and supremacy!/Glorify the Creator of the heavens.' This is followed by a lot of chest beating and the sort of overblown declarations one hears all too often in anthems, although at least this one's mercifully short. **[6]**

SCORE: 27 **WORLD RANKING: 195**

Scotland (*see also* United Kingdom) EUROPE

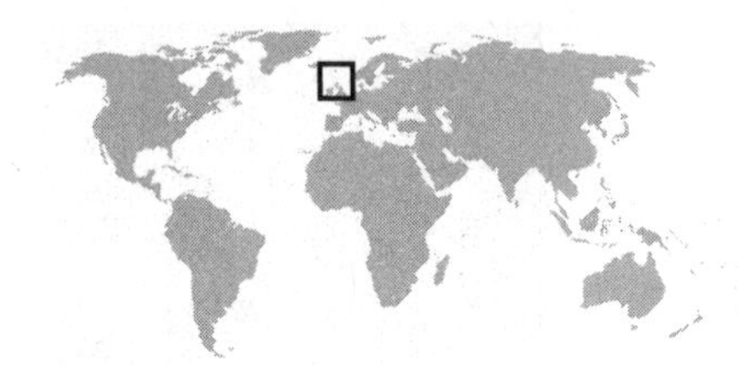

Name on driving licence: Scotland
Capital: Edinburgh (Dùn Èideann)
Population: 5.1 million
Dosh: Pound sterling = 100 pence
Size: 78,800 km² (3.8 Wales)

Complete history: Speak not the dread words '1 May 1707' to the Scots, for it was then that they were joined in unhealthy matrimony with the English (and the Welsh, incidentally, who were already in harness) to form something called the United Kingdom of Great Britain. Although some Scottish folk are now resigned to this, there are many otherwise sane Scots who would rather starve free in a bothy than spend another minute enslaved in the gulag that the English have made of Scotland. Despite the prison-camp atmosphere, the country is still pretty popular with tourists.

Popular misconception: 'All Scots eat deep-fried Mars bars.' Dinna e'en thenk it. Many Scottish people do not partake of this delight due to batter-related allergies. **[14]**

Where to avoid: Gretna, officially the world's most depressing town, and not just due to its proximity to England. **[3]**

Pub fact: Scotland possesses the largest wilderness in Western Europe. **[17]**

Made in Scotland: Three of the scariest people to stalk the planet: William Wallace (c.1272–1305), Robert the Bruce (1274–1329) and Ronnie Corbett (1930–). **[15]**

Anthem: Another country with competing claims to the title of official anthem. The front runners would have to be 'Flower of Scotland', a happy modern ditty about all the Scots who died at Bannockburn (a rare battle the Scots actually won, so they're rather fond of it); 'Scotland the Brave', a sugary sweet blend of 'misty Highlands' and 'purple islands'; and Robert Burns' 'Scots Wha' Hae', which unfortunately does nothing to dispel the Sassenach idea that the Scots are forever littering their conversation with this expression. **[10]**

SCORE: 59 **WORLD RANKING:** 85

AFRICA Senegal

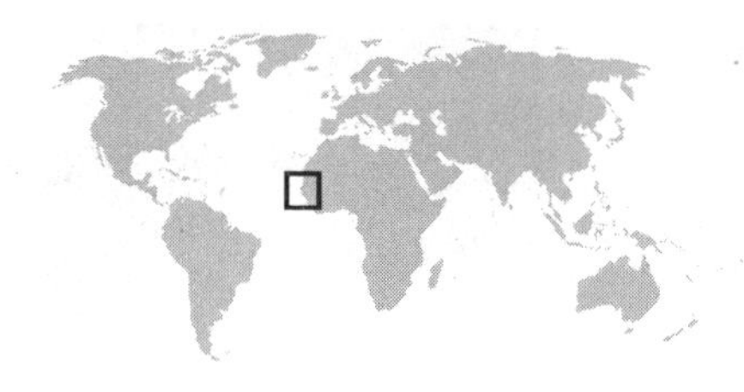

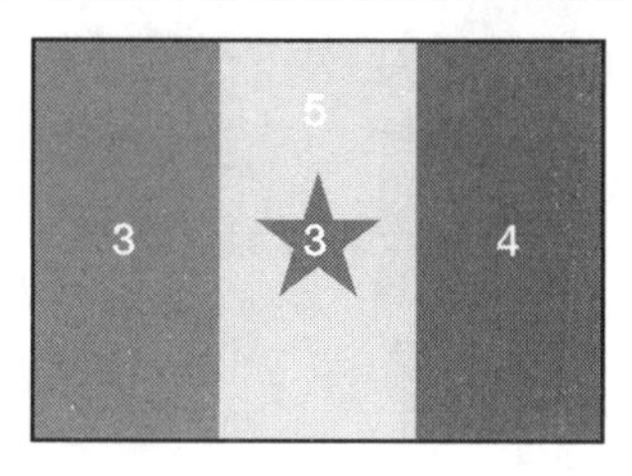

Name on driving licence: Republic of Senegal
Capital: Dakar
Population: 11 million
Dosh: CFA franc = 100 centimes
Size: 196,200 km^2 (9.5 Wales)

Complete history: In 1960 Senegal stopped being part of Mali and became the independent republic of Wartorn Senegal. Before all this, there was the usual awfulness of the slave trade run by the French, though to be fair, it was also run at times by the Portuguese, the Dutch and the British as each in turn darkened this stretch of coastline.

Crowning achievement: Hospitality. The Senegalese are so good at it that the word for it – *Teranga* – has taken on a life of its own and can no more be uttered without at the same time mentioning its sheer *legendariness* than a man can look at a blancmange falling from a train without imagining his own brain being dashed against the sleepers. **[16]**

Top spot: Dakar, which – like Côte d'Ivoire's Abidjan – claims to be the 'Paris of Africa'. Frankly, both assertions are about as plausible as Edinburgh's pretensions to being the 'Athens of the North'. Now, if Dakar lowered its sights and declared itself, say, the 'Amiens of Senegal', it might get a few more listeners. **[6]**

Made in Senegal: Youssou N'Dour, poster boy for world music and bringer of mbalax to the masses, though of course the earlier non-Westernised albums are the best ones. **[15]**

Pub fact: Senegal is the sixth-largest exporter of groundnuts. Ironically, these are usually consumed whole. **[11]**

Opening lines of national anthem: 'Pluck your koras, everyone/Beat the drums.' Upsettingly, it turns out that koras are some sort of harp–lute constructions rather than eyebrows. Still, the tune is very dancy. **[12]**

SCORE: 60 **WORLD RANKING:** 76

Serbia

EUROPE

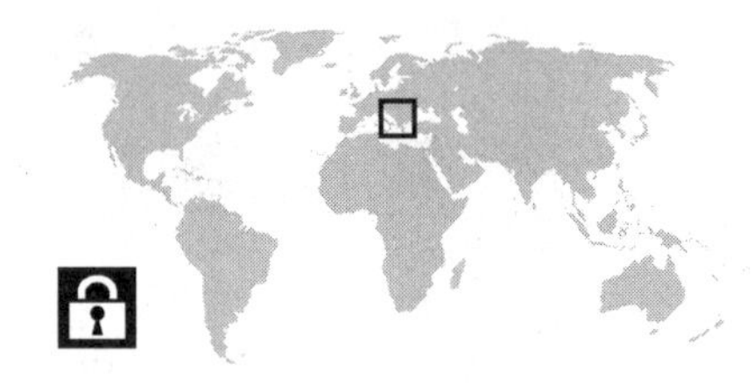

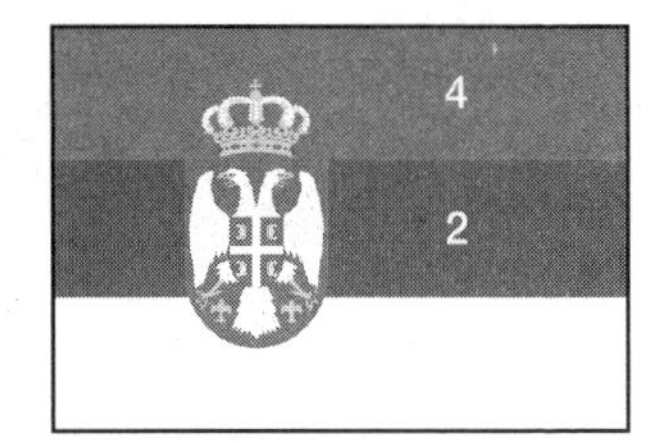

Name on driving licence: Republic of Serbia
Capital: Belgrade
Population: 9.4 million
Dosh: Serbian dinar = 100 para
Size: 88,361 km² (4 Wales)

Complete history: Serbian history goes back to a 7th-century leader spectacularly known as the Unknown Archont. However, since picking the right side in World War II, the Serbs have distinctly blotted their copybook.

Crowning achievement: Technically, Serbia is the world's newest country. Although Montenegro seceded from the State Union of Serbia and Montenegro on 3 June 2006, the Serbians only got around to counter-declaring their own independence, which they'd wanted 'for, like, ages' (it's difficult to imagine who they thought they were fooling), two days later. **[13]**

Gift to the world: Electrical physicist Nikola Tesla (1856–1943), born to Serbian parents (remember this for later) in the village of Smiljan, and who 'invented the 20th century'. In true Balkan fashion, there are many counter-claims to Tesla: by Austria (in 1856 Smiljan was still in the Austrian Empire), Croatia (where Smiljan is now), Hungary and France (where Tesla worked) and the US (where he ended up), but unusually, this hasn't led to war between the quarrelling parties. **[13]**

Popular misconception: 'It's always raining frogs in Serbia.' *Au contraire*, this hasn't occurred since the village of Odzaci was so afflicted in 2005. **[10]**

Pub fact: The Serbian form of *slivovitz* (every European country whose language uses a profusion of wž, šj or ćz sounds is bound by law to distil its own version) accounts for a cool 70 per cent of the country's 400,000-tonne plum harvest. **[8]**

National anthem cementing Serbia's reputation as a country of fun-loving jokers: 'God of Justice/Thou who saved us when in deepest bondage cast/Hear Thy Serbian children's voices/Be our help as in the past.' **[5]**

SCORE: 49 **WORLD RANKING:** 140

colour key: 1 = turquoise 2 = blue 3 = green 4 = red 5 = yellow 6 = orange 7 = pink 8 = purple 9 = brown

AFRICA Seychelles

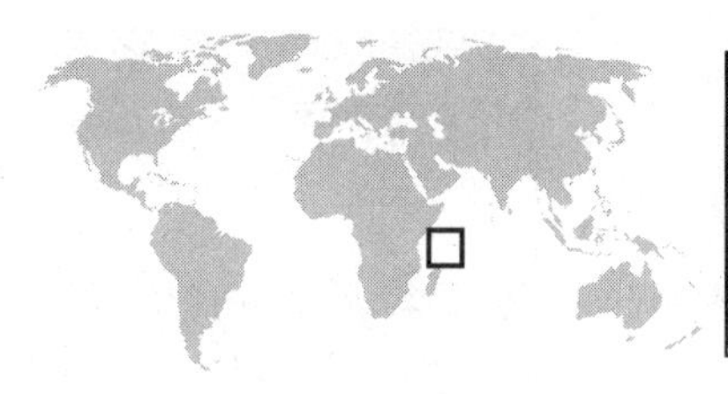

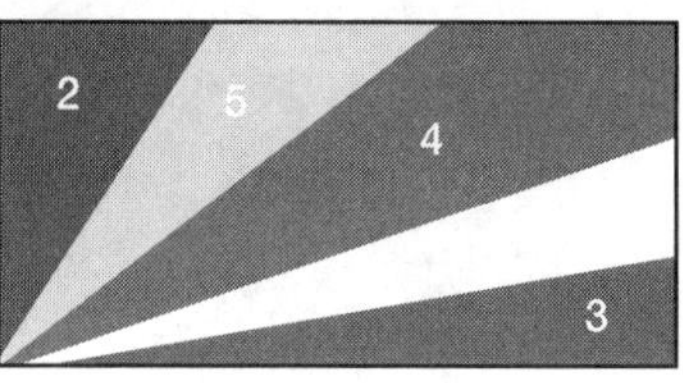

Name on driving licence: Republic of Seychelles
Capital: Victoria
Population: 81,000
Dosh: Seychelles rupee = 100 cents
Size: 450 km² (0.02 Wales)

Complete history: Take your pick between Austronesians and Arabs as the first visitors to Seychelles. It was definitely the Portuguese, however, who first reported the existence of an appealing archipelago in the Indian Ocean. This was a mistake because the French duly nabbed it, but lost it to the English, who lost it to the prevailing winds of progress. The islands are named after a French finance minister, which is about as dreary as it can possibly get.

Top spot: Aldabra, the world's largest atoll, is home to roughly five times as many giant tortoises as live on the Galapagos Islands. **[13]**

Popular misconception: 'The Seychelles is huge, which is why they're such a dominant world power.' An easy mistake to make but, astonishingly, Seychelles is Africa's smallest nation both by size and population. It's just the fact that the people are larger than life that makes it seem as if the fate of the world is largely in their hands. **[10]**

Pub fact: The largest seed known to humankind is found only in Seychelles. It's called the *coco de mer* (lit. 'clown of the sea'), it can weigh up to 45 pounds, and it grows into a giant palm fan if left to its own devices. **[12]**

Made in Seychelles: The Seychelles toc-toc, a bird famed throughout the isles for its ability to imitate the latter 50 per cent of a clock. **[10]**

Opening lines of national anthem: 'She sells Seychelles on the sea shore.' No, sadly not, it's the rather duller: 'Seychelles, our homeland/Where we live in harmony.' **[3]**

SCORE: 48 **WORLD RANKING: 146**

Sierra Leone

AFRICA

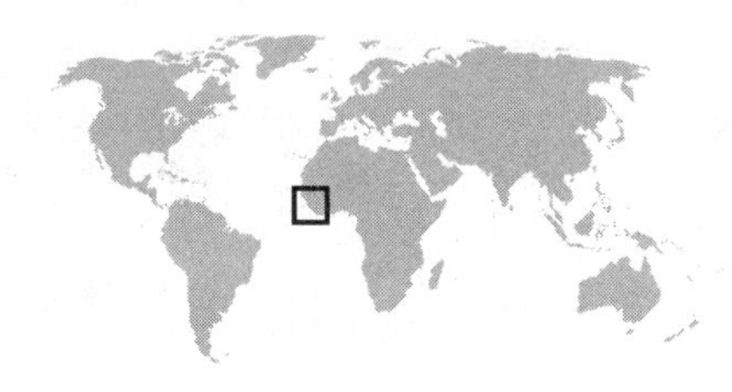

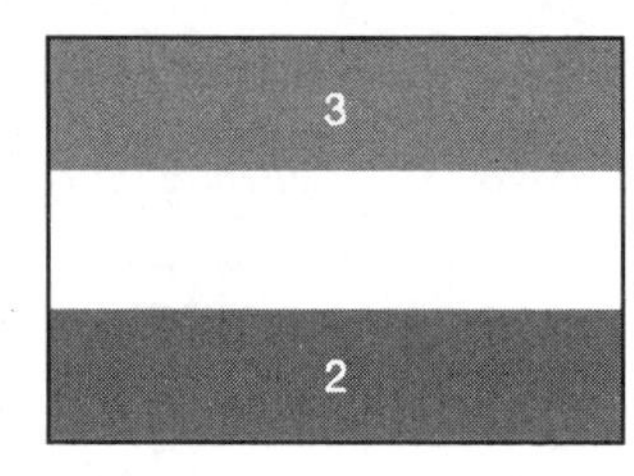

Name on driving licence: Republic of Sierra Leone
Capital: Freetown
Population: 5.9 million
Dosh: Leone = 100 cents
Size: 71,700 km^2 (3.5 Wales)

Complete history: The slaves have been freed, but the people are still in chains. Discuss.

Sad fact: After making that coveted last place in the United Nations' Human Development Index its own, the 2006 UN report disastrously placed Sierra Leone above Niger, making it only the second-worst place in the world to live. The survey also missed out some of the world's smaller countries as well as Afghanistan, Somalia and Iraq, so there's a chance that Sierra Leone is really only bottom-five material rather than out and out world champion. **[3]**

Pub fact: Two-thirds of the team representing Sierra Leone at the 2006 Commonwealth Games in Melbourne went missing. All fourteen athletes later appealed for political asylum, claiming they'd even rather endure the whole going-up-at-the-end-of-sentences-even-when-not-asking-a-question thing than go back to Freetown. **[5]**

Made in Sierra Leone: Graham Greene's *The Heart of the Matter*, which is based on his experiences as an MI6 agent in Freetown during World War II, but, like all his novels, is essentially about Catholicism, guilt, failure and the impossibility of truly knowing another person. **[10]**

Popular misconception: 'You can't find good rutile nowadays.' Yes, you can, there's some in Sierra Leone, although it's underground and digging it up is a ghastly business. Try settling for one of the other three forms of titanium dioxide found in nature. **[12]**

National anthem: 'So may we serve thee ever alone/Land that we love, our Sierra Leone.' Alone? Is that really the best rhyme they could come up with for Leone? What's wrong with 'bemoan' or 'overblown'? **[6]**

SCORE: 36 **WORLD RANKING: 181**

ASIA

Singapore

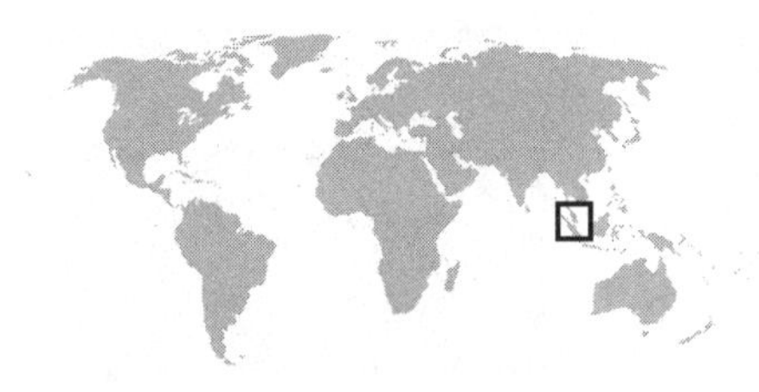

Name on driving licence: Republic of Singapore
Capital: Singapore City
Population: 4.4 million
Dosh: Singaporean dollar = 100 cents
Size: 650 km^2 (0.03 Wales)

Complete history: Legend has it that Singapore came into being in 1299, though presumably the island was there before that. Nothing much happened until Sir Thomas Stamford Raffles built a settlement there in 1819.

Top spot: Raffles Hotel, home of the Singapore sling. **[17]**

Customs to treasure: Chewing gum has been outlawed since 1992, on the grounds that it makes the pavements all sticky. The Americans, great believers in the right of free people everywhere to get stuck to pavements, pressurised Singapore into partially lifting the ban so that it no longer includes gums that help people to give up smoking. Still, anyone caught smuggling in 'normal' chewing gum faces a twelve-month prison term, which should give would-be miscreants something to chew over. Arf arf. **[14]**

Pub fact: In 1997, Singapore built itself the world's biggest fountain, the Suntec City Fountain of Wealth, an act that displayed an arguably unhealthy ignorance of the concept of hubris. **[7]**

Popular misconception: 'Singapore's national symbol, the Merlion (a sort of mermaid in which a lion deputises for the maid), comes from an ancient story involving a lion that came out of the sea to save someone from something or other, probably.' Oh that it were so. It was designed by a chap called Fraser Brunner in 1964 for the Singapore Tourism Board. The only lion to feature in Singaporean folklore was one apparently seen on land by some prince. **[6]**

Opening lines of national anthem: 'Come, fellow Singaporeans/Let us progress towards happiness together.' Sweeeet. Lots of holding hands and everything. **[16]**

SCORE: 60 **WORLD RANKING:** 76

Slovakia

EUROPE

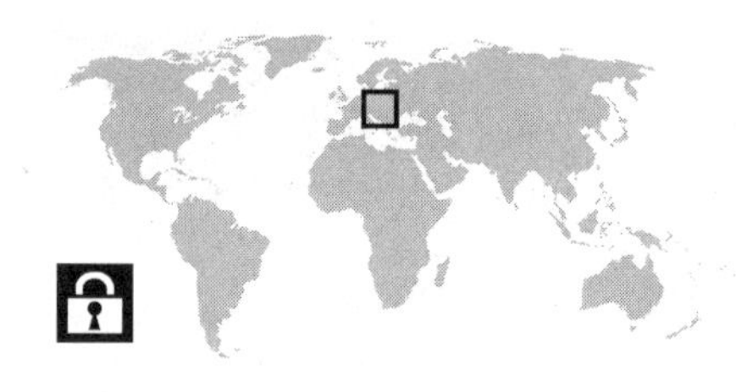

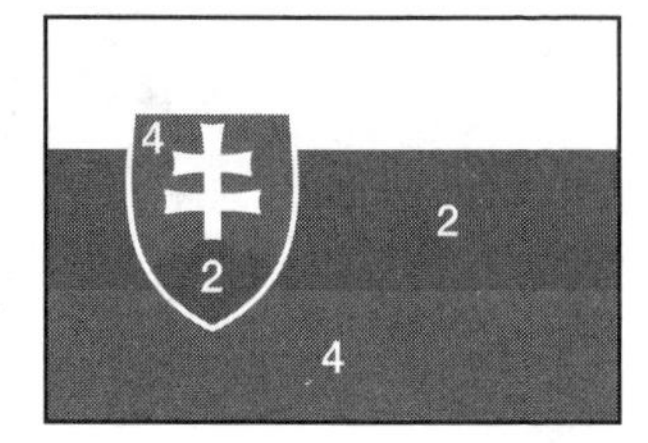

colour key: 1 = turquoise 2 = blue 3 = green 4 = red 5 = yellow 6 = orange 7 = pink 8 = purple 9 = brown

Name on driving licence: Slovak Republic
Capital: Bratislava
Population: 5.4 million
Dosh: Slovak koruna = 100 halierov
Size: 48,800 km^2 (2.4 Wales)

Complete history: The Slavs around here had to put up with being Hungarian for about a thousand years, and then a lowly suffix to Czecho-. They're understandably a bit bitter.

Crowning achievement: The summit of a hill called Krahule is the geographical centre of Europe. Block your ears to the counterclaims from other pretenders to the throne such as the Belarusians, Estonians, Lithuanians, Ukrainians and, especially, the Poles, who confusingly claim they have two places that are both the centre of Europe, for these people are all venal liars who would sell their grandmothers for some dodgy geographical 'evidence' to back up their wafer-thin cases. **[15]**

Top spot: Spiis, the largest castle in central Europe, wherever that is. **[10]**

Gifts to the world: The wireless telegraph, forefather of the radio (Jozef Murgaš, 1864–1929), and the water pillar, as used to pump water from mines (Jozef Karol Hell, 1713–89). Without these two Slovaks, we would still have to run down the valley to the pit owner to tell him the mine's been flooded. **[13]**

Pub fact: Bratislava has been the capital of two different countries. Before doing service as the top town of Slovakia, it did a 250-year stint as Pozsony, capital of Hungary. **[14]**

Entire national anthem: 'Lightning flashes over the Tatra, the thunder pounds wildly/Let them pause, brothers, they will surely disappear, the Slovaks will revive/This Slovakia of ours has been fast asleep until now/But the thunder and lightning are encouraging it to come alive.' For sheer out-and-out oddness, it's hard to imagine how this could be beaten. **[18]**

SCORE: 70 **WORLD RANKING:** 22

Slovenia

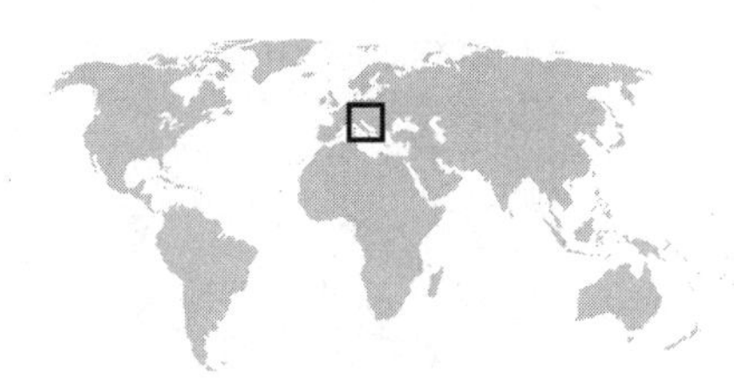

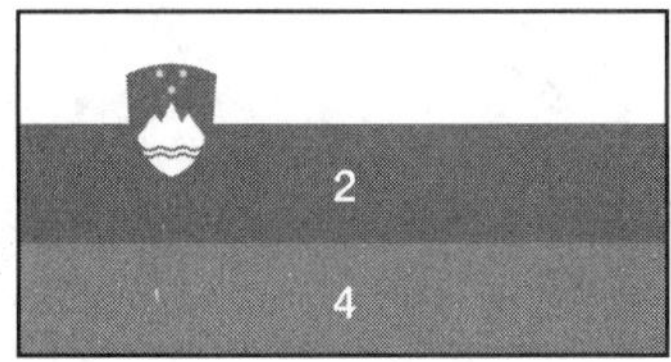

Name on driving licence: Republic of Slovenia
Capital: Ljubljana
Population: 2 million
Dosh: Tolar = 100 stotin
Size: 20,250 km^2 (1 Wale)

Complete history: The name Slovenia comes from *Slovnci* ('famous people'), the Slavs' rather self-aggrandising name for themselves. However, little were they to know that their fame would eventually derive from the misfortune of giving the English language the word for 'slave' (due to their being forced into slavery en masse by the Holy Roman Empire). The irony seeps from every pore.

Crowning achievement: Despite its lamentable history of enslavedness, Slovenia has contrived to become the best off post-Soviet nation, largely through being pretty and just sort of getting on with things. **[12]**

Top spot: It rather depends on who you are. Bled (mountains, lake, church on island, castle up above) for those who enjoy scenes from chocolate boxes customarily on sale at airports; Bohinj (mountains, lake, bridge) for cooler types who claim to prefer Nice over Cannes. Something for everyone, then. **[16]**

Made in Slovenia: The Lipizzaner horse – originally bred in Lipica to bear the substantial burden of the Austro-Hungarian nobility – is considered the Rolls-Royce of horse breeds now that they have been specially retrained to be ridden into swimming pools by rock stars. **[12]**

Pub fact: The Škocjan Caves form the world's largest underground system of canyons, the highest one of which can accommodate 33 double-decker buses on top of one other, though it does so only on special occasions. **[17]**

National anthem hearteningly free of jingoism: 'God's blessing on all nations/Who long and work for that bright day/When o'er earth's habitations/No war, no strife shall hold their sway/Who long to see/That all men free/No more shall foes, but neighbours be.' **[19]**

SCORE: 76 **WORLD RANKING:** 6

Solomon Islands AUSTRALASIA

colour key: 1 = turquoise 2 = blue 3 = green 4 = red 5 = yellow 6 = orange 7 = pink 8 = purple 9 = brown

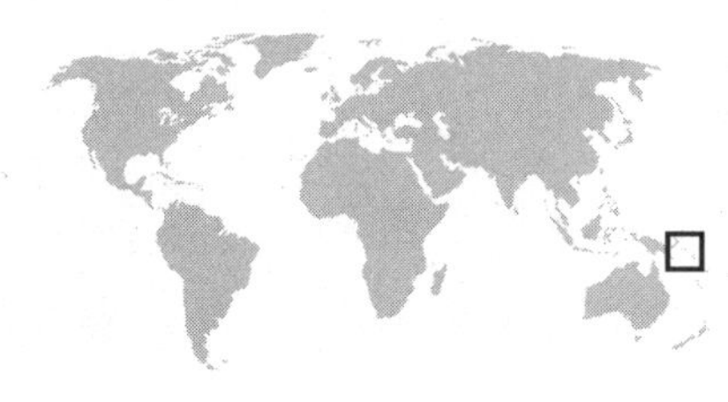

Name on driving licence: Solomon Islands
Capital: Honiara
Population: 524,000
Dosh: Solomon Islands dollar = 100 cents
Size: 28,450 km² (1.4 Wales)

Complete history: The Solomon Islands consists largely of hundreds of uninhabitable coral reefs, so it's likely that the Melanesians whom Pedro Sarmiento de Gamboa saw when he dropped by in 1568 were on the larger volcanic islands. The Solomons were British by the time of the Allies' successful Guadalcanal campaign in World War II, but for the minor detail of the Japanese occupation. Peace sort of reigned until about 2002 when there started an almighty scuffle between Guadalcanalians and Malaitans. The attendant chaos and wretched government response raised questions as to whether the Solomon Islands had become a so-called 'failed state'. The next year, a peacekeeping mission was sent in to find out.

Crowning achievement: Unlike the Bhutanese, Solomon Islanders steadfastly refuse to set up a television service, and are somewhat wary of the devil's lantern per se on the quite justifiable grounds that it introduces a dilution of home-grown culture for which the ability to watch endless repeats of *Friends* is scant recompense. **[18]**

Made in the Solomon Islands: Copra, aka the white bit of coconuts. **[6]**

Customs to treasure: Swearing is an imprisonable offence, making the Solomon Islands an understandably unpopular destination for sufferers of Tourette's syndrome. **[5]**

Popular misconception: 'The gold deposits on the Solomon Islands were the source of King Solomon's great wealth.' Sadly not, but the touchingly deluded Alvaro de Mendaña de Neira believed so when thinking up a name for the islands. **[3]**

Opening lines of national anthem: 'God save our Solomon Islands from shore to shore/Bless all her people and her lands.' **[12]**

SCORE: 44 **WORLD RANKING:** 167

AFRICA

Somalia

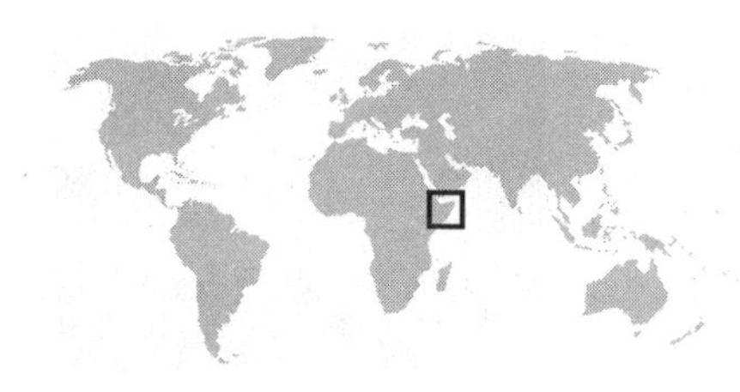

Name on driving licence: Somalia
Capital: Mogadishu
Population: 8.3 million
Dosh: CFA franc = 100 centimes
Size: 637,700 km^2 (31 Wales)

Complete history: Regrettably, Somalia does possibly the least convincing impersonation of a nation state of any country in the world. There has been no central government to speak of since 1991 and power has resided in the hands of sundry warlords, the self-proclaimed governments of Somaliland and Puntland, and the Islamic Courts Union. The only people who don't seem to have any power are the members of the interim transitional government which, of course, is the only authority recognised by the UN. In the past, Somalis have been blessed by the patronage of Fascist Italy, Britain, France, Egypt and Ethiopia, all of whom have kept the peace by dint of killing the locals.

Crowning achievement: It's an ill war that blows nobody any good. Somalia has one of the lowest HIV/AIDS infection rates in Africa. **[15]**

Pub fact: The country also has Africa's cheapest mobile call rates, so at least Somalis can chat to their friends about the anarchy surrounding them without stretching their finances unduly. **[11]**

Sad fact: With the advent of the UN peacekeeping mission in December 1992, Somalia unwittingly provided a seminal moment in news-gathering history when US Marines arrived on the beach to be confronted not by militiamen but television cameras. **[7]**

Made in Somalia: The Somalian wild ass, a grey donkey-esque creature racing headlong into extinction on its zippy zebra-lite legs. **[14]**

Chorus of national anthem drowned out by the noise of Kalashnikovs: 'Somalia wake up/Wake up and join hands together/We must help the weakest of our people/All of the time.' **[17]**

SCORE: 64 **WORLD RANKING:** 54

South Africa AFRICA

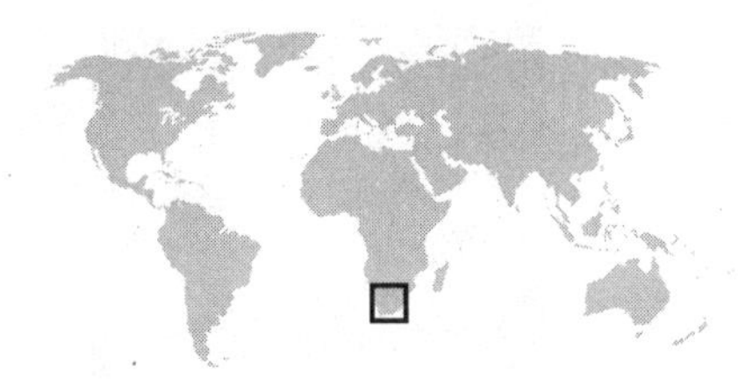

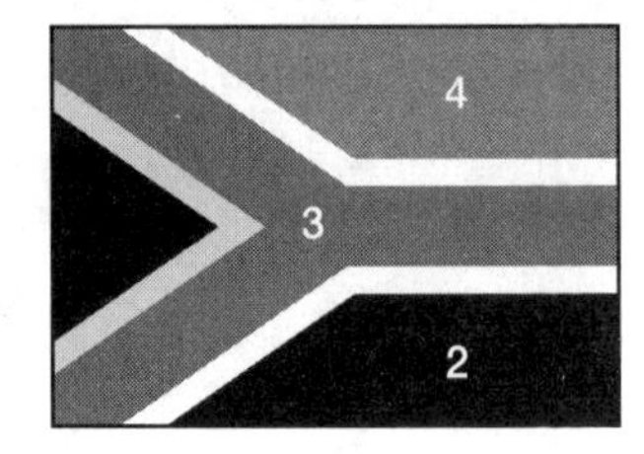

Name on driving licence: Republic of South Africa
Capital: Various
Population: 43 million
Dosh: Rand = 100 cents
Size: 1,219,900 km^2 (59 Wales)

Complete history: Whatever the Bantu-speaking folk did to merit such a virulent visitation by the Dutch and British, it must have been pretty bad. It all kicked off with the Great Trek when Dutch Boers (aka farmers), resenting British rule and anti-slavery laws, invaded Zululand. Before you knew it there were two Boer Wars, two Zulu Wars and apartheid. Thankfully, everything was put right by President 'Free' Nelson Mandela. However, it's all going to pot again thanks to his successor, occasional HIV-denier Thabo Mbeki.

Customs to treasure: Members of South African sports teams are called 'proteas'. This may make them sound hard as nails but the *Protea* is actually a rather pretty flower genus found uniquely in the land of the rand. Perhaps if Australian sportsmen were called 'acacias' we'd learn to like them a bit more. **[13]**

Pub fact: Of which, no English person attempting an Afrikaans accent can go for more than thirty seconds without it drifting into Australian. **[8]**

Made in South Africa: More than half the world's gold supplies over the last hundred years or so, which is why everyone in South Africa over the age of eighteen has their own Porsche. **[5]**

Sad fact: How many capitals does a country need? South Africans own three: Pretoria/Tshwane (administrative capital), Cape Town (legislative capital) and Bloemfontein (judicial capital). Naturally enough, none of these is South Africa's biggest city, which is Johannesburg. Nothing like sharing out the love. **[12]**

Opening lines of national anthem: 'God bless Africa/Lift her horn on high.' Indeed. **[16]**

SCORE: 54 **WORLD RANKING:** 114

EUROPE Spain

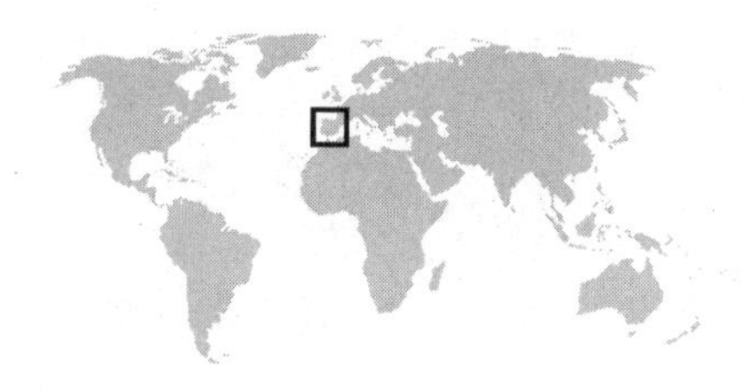

Name on driving licence: Kingdom of Spain
Capital: Madrid
Population: 40 million
Dosh: Euro = 100 cents
Size: 504,800 km^2 (24 Wales)

Complete history: The Spanish have devoted many centuries of their existence to being simply awful. Initially Iberians, they were joined by a mixture of Phoenicians, Celts and Greeks. The doughty Carthaginians eventually conquered the country but lost it to the might of Rome after the Second Punic War. Next came the Visigoths, followed by the Moors, who were at length seen off by the Christians. All fairly standard stuff so far. It all went wrong in the 15th century when they had the idea that their silly hats and lisps somehow made them into some kind of master race and that by rights they should take over the world, preferably killing or enslaving its inhabitants wherever possible. Still, no one can say that they have not paid for their misdeeds: 1.5 million Britons now live in Spain, their copies of the international version of the *Daily Mail* flapping like vituperative seagulls on the soft Mediterranean breeze. What goes round, comes round, and all that.

Crowning achievement: In 2002, Spain contrived to fight a miniature war with Morocco over the microscopic Isla Perejil, despite the fact that only goats live there. **[3]**

Top spot: The Alhambra, as palaces go, very more-ish. **[15]**

Sad fact: Bull fighting, what's *that* all about? **[0]**

Gifts to the world: Gazpacho, the acoustic guitar, Don Quixote, sherry, the steam-powered submarine, Manuel. **[13]**

National anthem: Like the noble Sammarinese, Spaniards stand mute at the playing of their national anthem. This is not because, like John Redwood, they do not know the words, but because there are none. **[18]**

SCORE: 49 **WORLD RANKING:** 140

Sri Lanka

ASIA

colour key: 1 = turquoise 2 = blue 3 = green 4 = red 5 = yellow 6 = orange 7 = pink 8 = purple 9 = brown

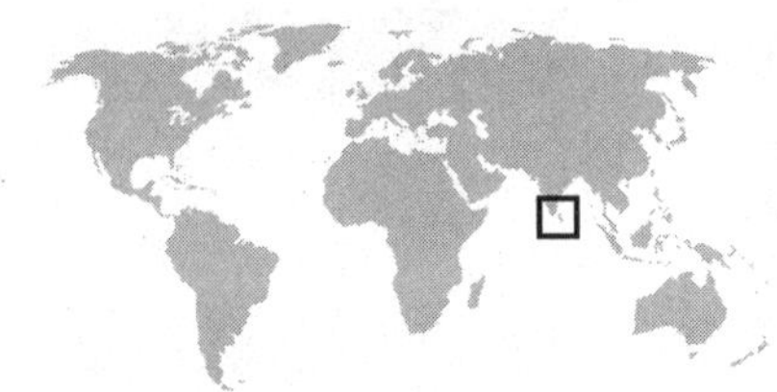

Name on driving licence: Democratic Socialist Republic of Sri Lanka
Capital: Colombo
Population: 20 million
Dosh: Sri Lankan rupee = 100 cents
Size: 65,600 km^2 (3.2 Wales)

Complete history: The Sinhalese and Tamils were unlucky enough to be assailed by the Portuguese, Dutch and British. Since independence there's been a spot of bother with the Tamil Tigers, a group that turn out not to be tigers at all but freedom fighters and/or terrorists, depending on your politics.

Crowning achievement: It's a close-run thing between electing the world's first female prime minister (Sirimavo Bandaranaike, 1960) and running the world's largest elephant orphanage. **[15]**

Top spot: Kandy, which was never taken by force by any of the colonial powers and whose Temple of the Sacred Tooth Relic houses Buddha's tooth. **[12]**

Sad fact: No one quite knows how to refer to Sri Lanka since the official designation is constantly undermined by reports smuggled off the island on bits of bark that the locals still call their country Ceylon. **[8]**

Made in Sri Lanka: Suresh Joachim, serial holder of Guinness World Records from 'Balancing on one foot' (76 hours, 40 minutes), to the somewhat technical 'Distance run while carrying a 4.5-kg brick in a nominated ungloved hand in an uncradled downward position' (126 km), to the merely asinine 'Longest time spent watching television' (69 hours, 48 minutes). It's all done in the name of world peace, apparently. **[8]**

National anthem that some might say strays over that all-important line that divides 'fondness' and 'blind adoration': 'Receive our grateful praise sublime, Lanka! We worship Thee/Thou gavest us Knowledge and Truth/Thou art our strength and inward faith/Our light divine and sentient being/Breath of life and liberation.' **[3]**

SCORE: 46 **WORLD RANKING:** 157

Sudan

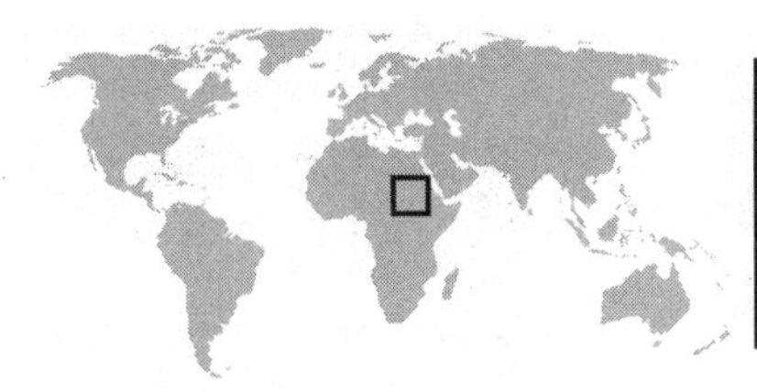

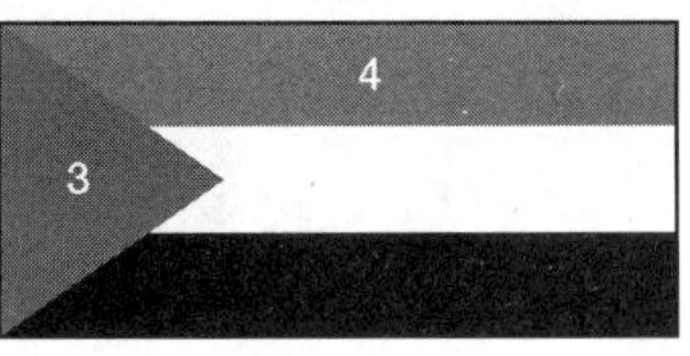

Name on driving licence: Republic of the Sudan
Capital: Khartoum
Population: 36 million
Dosh: Sudanese pound/dinar = 100 piastres
Size: 2,505,800 km² (121 Wales)

Complete history: The Sudanese haven't always had a lovely time of it, you know. For a start there was a lot of confusion about where Sudan actually was, since most of it had gone by the name of Nubia for as long as anyone could remember. By the time everything got sorted out, the blessed Egyptians invaded, followed by the British. When they left, everyone decided they would probably like to secede from everyone else, and violence has reigned ever since.

Pub fact: Sudan is the largest country in Africa. If you had a mind to, and asked everyone's permission, you could fit 390 million football pitches in it. Indeed, if only the country weren't perpetually at war no doubt someone would have got round to doing so by now. **[14]**

Sad fact: An estimated 400,000 people have died in the Darfur region as a direct or indirect result of the conflict up to the end of 2006. **[0]**

Popular misconception: 'Sudan is a dry country where only dry things happen.' Bogging well isn't, you know – it's the site of Sudd, the world's second-largest swamp. **[12]**

Made in Sudan: Manute Bol who, at 2.31 m (7'7"), is tall even by basketball standards. During his NBA career he earned several million dollars which he promptly spent supporting the Sudan People's Liberation Army, before becoming a peace activist. **[9]**

Opening lines of national anthem: 'We are the army of God and of our land/We shall never fail when called to sacrifice/ … /We give our lives as the price of glory.' Death, death, death. **[0]**

SCORE: 35 **WORLD RANKING: 184**

Suriname

AMERICAS

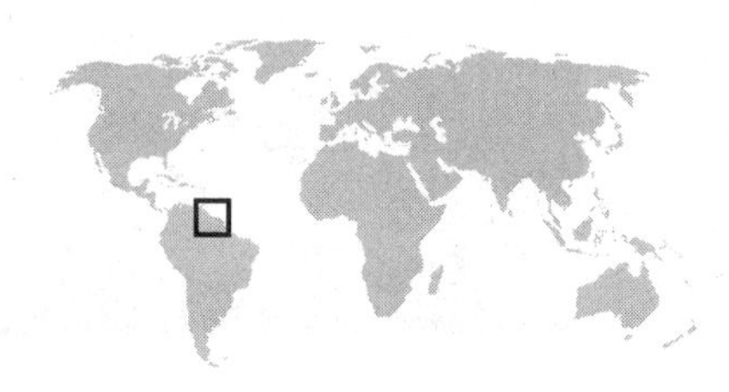

Name on driving licence: Republic of Suriname
Capital: Paramaribo
Population: 437,000
Dosh: Suriname guilder = 100 cents
Size: 163,300 km^2 (7.9 Wales)

Complete history: We may have forgotten it now that we're all such great chums, but there was a time when the Dutch and the British didn't get on at all, despite the whole William of Orange thing. For hundreds of years, the two fought churlishly over this bit of South America that the Spanish and Portuguese considered so scraggy they couldn't be doing with it. The Hollanders won in the end, but then lost to the Surinamese themselves in 1975.

Top spot: Paramaribo, possibly the most ethnically diverse city in the Americas, and home to birdsong competitions each Sunday. **[12]**

Made in Suriname: Aluminium, the metal from which the Statue of Eros in Piccadilly Circus is cast, and a word so difficult for the American tongue to pronounce that the second i was secretly melted down to stop further embarrassment. **[7]**

Popular misconception: 'It's spelt Surinam.' No it's not, otherwise it wouldn't be an anagram of aneurism. Which it is. **[14]**

Pub fact: The Treaty of Breda in 1667 gave the Dutch the British bit of Suriname in return for the island of Manhattan. This might seem like a poor deal for the Orange folk but for the fact that they got to hold on to Suriname for just over 300 years, whereas the British lost Manhattan to the locals as early as 1783. Or at least, not so much 'lost it' as made a tactical withdrawal. **[12]**

Opening lines of national anthem that suggest someone's been hiding loudspeakers in the ground again: 'Rise countrymen, rise/The soil of Suriname is calling you.' **[12]**

SCORE: 57 **WORLD RANKING:** 97

AFRICA

Swaziland

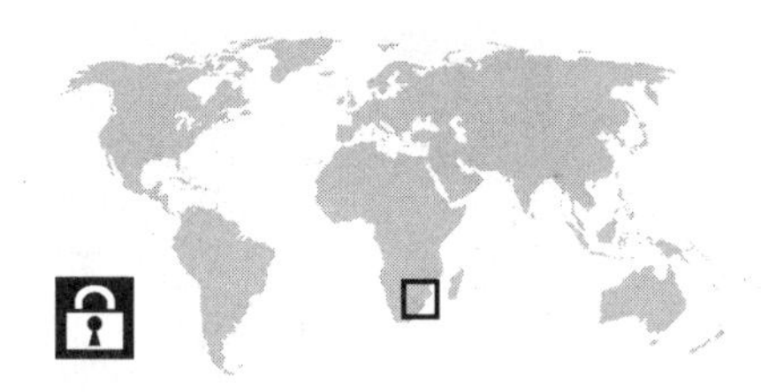

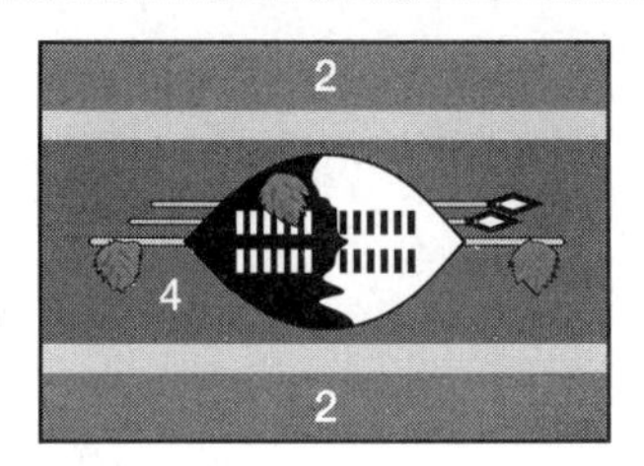

Name on driving licence: Kingdom of Swaziland
Capital: Mbabane
Population: 1.2 million
Dosh: Lilangeni = 100 cents
Size: 17,400 km^2 (0.8 Wales)

Complete history: Tradition has it that Ngwane II formed the Swazi nation sometime in the 18th century. When the Zulus attacked, the Swazis asked the British to defend them, which they did, before deciding they fancied ruling them for a bit too.

Popular misconception: 'Swaziland is a nation that used to exist during the days of the British Empire but which is now part of some country with a more modern-sounding name, like Angola.' By rights, it should be but several years ago it got caught in a stray time envelope and was sent crashing back to 1850. It is therefore only due to join late-20th-century naming schemes in about 100 years' time. **[6]**

Pub fact: The king is not only above the Constitution but above public criticism too (though you are allowed to mutter about him under your breath as long as no one can make out the words). **[4]**

Sad fact: The current head honcho, King Mswati III, is one of over 600 children squired by his late father, the 100-spoused King Sobhuza. **[2]**

Flag fact: There really aren't enough flags with oxhide shields on them, so well done, Swaziland. The frilly tassels symbolise the Swazi monarchy, which possibly says more about them than they'd care to admit. **[12]**

Opening lines of national anthem that sound suspiciously as if a royal hand was involved in the writing thereof: 'O Lord our God, bestower of the blessings of the Swazi/We give Thee thanks for all our good fortune/We offer thanks and praise for our King.' **[3]**

SCORE: 27 WORLD RANKING: 195

Sweden

EUROPE

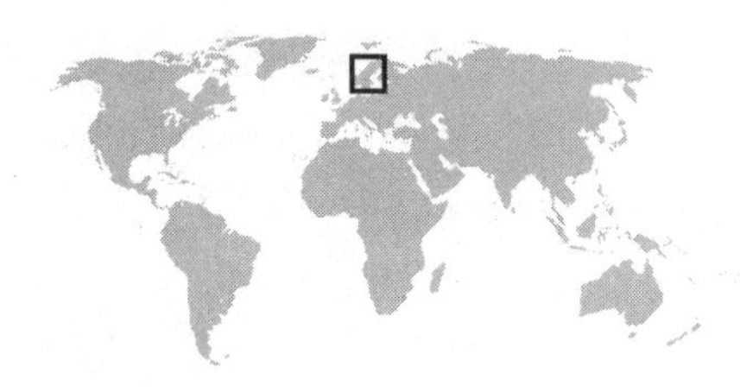

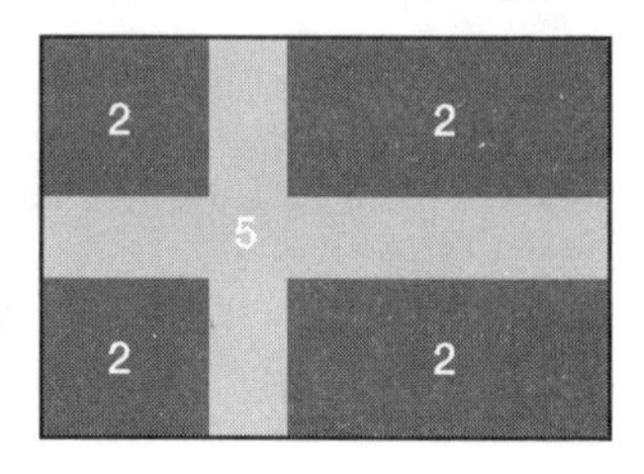

Name on driving licence: Kingdom of Sweden
Capital: Stockholm
Population: 9 million
Dosh: Swedish krona = 100 öre
Size: 450,000 km^2 (22 Wales)

Complete history: Sweden has always been an exciting country on the cutting edge of everything. Sometime victors over the Germans and Danes, and past possessors of quite large chunks of Europe, today's Swedes can rest knowing in their hearts that they will always be slightly more neutral than the Swiss.

Crowning achievement: Sweden's extraordinary welfare state? Its world-leading stance on the rights of women? Or bringing Billy bookcases to the masses? **[15]**

Gifts to the world: Dynamite (Alfred Nobel, 1866), the blowtorch (Carl Rickard Nyberg, 1881) and, for when the two combine, the pacemaker (Rune Elmqvist, 1958). **[11]**

Customs to treasure: The Thing of All Swedes. Although many people over the ages have hypothesised as to the nature of the Thing – blondness, perhaps, or a taste for early Black Sabbath – the dreary truth is that it was a general assembly held each spring up until the Middle Ages, when the Swedes presumably discovered a more interesting thing. **[10]**

Gifts to the world: Abba, of whom everyone has heard; The Cardigans, of whom most people have heard; and Komeda, of whom almost no one has heard, which is one of the great tragedies of our times. **[18]**

Anthems: Like the Danes, Norwegians and Thais, the Swedes luxuriate in the balmy waters of both a royal and a national anthem. The national anthem goes for the epic sweep: 'Thou ancient, thou unbound, thou north of the far'; while the royal anthem has more of the hey-nonny-nonny about it: 'Once from the depths of Swedish hearts/A full and artless song did start.' **[14]**

SCORE: 68 **WORLD RANKING: 31**

colour key: 1 = turquoise 2 = blue 3 = green 4 = red 5 = yellow 6 = orange 7 = pink 8 = purple 9 = brown

EUROPE **Switzerland**

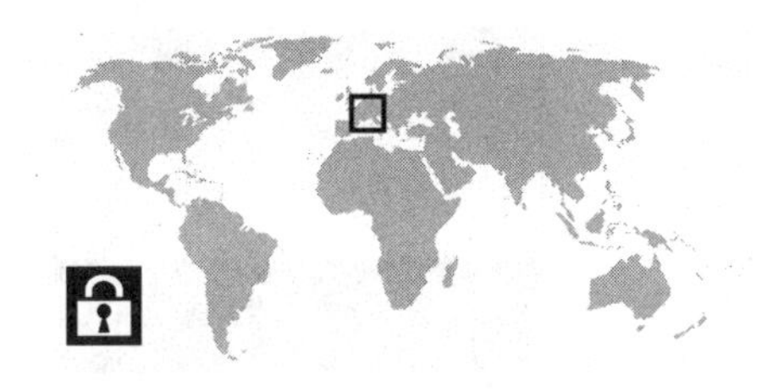

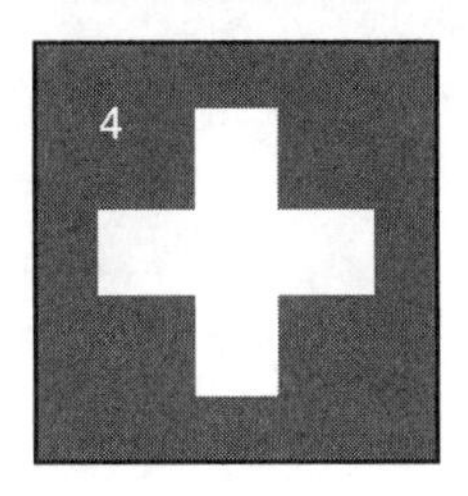

Name on driving licence: Swiss Confederation
Capital: Bern
Population: 7.5 million
Dosh: Swiss franc = 100 centimes
Size: 41,300 km^2 (2 Wales)

Complete history: Try to forget for a moment that the Reverend Sydney Smith wrote 'I look on Switzerland as an inferior sort of Scotland', for that was in 1815, and Scotland's got a lot worse since then. Furthermore, Switzerland has gone from a blur of perpetually warring cantons to a picture-perfect jigsaw puzzle of peaceful tedium, meaning today's Swiss can rest knowing in their hearts that they will always do slightly more national service than the Swedes.

Crowning achievement: The invention of Smell-o-Vision (as used in the 1960 film *Scent of Mystery* starring Denholm Elliot and Peter Lorre) must come pretty high up the rankings, although the Swiss do also claim to own the world's oldest-named cheese, the 7th-century Sbrinz. **[12]**

Made in Switzerland: Carl Jung, of whom we are all collectively conscious, whether we like it or not. **[11]**

Pub fact: For all the hilarious jokes about their navy, the Swiss do actually possess a number of military patrol boats for swanning around various large lakes, and the Swiss Navy brand of peppermints, which must scare the living daylights out of potential invaders. **[14]**

Gifts to the world: The Swiss army knife, Gruyère cheese, cellophane, cowbells, alphorns, the Rorschach ink-blot test. All the ingredients for a perfect picnic. **[17]**

Opening lines of national anthem: 'When the morning skies grow red/And o'er their radiance shed.' Sheds are not mentioned nearly enough in national anthems so one can only applaud the Swiss and move on quickly. **[16]**

SCORE: **70** WORLD RANKING: **22**

Syria

MIDDLE EAST

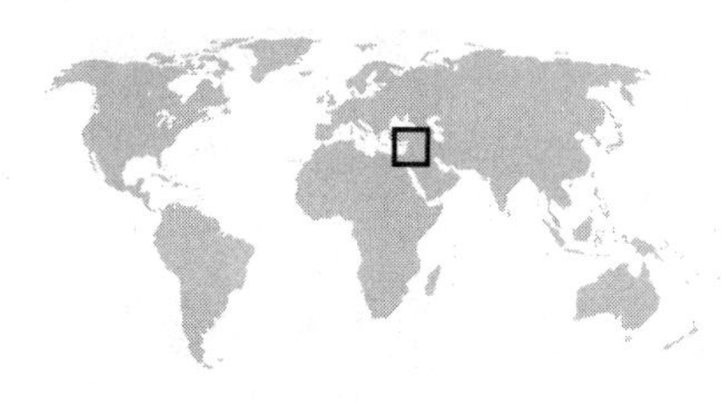

Name on driving licence: Syrian Arab Republic
Capital: Damascus
Population: 18 million
Dosh: Syrian pound = 100 piastres
Size: 184,200 km^2 (8.9 Wales)

Complete history: 'What could be sillier/Than Syria?' as poet Stonking Ralph Kierkegaard once enquired. While his question was probably rhetorical, it behoves us to confirm that there are sillier things, one of them being that no one outside the realm of viticulture has ever learnt precisely what the difference is between a raisin and a currant. Anyway, it's the French who are to blame because they invented Syria, or at least its borders, which bear little resemblance to the nation of antiquity. This explains to some degree why the Syrians have used Lebanon as a sort of puppet nation. For casual conversation in pubs, this is probably all you need to know – say anything further and you will be bluffing.

Crowning achievement: Syria claims that in Damascus it has the world's oldest continuously inhabited capital city, and until anyone else can come up with one that was founded before 2500 BC they've got a point. **[14]**

Gift to the world: *Ugarit cuneiform*, a terribly early alphabet. **[15]**

Customs to treasure: Throne hopping. The French toppled Greater Syria's King Faisal in 1920. Undeterred, he popped across to Iraq and became their king instead. **[13]**

Made in Syria: Queen Zenobia, who successfully repelled the might of Rome, conquered Egypt, then poisoned herself when she at last fell into the hands of the Romans. You go girl, as her subjects no doubt encouraged from behind a war elephant. **[17]**

National anthem proving that even heavenly bodies need to take the weight off their feet sometimes: 'A hallowed sanctuary/The seat of the stars.' **[16]**

SCORE: 75 WORLD RANKING: 8

ASIA Taiwan

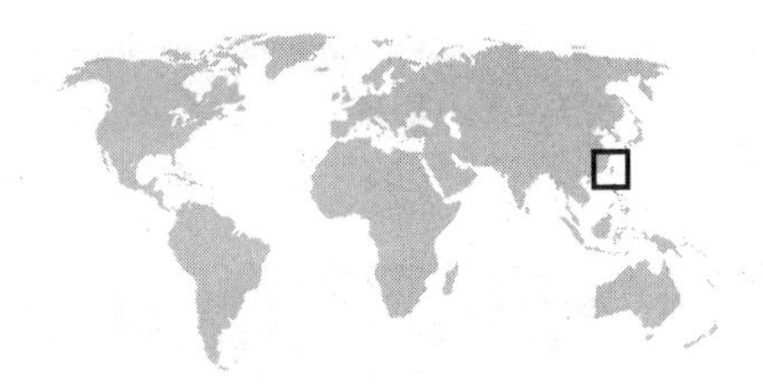

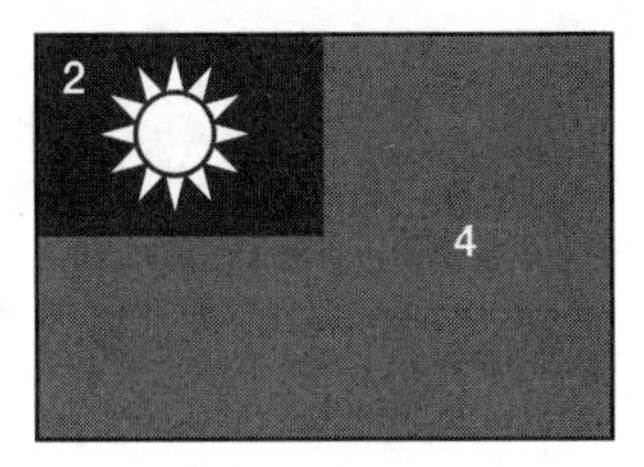

Name on driving licence: Republic of China
Capital: Taipei
Population: 23 million
Dosh: Taiwan dollar = 100 cents
Size: 36,000 km^2 (1.7 Wales)

Complete history: The island has been inhabited since at least 8000 BC but only really came into its own when Chiang Kai-shek's Kuomintang had to flee there in 1949. Since then, the world has been blessed with two countries calling themselves China, which fair cheers one up of a gloomy afternoon.

Crowning achievement: Finding itself included in university slang signifying the various classes of degree. Thus, while a First is a 'Geoff' (i.e. Geoff Hurst), a 2.2 is a 'Desmond' (in honour of South Africa's leading cleric) or a 'Trevor' (South Africa's leading cleric's brother) and a Third is a 'Thora' (Hird), a 2.1 is a 'Maiden' (aka 'Made in Taiwan'). Doesn't really get better than that, does it? **[4]**

Made in Taiwan: Everything. **[15]**

Customs to treasure: Hunter-gatherering. The Yami, one of the last peoples on the planet to survive by this means, live on the island of Lanyu, hardly bothering to contribute to global warming. **[17]**

Pub fact: Taiwan used to be known rather more lyrically as Formosa, which means 'beautiful' (not to be confused with Formica, which is an antonym). **[12]**

Opening lines of national anthem: 'The Three Principles of the People: Our aim shall be/To found a free land, world peace be our stand.' Taken from a riveting speech by the country's first president, the anthem is, naturally enough, banned in the People's Republic. Indeed, at global events such as the Olympics, Taiwan has to use a different anthem called 'The Flag Raising Song', which is rather less inspired and goes on about different colours a lot. **[10]**

SCORE: 58 **WORLD RANKING:** 91

Tajikistan

ASIA

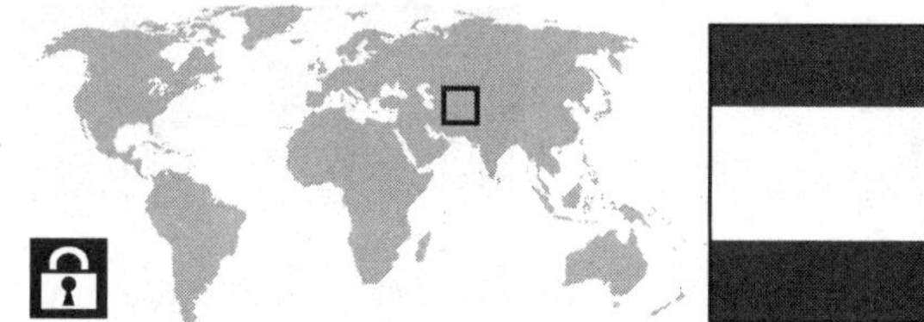

Name on driving licence: Republic of Tajikistan
Capital: Dushanbe
Population: 7 million
Dosh: Somoni = 100 diram
Size: 143,100 km^2 (6.9 Wales)

Complete history: Once safely out of the Soviet Union, the Tajiks immediately arranged a civil war for themselves but got bored after a few years and went back to farming cotton.

Top spot: The Pamirs, the mountain range known to the locals as the 'Roof of the World' (though, of course, if it were the roof, it would be on top, like the sky, but don't mention this). The Yeti is known to live here. If nothing else, he gets around. **[10]**

Pub fact: The Silk Road – as used by traders going back and forth from China to the Med – passed through Tajikistan and brought with it the seventeen-year-old Marco Polo. It is rumoured his experiences inspired him to create 'a new kind of sporting recreation, a species of hockey played on horseback'. **[15]**

Sad fact: Tajikistan suffers along with the other '–stans' in that no one outside of their region has much of an idea which one is which or where they might be found. Tajiks therefore live in constant fear of being mistaken for Afghanistan and summarily carpet bombed. This would be a blow to the country's economy since any such dumping would seriously jeopardise Tajikistan's home-grown carpet industry, which is more or less the only one they've got. **[4]**

Made in Tajikistan: Carpets. See, told you. **[13]**

Opening lines of national anthem: 'Our beloved country/We are happy to see your pride.' The rest of the ditty flirts with the possibility of venturing into a land of mush without quite making it through customs. **[9]**

SCORE: 51 WORLD RANKING: 130

colour key: 1 = turquoise 2 = blue 3 = green 4 = red 5 = yellow 6 = orange 7 = pink 8 = purple 9 = brown

AFRICA Tanzania

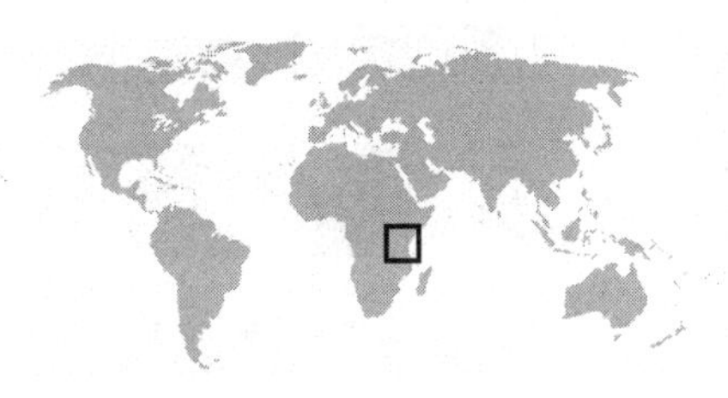

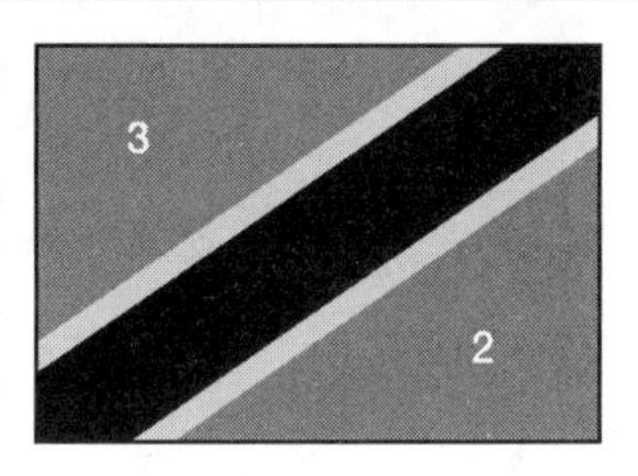

Name on driving licence: United Republic of Tanzania
Capital: Dodoma
Population: 37 million
Dosh: Tanzanian shilling = 100 cents
Size: 945,100 km^2 (46 Wales)

Complete history: Tanzania became itself when mainland Tanganyika came together in mystical union with the island of Zanzibar in 1964. This was very good news for Zanzibar because it's no more than two pathetic scraps of island and yet it got just as many letters in the new name as the proper country-sized Tanganyika. No doubt this was largely down to the Zanzibar-based birth of Farookh Bulsara some eighteen years beforehand, although admittedly his ability to influence the naming of nations became greater when he became the frontman of block-harmony specialists Queen.

Top spot: The Serengeti is the largest game park on the planet, as well as being the world's oldest ecosystem, though how they work that out is a mystery life is too brief to unravel. Anyway, it's famed for its migrations of huge herds of things. If you *could* see wildebeest sweeping majestically across the plain from a Torquay hotel bedroom window, this is the plain they would be on. **[19]**

Pub fact: Tanzania owns part of Africa's largest lake (Victoria) and highest mountain (Kilimanjaro), which some might consider a shade on the greedy side. **[15]**

Made in Tanzania: The deeply endangered black rhino, which can be found running around Ngorongoro Crater looking worried. **[14]**

Sad fact: More slaves passed through Zanzibar than any other trading centre on Earth. Which is quite a past to live down. **[0]**

Opening lines of national anthem: 'God Bless Africa/Bless its leaders.' The anthem doesn't get around to asking for a blessing on Tanzania until after the chorus, which is rather sweet. **[17]**

SCORE: 65 **WORLD RANKING: 50**

Thailand

ASIA

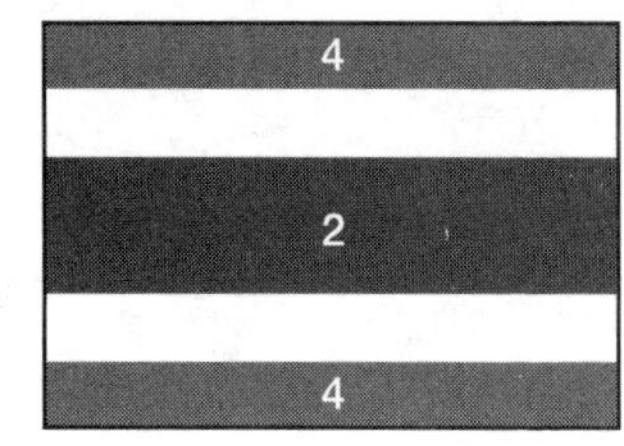

Name on driving licence: Kingdom of Thailand
Capital: Bangkok
Population: 65 million
Dosh: Baht = 100 satang
Size: 514,000 km² (25 Wales)

Complete history: The Thais arrived. They're still there. They've attempted to spice up this lacklustre attempt at a history by having the odd coup, but they're fooling no one.

Crowning achievement: Every nation in South East Asia has been colonised by Europeans at some time or other except Thailand. Whether this was an oversight by the Europeans or a deliberate snub has been the stuff of heated debate for decades between historians otherwise unable to get lives. **[16]**

Made in Thailand: The tuk-tuk, a trip round Bangkok in which is a very cheap way of discovering whether or not you have lost the desire to go on living. **[12]**

Customs to treasure: The *wai*, a sneaky method of shaking hands without actually shaking hands. **[11]**

Sad fact: At the annual Surin Elephant Round-up Show, 250 or so elephants get through about 50 tonnes of fruit and veg. As befits such a noble beast, the food is served on silk tablecloths. **[10]**

Anthems: Thailand is the fourth in the quartet of states that foist both a national and a royal anthem on their weary people. The national anthem is a far from generous offering that suggests that Thais are not taught to share their toys in the nursery: 'Thailand embraces in its bosom all people of Thai blood/Every inch of Thailand belongs to the Thais.' The royal anthem, meanwhile, is so sycophantic and toady it fair churns the stomach: 'To the supreme Protector of the Realm/The mightiest of monarchs complete with transcendent virtues/ … /We wish that whatsoever Your Majesty desires/The same may be fulfilled.' Bleah. **[0]**

SCORE: 49 **WORLD RANKING:** 140

AFRICA Togo

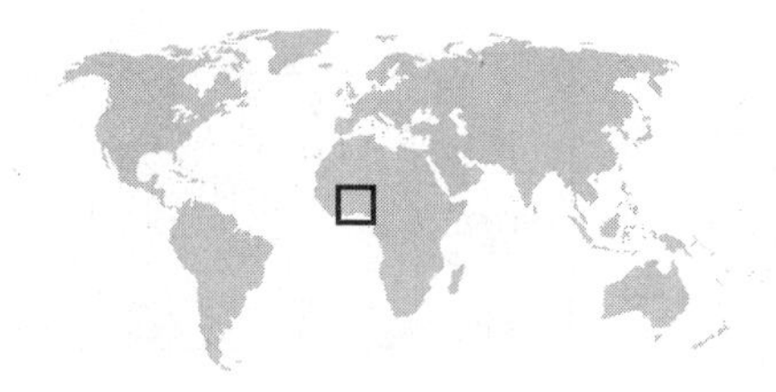

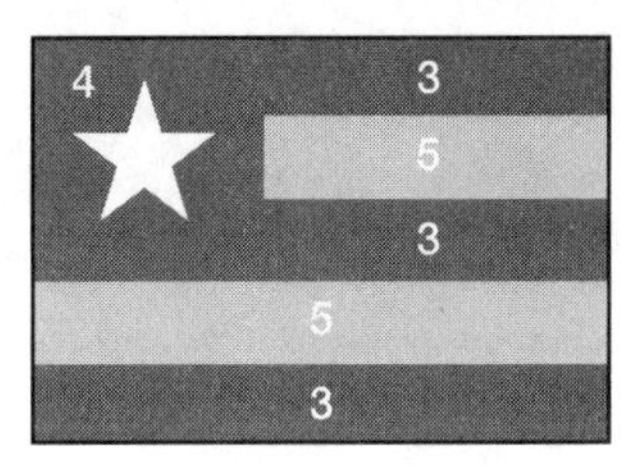

Name on driving licence: Republic of Togo
Capital: Lomé
Population: 5.6 million
Dosh: CFA franc = 100 centimes
Size: 56,800 km^2 (2.7 Wales)

Complete history: First they became part of the 'Slave Coast', then they got taken over by Germans. Neither big nor clever.

Sad fact: From 1967, Gnassingbé Eyadéma held power for a largely uninterrupted 38 years – always a healthy state of affairs. Unusually for an African leader he didn't die in a mysterious plane crash, though he did at least manage to be on a plane when he succumbed to a heart condition. This came as a surprise to the Togolese, who had no idea that he had a heart. **[3]**

Customs to treasure: September's Agbogboza festival, which heralds the new year, the new yam and the new you. **[11]**

Made in Togo: *Fufu*, a food prepared exclusively by women (what would happen if a man were to try his hand at it is unclear – perhaps he would implode into a fluffy ball of metrosexuality). Yams are boiled until soft then smashed to pieces. Somehow, this makes them into a dough which is then eaten with a sauce in an attempt to disguise the flavour. **[9]**

Popular misconception: 'Togo was named after the American way of saying take-away.' Thankfully not. Togo is the Ewé word for 'behind the sea', which is not quite as descriptive as perhaps it might have been. They're probably a bit embarrassed about it now and wish someone *would* take it away. **[6]**

Opening lines of national anthem that have provoked a storm of compensation claims from weak, troubled and unhappy forefathers: 'Hail to thee, land of our forefathers/Thou who made them strong, peaceful and happy.' **[7]**

SCORE: 36 **WORLD RANKING: 181**

Tonga

OCEANIA

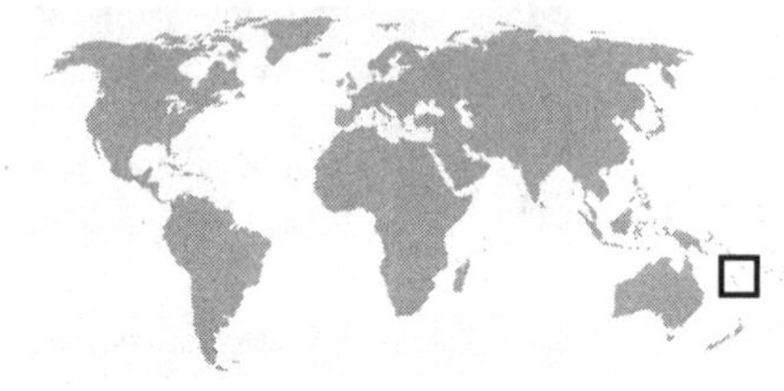

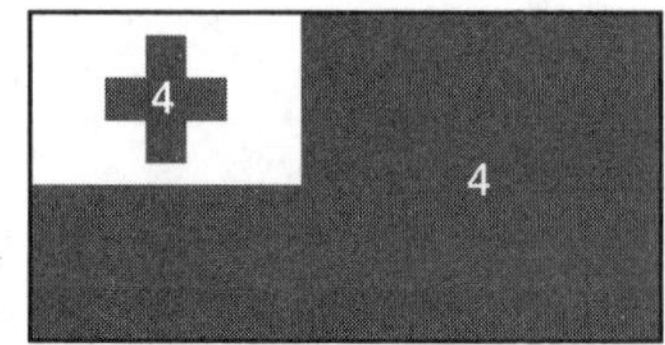

Name on driving licence: Kingdom of Tonga
Capital: Nuku'alofa
Population: 110,000
Dosh: Pa'anga = 100 seniti
Size: 750 km^2 (0.03 Wales)

Complete history: What drove first the Dutch then the British not to interfere too alarmingly with these particular Polynesians is a mystery easily solved by the fact that there's really not much to exploit in Tonga, save the generosity of the people.

Pub fact: Captain Cook named Tonga the 'Friendly Islands' on the grounds that no one killed him when he visited. Cook was (generally) a good egg. The Tongans recognised this and so spared his life by not being able to decide how they were going to murder him until after he'd left. **[12]**

Crowning achievement: Featuring so heroically in the mutiny on the *Bounty*. In 1789, Captain Bligh and eighteen loyal crewmen were set adrift near the Tongan isle of Tofua. Bligh was (generally) a bad egg. The Tofuans recognised this and so attacked his party, killing one of them and sending the rest packing. Rah! **[14]**

Made in Tonga: King George Tupou V, owner of a brewery, an airline, a mobile-phone company and a London taxi. **[2]**

Sad fact: When Tonga came into US$28 million by selling citizenship to Hong Kongers prior to the Chinese takeover, they gave it to a man named Jesse Bogdonoff, Tonga's official court jester, to invest for them. This may seem like a foolish move and indeed it transpired, for the money was never seen again. **[3]**

Opening lines of national anthem: 'Oh Almighty God above/Thou art our Lord and sure defence.' The song continues in a touchingly humble manner heard all too rarely in the bombastic world of the state anthem. **[15]**

SCORE: 46 **WORLD RANKING:** 157

colour key: 1 = turquoise 2 = blue 3 = green 4 = red 5 = yellow 6 = orange 7 = pink 8 = purple 9 = brown

AMERICAS

Trinidad & Tobago

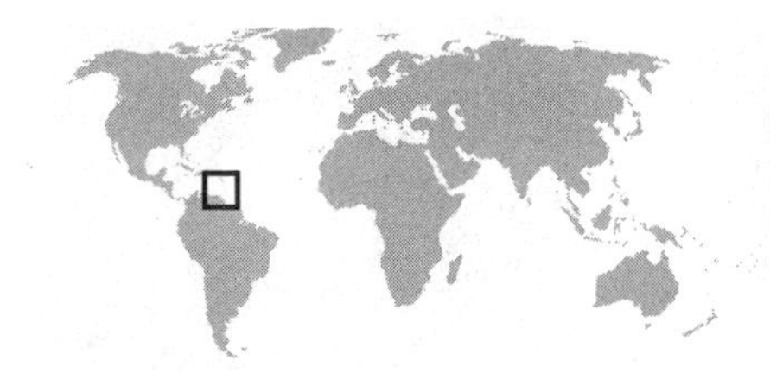

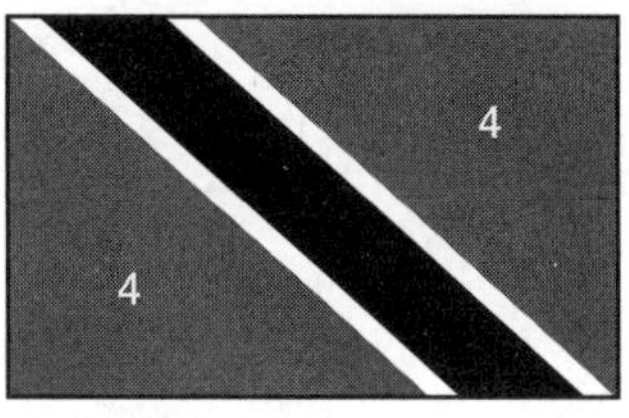

Name on driving licence: Republic of Trinidad and Tobago
Capital: Port of Spain
Population: 1.1 million
Dosh: Trinidad and Tobago dollar = 100 cents
Size: 5,100 km^2 (0.25 Wales)

Complete history: It's the usual sordid story – the Spanish did for the Arawaks and Caribs. The British did for the Spanish. The French did for the British. The British did for the French. For hundreds of years the beaches were strewn with nothing but handbags. Eventually, the British handed the islands back to their original owners, or would have done, had there been any left, so they gave them to the descendents of the slaves deposited therein. Expected to be thanked too, I shouldn't wonder. Trinidad is right next to Venezuela, by the way, which surprises some people.

Top spot: Port-of-Spain, if only for the wonderful jest of renaming what was Puerto de España in such a literal fashion to remind the Spanish of what they once owned. All right, perhaps it's a bit childish. **[7]**

Gifts to the world: Calypso music, steel bands, limbo dancing, the acknowledgement that an entire nation can be reduced to a stereotype if needs be. **[8]**

Customs to treasure: The Buccoo Goat Race Festival. The jockeys get the worst of it as goats slam themselves up the racetrack hell bent on retribution for the last twelve months of indifferent treatment. **[16]**

Pub fact: Pitch Lake is the world's largest bitumen lake. This makes it harder work to swim in than your ordinary lake, but ideal if you've always yearned to set fire to yourself while doing so. **[13]**

National anthem: 'Here every creed and race/Find an equal place.' Amen to that, kids. Are you listening, Thailand? **[20]**

SCORE: 64 WORLD RANKING: 54

Tunisia

AFRICA

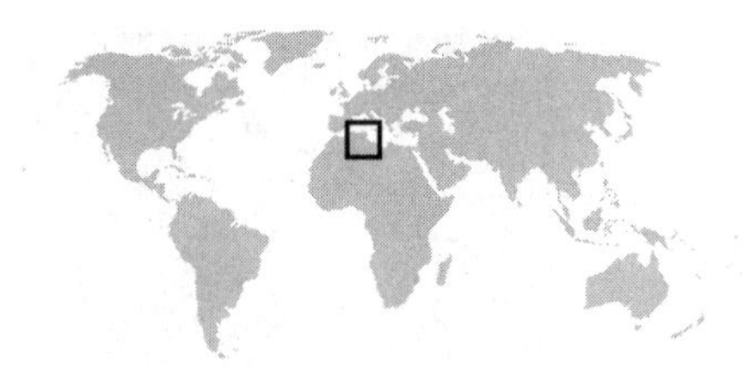

Name on driving licence: Republic of Tunisia
Capital: Tunis
Population: 10 million
Dosh: Tunisian dinar = 1,000 millimes
Size: 163,600 km² (7.9 Wales)

Complete history: Imagine Tunisia as a suitcase and history as clothes and there's absolutely no way you're going to get all the latter into the former and close it, no matter how many eighteen-stone men you employ to jump on the lid. The wandering Berbers came first, followed by a succession of Phoenicians, Romans, Vandals, Arabs, Normans (from Sicily, confusingly), Almohads, Berbers again (huzzah!), Spaniards, Ottoman-Turks, French, Germans and, finally, the Allies, but they've left now.

Top spot: The Antonine Baths, Africa's largest, although since it was only the Romans who built baths, this is not quite as impressive as first it might seem. **[8]**

Customs to treasure: Imprisoning cyber-dissidents. **[0]**

Pub fact: *Tunis* either comes from the Berber word for a promontory – an explanation that stretches beyond the confines of mere dullness into a dark land of unmitigated ennui – or from the phrase for 'spend the night', which is infinitely preferable because it conjures up images of a sofa bed, a far friendlier prospect after a hard day's imprisoning than some sort of headland. **[10]**

Made in Tunisia: *The Life of Brian*. **[13]**

National anthem: 'We are ready to die, if it is necessary, die so that our country will live!/ ... /There is nobody in our country who refuses to be in the ranks of its soldiers!' Yet another deathwish anthem. This is surely the reason that in the future, if there is one, folk will look back on our system of nation states and tut and shake their heads sadly, and tut again. **[0]**

SCORE: 31 WORLD RANKING: 189

EUROPE/ASIA

Turkey

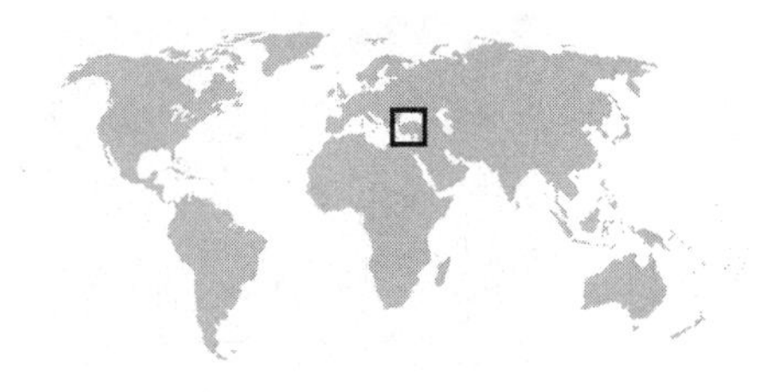

Name on driving licence: Republic of Turkey
Capital: Ankara
Population: 69 million
Dosh: New Turkish lira = 100 kurus
Size: 780,600 km^2 (37 Wales)

Complete history: Turkey Turkey Turkey – bridge between Asia and Europe, Muslim and Christian, East and West. So goeth the old saw. However, as the kerfuffle over the country's proposed European Union membership demonstrates, it's not really a bridge between anything much, although it is admittedly a good means of getting from Syria to Bulgaria without getting your feet unduly wet. Meanwhile, the only person to happen in Turkish history was Ataturk, founder and first president of the Turkish republic, and now a major divinity.

Top spot: Depending on your religion, either the Hagia Sophia (a cathedral converted into a Mosque by the Ottomans) or the Grand Bazaar, purportedly the world's oldest surviving temple to Commerce. **[10]**

Gifts to the world: Turkish denial, Turkish anger, Turkish depression, Turkish acceptance, Turkish delight. **[12]**

Made in Turkey: The kaftan, although it takes a special kind of person to look 'cool in the kaftan'. **[13]**

Customs to treasure: 'Insulting Turkishness' or even besmirching the good name of Ataturk (may he be forever lauded) can get you a prison sentence. However, the law is a bit blurry with respect to praise heaped on Turkishness or Ataturk in what might be perceived as a sarcastic manner, so next time you're visiting you might want to give it a go in the name of free speech. **[0]**

Opening lines of national anthem bemoaning the phasing out of real fires throughout the Turkish republic: 'Fear not and be not dismayed, this crimson flag will never fade/It is the last hearth that is burning for my nation.' **[11]**

SCORE: 46 WORLD RANKING: 157

Turkmenistan ASIA

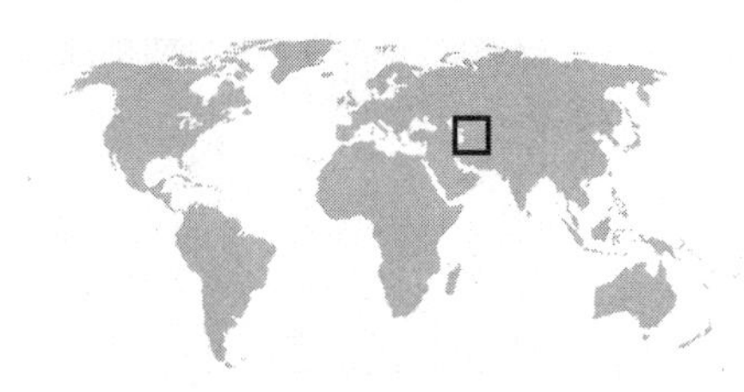

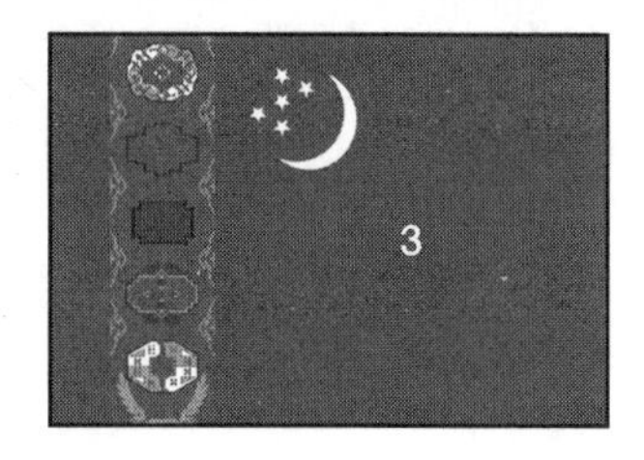

Name on driving licence: Republic of Turkmenistan
Capital: (Variants of) Ashgabat
Population: 4.9 million
Dosh: Turkmen manat = 100 tenesi
Size: 488,100 km^2 (24 Wales)

Complete history: It was about a millennium ago that Turkic people arrived on the scene and it was not long before Merv – as unlikely a name as you'd wish on a megalopolis – was the largest city in the world. Genghis Khan popped over for the weekend with his merry Mongol hordes, passing the baton eventually to Tamerlane, the man responsible for Edgar Allan Poe.

Crowning achievement: Not knowing how to spell its own capital. Choose one from Ashgabat, Aşgabat, Ashkhabád, Ashkabat, Ashgabad and Ishqábád. **[10]**

Made in Turkmenistan: Self-styled Turkmenbashi ('Father of all Turkmen') and, naturally, president for life, Saparmurat Niyazov (1940–2006) was a real mad despot's mad despot. He decreed that all hospitals and libraries outside the capital be closed (on the grounds that ill people should visit Ashgabat and that 'Turkmen don't read books anyway'); renamed almost everything after himself, including January; declared an annual 'Turkmen Melon Day' in honour of the winsome spherical fruit; and imprisoned any-one whose thinking even appeared to be out of step with his own, while his country slowly starved. His undoing was in not banning fatal heart attacks. The nation was not plunged into mourning. **[0]**

Made in Turkmenistan: Fermented camel's milk, even tastier than fermented camel, apparently. **[5]**

Customs to treasure: *Lipioshka* (an unleavened bread) should always be placed upper side up (which is fine if you can work out which side that is) and never put on the ground. Even in a bag. **[13]**

Opening lines of national anthem: 'The great creation of Turkmenbashi/Native land, sovereign state.' **[0]**

SCORE: 28 **WORLD RANKING:** 193

Tuvalu

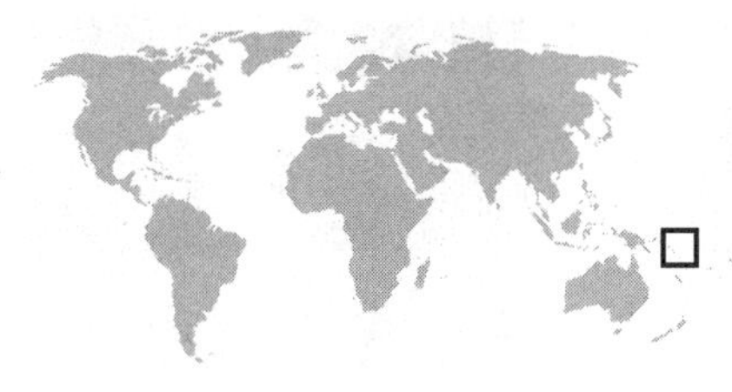

Name on driving licence: Tuvalu
Capital: Fongafale
Population: 11,000
Dosh: Tuvaluan dollar/Australian dollar = 100 cents
Size: 26 km² (0.001 Wales)

Complete history: The United Nations' smallest member by population is inhabited by Polynesians; was ravaged by Peruvian slave traders in the 1860s; became the Ellice Islands and then part of the Gilbert and Ellice Islands under the British; gained its independence in 1975; and sank beneath the waves in about 2050.

Crowning achievement: Tuvalu periodically declares that it will take the US and Australia to the International Court of Justice for precipitating climate change and not signing up to the Kyoto Protocol – a brave but, one can't help feeling, vain gesture. **[14]**

Pub fact: Tuvalu's most profitable export in recent years has been its 'tv' internet country code which it has sold to many of the world's hip hop and happening television stations (and MTV). **[14]**

Popular misconception: 'Tuvaluans don't marry dolphins.' Ah, but they do. Admittedly, it hasn't happened for a while now, but a man was once joined in holy matrimony with a woman who turned out to be a bottlenose dolphin and who returned to her own after giving her husband two sons, so there. **[12]**

Customs to treasure: Rain-making. The last woman on the island of Niutao to be able to do this was called Taia Teuai, who died in 1892, so they're overdue a successor. Applications for the post are being encouraged from native Mancunians, all of whom are believed to possess this power. **[11]**

Opening lines of national anthem: '"Tuvalu for the Almighty"/Are the words we hold most dear.' One in the eye there for people who claim that Tuvalu hasn't a prayer. **[15]**

SCORE: 66 **WORLD RANKING:** 44

Uganda

AFRICA

colour key: 1 = turquoise 2 = blue 3 = green 4 = red 5 = yellow 6 = orange 7 = pink 8 = purple 9 = brown

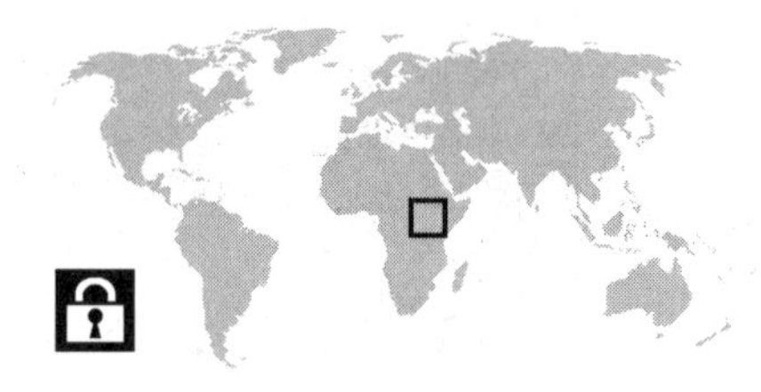

Name on driving licence: Republic of Uganda
Capital: Kampala
Population: 26 million
Dosh: Ugandan shilling = 100 cents
Size: 236,000 km² (11 Wales)

Complete history: If one must use a language, one might as well be Nilotic-speaking, as Uganda's earliest inhabitants were. More recently there have been the British and Idi Amin and military coups, but everything's all right now. Sort of.

Crowning achievement: Uganda is hailed as a rare success story in the fight against HIV/AIDS, although since it was one of the African nations most ravaged by the disease, it's probably just as well. **[13]**

Made in Uganda: *Waragi*, a gin made from bananas; and *waraba*, a banana made from gin. **[14]**

Sad fact: In 1987, Joseph Kony declared he was a medium channelling the spirits of Juma Oris (a government minister under Idi Amin) and a Chinese general. This did not auger well and the Lord's Resistance Army – for whom no act of violence appears beyond the pale – is the result. **[0]**

Flag fact: The black, yellow and red of the Ugandan flag represent the people, the sun and brotherhood respectively, and have been arranged to resemble a badly chosen jumper given to an unmarried uncle who would have been happy with vouchers and is now desperately trying to keep the disappointment from showing in those tawny eyes that once drew soft glances from many a secret admirer but which have long since lost their lustre. **[8]**

National anthem: 'Oh Uganda! The land of freedom/Our love and labour we give/And with neighbours all/At our country's call/In peace and friendship we'll live.' Yay! Ugandans immerse their souls in love to a tricky A-B-C-C-B rhyming scheme. **[17]**

SCORE: 52 WORLD RANKING: 125

Ukraine

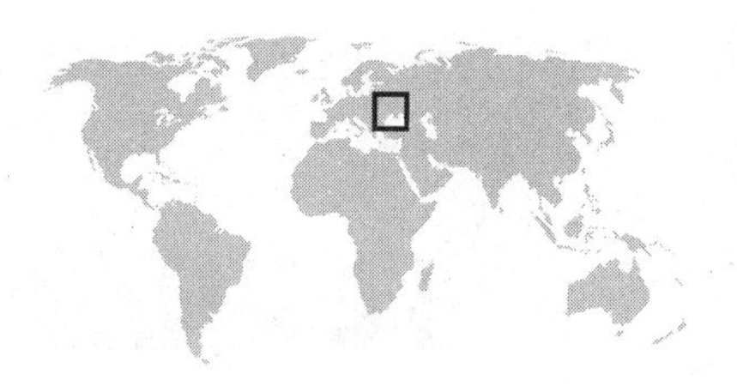

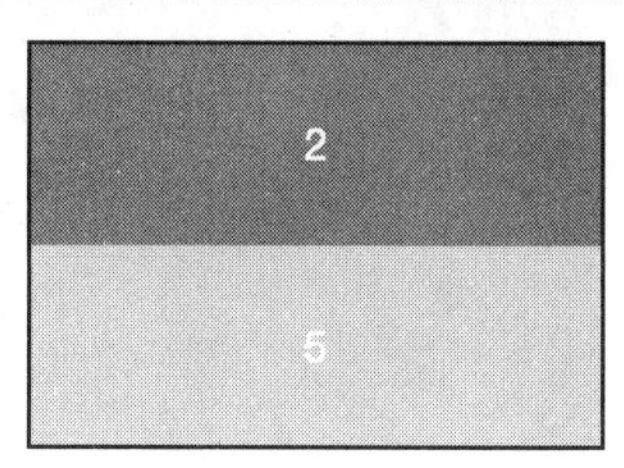

Name on driving licence: Ukraine
Capital: Kiev
Population: 48 million
Dosh: Hryvnia = 100 kopiykas
Size: 603,700 km^2 (29 Wales)

Complete history: Neolithic chaps were freezing their little beards off around here and using them as chisels to carve illustrations of their pets onto the sides of caves long before the Orange Revolution was thought of, but they would doubtless have approved of the latter once they'd grasped the concepts of civil disobedience and democracy.

Top spot: Kiev, mostly for the astonishingly exciting legend of its foundation. Apparently, four siblings – Kiy, Khoriv, Shchek and Lybid – named the city after their older brother Kyiv. Only they evidently spelt it wrong, so presumably they weren't that close. And anyway, why wasn't Kyiv there to found the city with them? Nobody seems prepared to come up with an answer for that one and, until they do, Ukrainians will always be treated with suspicion. **[10]**

Pub fact: Both 'Ukraine' and U2's permahatted guitarist 'The Edge' derive their names from words meaning 'the edge', although only one of them also answers to the name 'Dave Evans'. **[15]**

Popular misconception: 'A third of Ukrainians live below the UN poverty line (US$2/day).' Apparently not. Everyone's really well off. **[10]**

Gift to the world: Fallout. **[0]**

Opening line of national anthem: 'Ukraine's glory has not perished, nor her freedom.' This was revised in 2003 from the original 'Ukraine has not perished, neither her glory, nor freedom', which suggests that Ukraine perished at some point in this year, possibly after a lingering post-Chernobyl illness, but that her glory lives on, or at least can still be picked up by a Geiger counter. **[5]**

SCORE: 40 **WORLD RANKING:** 176

United Arab Emirates MIDDLE EAST

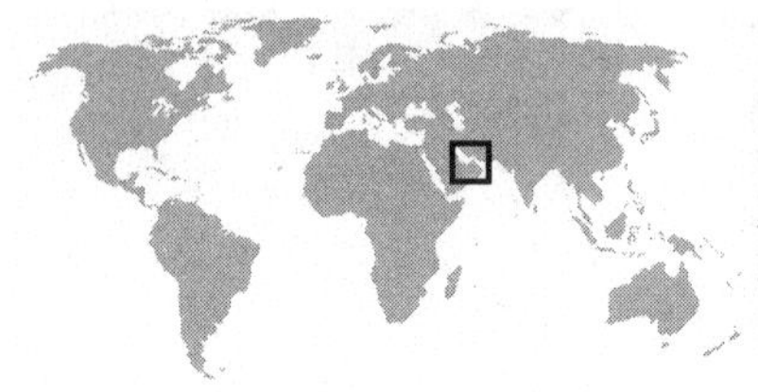

colour key: 1 = turquoise 2 = blue 3 = green 4 = red 5 = yellow 6 = orange 7 = pink 8 = purple 9 = brown

Name on driving licence: United Arab Emirates
Capital: Abu Dhabi
Population: 2.5 million
Dosh: UAE dirham = 100 fils
Size: 82,900 km² (4 Wales)

Complete history: There was a lot of rootlessness around these parts before the emirates took shape in the 18th century. However, such was their bickering that Britain was forced to step in and make them all sign a truce. They thus became known as the Trucial States, a name so ridiculous that they united almost immediately in order to change it.

Crowning achievement: Owning an airline that is better known than the country from which it hails. **[10]**

Pub fact: The UAE possesses the world's largest artificial harbour, so there'll be plenty of room for your artificial yacht next time you're cruising in the district. **[8]**

Made in the UAE: Oil. Erm. That's it. **[5]**

Sad fact: The seven-star sail-shaped Burj Al Arab in Dubai is the world's tallest hotel (321 metres). Keen to outdo themselves, Dubai's Rose Tower is due to open in 2007 and is 10 metres taller. Both hotels have been designed for guests who have become so disconnected from reality that they no longer know the value of money, but do like the idea of plasma TV screens with goldleaf frames. **[0]**

National anthem: The anthem had no words at all until 1996 when suddenly it was decided that the good burghers of the UAE could not go a minute longer without recourse to such lines as: 'Work sincerely, work sincerely', and, 'My country, my country, my country, my country'. There's also the obligatory 'We all sacrifice for you, we supply you with our blood', which is an unpleasant image. **[4]**

SCORE: 27 **WORLD RANKING: 195**

EUROPE United Kingdom

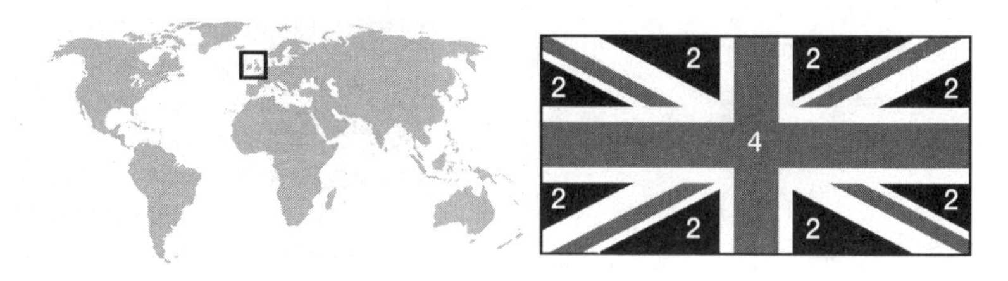

Name on driving licence: United Kingdom of Great Britain and Northern Ireland
Capital: London
Population: 60 million
Dosh: Pound sterling = 100 pence
Size: 244,800 km^2 (11 Wales + Wales)

Complete history: The United Kingdom comprises the English, who believe themselves British; the Welsh, who are just Welsh; the Scots, who would slit your throat for suggesting they were anything but Scots; and the Northern Irish, who never have been British and cry themselves to sleep every night on account of it (well, half of them). The first three of these nations used to run the world, but only the English are genuinely surprised that no one has thanked them yet.

Crowning achievement: 'Stand and Deliver' (Adam and the Ants, 1981). **[20]**

Pub fact: The UK operates more CCTV cameras per capita than any other country on Earth. This is not due to the fact that British citizens are likely to carry out a crime at any given moment, but because they're all extremely good-looking people whose every step is a fascinating study to the world. **[0]**

Sad fact: The UK has 217 words and phrases meaning 'to get drunk'. **[3]**

Made in the UK: Wilfred Owen, Robert Burns, Dylan Thomas, Seamus Heaney. What's the point of living if you're not a poet? **[20]**

National anthem: The British invented the national anthem, so it's all their fault really. Their turgid dirge has dragged their unwilling bodies from sedentary positions since 1745. There are five verses, but no one but hopelessly craven royalists knows any but the first. This is a shame because in verse two, rather than calling for the instant death of the enemies of the realm, the British ask God to 'frustrate their knavish tricks'. **[5]**

SCORE: 48 **WORLD RANKING: 146**

United States of America

AMERICAS

Name on driving licence: United States of America
Capital: Washington
Population: 300 million
Dosh: US dollar = 100 cents
Size: 9,626,100 km^2 (465 Wales)

Complete history: 'America had often been discovered before Columbus, but it had always been hushed up', as Oscar Wilde so rightly opined. Since then, the Land of the Free and the Home of the Brave Corp (est. 1776) has fought the British, Native Americans, Mexicans, Germans (somewhat belatedly, twice), World Communism (though the Chinese are now acceptable, apparently) and has bombed more countries than anyone else in the history of the planet. Currently it's engaged in something called the War on Terror™. This largely consists of imprisoning (preferably without trial) and/or killing Terrorists™ or, if this is not possible, any civilian engaged in un-American activities such as 'looking shifty' or 'being called Ahmed'. Happily, everyone in the world applauds these tactics, and even the Terrorists™, in their better moments, know them to be right. Back at home, any American caught not eating, driving or otherwise using up the world's resources is liable to arrest.

Crowning achievement: The Americans were the first to put men on the moon and, perhaps more laudably, bring them back again. **[15]**

Pub fact: Buzz Aldrin's mother's maiden name was Moon, but they still made him get off second. **[8]**

Made in America: Jimmy Hendrix; the glass harmonica; irony (oh, all right, not really). **[14]**

Top spot: New York. Every other city suffers from being too American. **[12]**

Opening lines of national anthem: 'Oh, say can you see by the dawn's early light/What so proudly we hailed at the twilight's last gleaming?' No, I can't quite make it out, what is it exactly? **[3]**

SCORE: 52 **WORLD RANKING:** 125

colour key: 1 = turquoise 2 = blue 3 = green 4 = red 5 = yellow 6 = orange 7 = pink 8 = purple 9 = brown

Uruguay

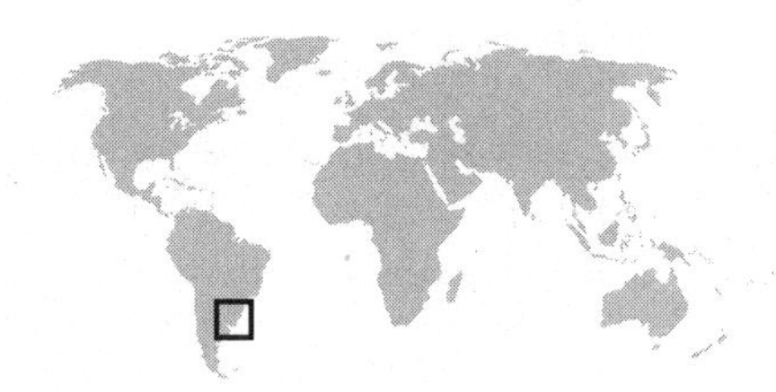

Name on driving licence: Eastern Republic of Uruguay
Capital: Montevideo
Population: 3.4 million
Dosh: Uruguayan peso = 100 centésimos
Size: 176,200 km^2 (8.5 Wales)

Complete history: The Guaraní were here first but you'll be lucky to find any now. The former 'Switzerland of South America' (ho-hum) is full of the descendents of Spaniards and Italians who live in towns called Fray Bentos. Weep, ye Amerindians.

Crowning achievement: Uruguay is by far the tiniest country (by population) to have won the World Cup. Indeed, they've done so twice, just to prove the first one wasn't a fluke. **[16]**

Top spot: Dispiritingly, there are no towns in Uruguay named after numbers except Treinta y Tres ('Thirty-three'). Even this turns out not to be an *hommage* to the integer but to the thirty-three Uruguayans who, on 19 April 1825, revived the struggle for independence against the Luso-Brazilians. Taken in that context, thirty-three doesn't really seem like quite enough. **[10]**

Customs to treasure: Uruguay was the first country in South America to create a welfare state, beating so-called modern countries such as Britain to the punch by thirty years. This makes the country a top choice should you feel a cerebral embolism coming on. **[15]**

Pub fact: No Uruguayan has ever knowingly eaten a vegetable. Do they not know that meat is murder? **[0]**

Opening lines of national anthem: 'Easterners [i.e. Uruguayans], our homeland or the grave!/Freedom, or die a glorious death.' There are eleven verses of this hearty nonsense and a chorus 'twixt each one, meaning that a complete rendering can easily run to five minutes. This is why Uruguayans have a phobia of state occasions, international sporting events, and tuba players generally. **[3]**

SCORE: 44 **WORLD RANKING:** 167

Uzbekistan ASIA

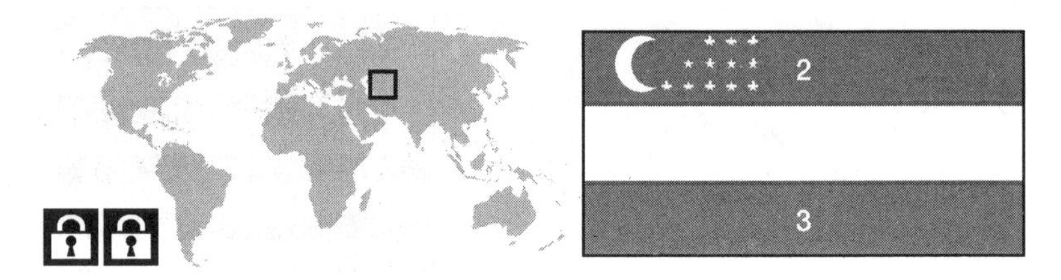

colour key: 1 = turquoise 2 = blue 3 = green 4 = red 5 = yellow 6 = orange 7 = pink 8 = purple 9 = brown

Name on driving licence: Republic of Uzbekistan
Capital: Tashkent
Population: 26 million
Dosh: Sum = 100 tiyn
Size: 447,400 km^2 (22 Wales)

Complete history: The usual Alexander the Great/Genghis Khan/Tamerlane stories abound, though the Russians got here surprisingly early (1860s), presumably to save the Soviets the trouble later on.

Top spot: Samarkand, the self-styled 'Centre of the Universe', became a byword for exoticism in 1913 with English poet James Elroy Flecker's *The Golden Journey to Samarkand*, and had its position in the popular consciousness cemented by a Monty Python appearance in the guise of the book *Thirty Days in the Samarkand Desert with the Duchess of Kent*. **[15]**

Where to avoid: Andijan, where a massacre carried out by the Uzbek army in May 2005 left up to 1,000 opposition demonstrators dead. According to government sources, only 200 people perished, and they were killed by the organisers of the demonstration. Like Hitler said, if you're going to tell a lie, make it a big one. **[0]**

Sad fact: Being a good communist, President Karimov is a dictator of the old school mass-arrests-purges-more-mass-arrests variety, and has garnered the country a European Union arms embargo and a UN report that describes the use of torture there as 'systematic'. He is, however, a lovely man who sends his mother flowers. **[0]**

Made in Uzbekistan: Cotton. Everyone's afraid to produce any other crop in case it's deemed not pro-Karimov enough. Sugar beet, for example, is well known for its dissident tendencies. **[3]**

Opening lines of national anthem: 'Stand tall, my free country, good fortune and salvation to you/You, yourself, a companion to friends, o loving one!' Contrary to popular belief, Uzbekistan is just one big hugfest. **[15]**

SCORE: 33 **WORLD RANKING:** 186

AUSTRALASIA

Vanuatu

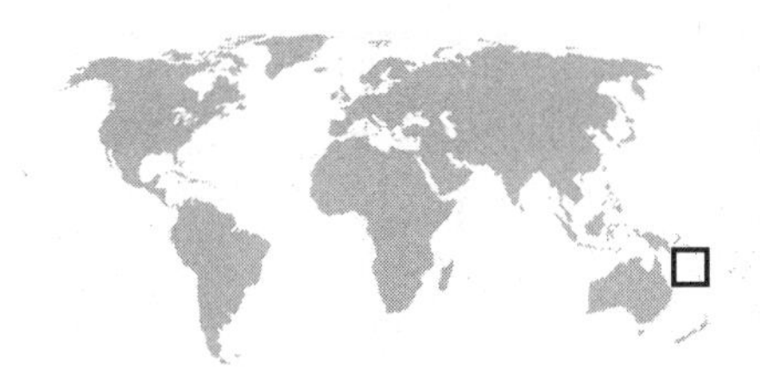

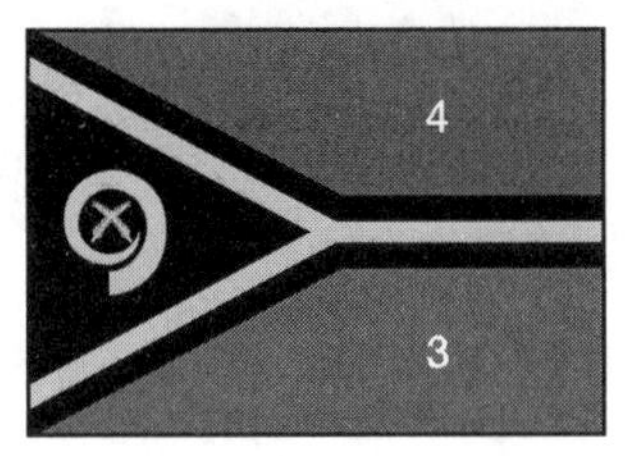

Name on driving licence: Republic of Vanuatu
Capital: Port Vila
Population: 207,000
Dosh: Vatu = 100 centimes
Size: 12,200 km² (0.6 Wales)

Complete history: How you recover from being called the Anglo-French Condominium of the New Hebrides (1906–80) is anyone's guess, but the doughty Melanesians of Vanuatu are doing their best.

Crowning achievement: Understanding each other. Vanuatans share 105 indigenous languages, the highest concentration per capita anywhere on Earth. **[15]**

Customs to treasure: In Vanuatu, a person's wealth is not calculated by how much they have but how much they have given away. **[20]**

Made in Vanuatu: *Kava*. Not to be confused with its weedy homophone cava, kava is narcotic rather than alcoholic and is traditionally drunk from coconut shells, although it apparently tastes no less disgusting out of a glass. **[10]**

Religious affiliations: The John Frum Cargo Cult, which stems from a vision in the late 1930s of a white man who came to rescue the Vanuatans from missionaries and colonial officials. When US troops arrived a few years later with their magical hospitals, jetties, corrugated steel huts and chewing gum, it was clearly the work of John Frum. As the troops left he appeared again, promising to return with loads of similar goodies (hence 'cargo cult'). Vanuatans duly built rudimentary airstrips, warehouses and bamboo control towers in readiness. They're still ready to this day, meeting weekly to worship John Frum in a service that lasts from dusk till dawn. **[10]**

Opening lines of national anthem: 'We are happy to proclaim/We are the People of Vanuatu!' Mind you, any nation whose anthem is entitled 'Yumi, yumi, yumi' can't help being happy really. **[16]**

SCORE: 71 **WORLD RANKING: 16**

Vatican City

EUROPE

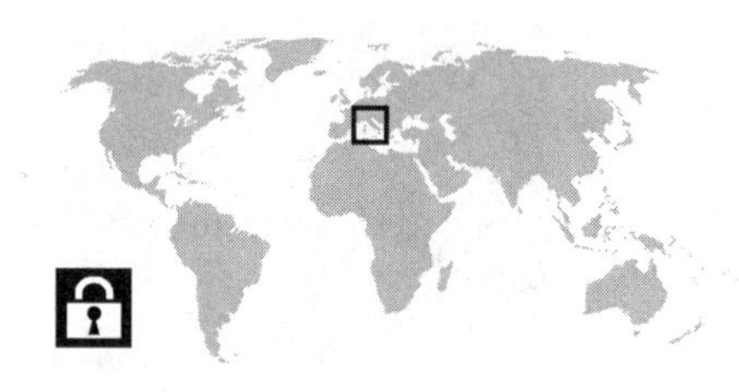

colour key: 1 = turquoise 2 = blue 3 = green 4 = red 5 = yellow 6 = orange 7 = pink 8 = purple 9 = brown

Name on driving licence: State of the Vatican City
Capital: Vatican City
Population: 900
Dosh: Euro = 100 cents
Size: 0.44 km^2 (0.00003 Wales)

Complete history: The popes used to own loads of Italy and so to be reduced to less than a fifth of a square mile is something of a come down. It's got a curious square flag which can't help either.

Crowning achievement: The Vatican City is the smallest independent state in the world and the only one to be entirely within the confines of another country's capital city. Furthermore, its population is entirely masculine: the last woman to live there was the (possibly legendary) first Pope John VIII (aka Pope Joan) who turned out not to be as male as is traditional for a pontiff. **[12]**

Gift to the world: Monsignor Hugh O'Flaherty, the 'Scarlet Pimpernel of the Vatican', who saved an estimated 4,000 Jews and Allied soldiers from the Nazis during World War II, and whose accent later went on to provide Gregory Peck with some awkward moments. **[20]**

Pub fact: To determine whether a pope is dead, a man named the Camerlengo has the duty of striking him three times on the head with a silver hammer while calling his name (the pope's, not his own, that wouldn't make sense). **[15]**

Religious affiliations: Catholicism is very popular, as is the phrase: 'Is the Pope a Catholic?' **[12]**

Opening lines of national anthem: 'O happy Rome, O noble Rome/You are the seat of Peter, whose blood was shed in Rome.' Peter is presumably unchuffed that his violent death is viewed as having bestowed happiness and nobility upon the city, but is probably secretly pleased to have got a mention anyway. **[8]**

SCORE: 67 WORLD RANKING: 37

Venezuela

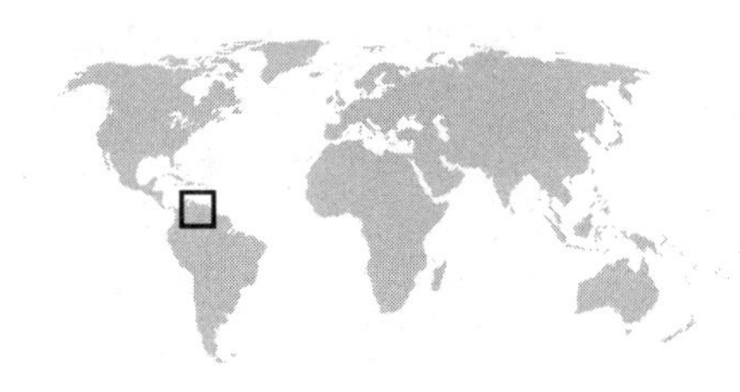

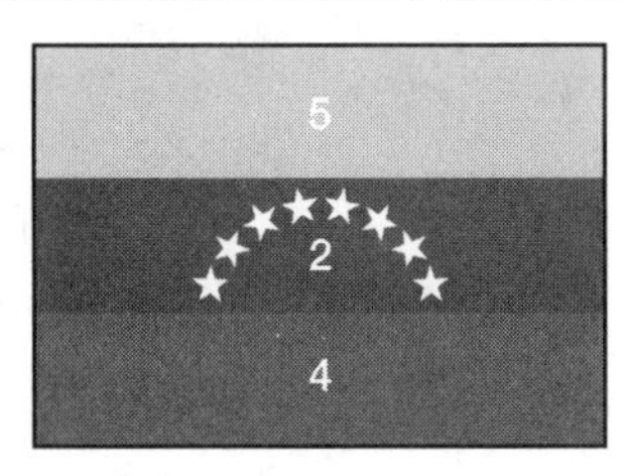

Name on driving licence: Bolivarian Republic of Venezuela
Capital: Caracas
Population: 25 million
Dosh: Bolívar = 100 céntimos
Size: 912,000 km^2 (44 Wales)

Complete history: Christianity was introduced to the Arawaks and Caribs by the conquistadores, though not quite in the sensitive way Christ would doubtless have wished. It was not until 1821 that Simón Bolívar finally expelled the Spaniards, but he had to wait until 1999 before President Hugo Chávez included his name in the country's title.

Customs to treasure: President Chávez's weekly telemarathon *Aló Presidente*, in which he goes on about stuff for ages. **[5]**

Top spot: The Angel Falls (979 metres) are the highest on Earth and are named not after some angelic visitation but to their 'discovery' in 1935 by an American aviator called Jimmy Angel. The Permone people, who didn't realise the falls hadn't been discovered, already had a name for them (Churun Meru) which doesn't involve any angels, or indeed pilots. **[16]**

Sad fact: Venezuela has won more Miss World and Miss Universe titles than anyone else. Thankfully, the rest of us know these events to be nonsense idiotheques of such shallowness that if they were puddles you'd have difficulty drowning dormice in them, not that you'd want to because dormice are a protected species. **[0]**

Made in Venezuela: Polarisation. Chávez is loved by the poor and reviled by the rich in equal measure, so he must be doing something right. **[12]**

National anthem: '"Off with the chains! Off with the chains!"/Cried the Lord, cried the Lord/And the poor man in his hovel/Implored freedom/At this holy name, there trembled/The vile selfishness that had triumphed.' Get us to nunneries! Some day all anthems will be like this. **[20]**

SCORE: 53 **WORLD RANKING: 119**

Vietnam

ASIA

colour key: 1 = turquoise 2 = blue 3 = green 4 = red 5 = yellow 6 = orange 7 = pink 8 = purple 9 = brown

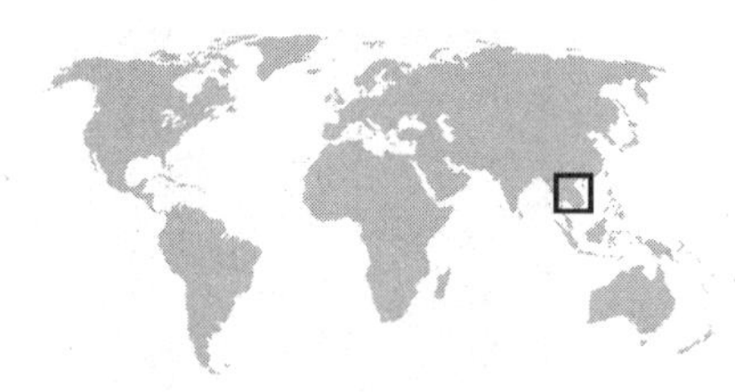

Name on driving licence: Socialist Republic of Vietnam
Capital: Hanoi
Population: 83 million
Dosh: Dông = 10 hao = 100 xu
Size: 329,600 km² (16 Wales)

Complete history: Annam endured over a thousand years under Chinese rule before loosing the chains in 939. It split apart 600 years later and was only glued back with the help of the dear old French in 1802, by which time it had become Vietnam. Beware Frenchies bearing gifts, however, for soon they wanted the place for themselves, only losing it to the Japanese in World War II. In 1954, Vietnam split again, and that's when all the trouble started.

Crowning achievement: Successfully fighting off the Americans, but then that's become somewhat trendy nowadays. **[14]**

Popular misconception: 'The average age of American soldiers fighting in the Vietnam war was n-n-n-n-nineteen.' Nope. It was tw-tw-tw-tw-twenty-two, though that hardly sounds so good, does it? **[3]**

Top spot: The Temple of Literature in Hanoi, which housed the first Vietnamese university back in 1076. **[13]**

Made in Vietnam: The Vietnamese pot-bellied pig, which is not rotund out of gluttony but because its backbone happens to dip a bit in the middle. It's also renowned for being a rather good-natured pig, as pigs go. Makes you ashamed for having made fun of it all these years, huh? **[15]**

Opening lines of national anthem: 'Soldiers of Vietnam, we go forward/With the one will to save our Fatherland.' All right, we know your country's named after a war and everything, but not *everyone* has to be a soldier, do they? What about hairdressers and theatrical designers? Surely, you could do with some of them? No? Oh, all right, have it your own way. **[2]**

SCORE: 47 WORLD RANKING: 150

EUROPE **Wales** (*see also* United Kingdom)

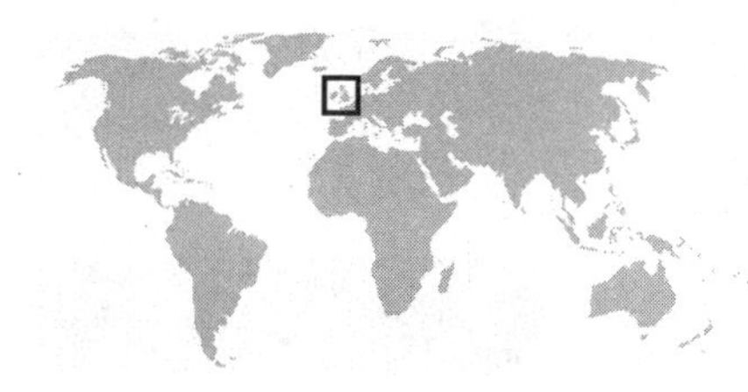

Name on driving licence: Principality of Wales
Capital: Cardiff (Caerdydd)
Population: 3 million
Dosh: Pound sterling = 100 pence
Size: 20,761 km² (Itself)

Complete history: Despite what you may have read elsewhere, the Welsh do not have a chip on their shoulder or live in conditions of filth and despair. They are a harried people, it is true, and have never enjoyed sovereignty over their entire nation, but they do not blame the English for this, and if they do, they do so in Welsh to spare the Englishman's feelings.

Customs to treasure: Unlike the English, the Welsh enjoy a good meal but, out of patriotic fervour, will permit of only five different foodstuffs to enter their bodies: *bara brith*, Welsh rarebit, *cawl cennin* (leek stew), Welsh cakes and laverbread. **[13]**

Made in Wales: Owain Glyndwr (1359–c.1421). The scourge of Henry IV very nearly made Wales an independent nation before it all went belly up around 1406. **[12]**

Top spot: Rhyl, a place of culture, fine architecture and Riviera-style beaches. **[16]**

Pub fact: The Welsh have more castles per head of population than any other people on Earth. You would think this would make them happy, especially since their ancestors didn't even have to go to the trouble of building most of them, but apparently not. **[8]**

Anthem: 'The land of my fathers is dear unto me/Old land where the minstrels are honoured and free.' 'Land of My Fathers' (also adopted by Celtic 'nations' Brittany and Cornwall) is the unofficial anthem of Wales, although 'God Save the Queen' doesn't have any official status either, so you can sing whatever takes your fancy really. 'Delilah', for example, would probably rouse the emotions just as efficiently as anything else. **[12]**

SCORE: 61 **WORLD RANKING: 70**

Yemen

MIDDLE EAST

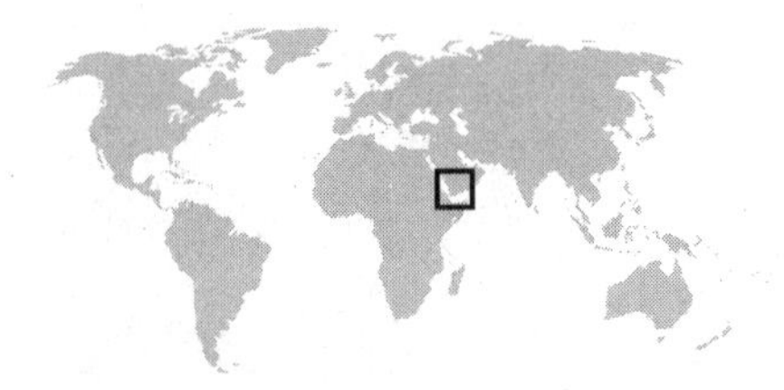

Name on driving licence: Republic of Yemen
Capital: Sana'a
Population: 20 million
Dosh: Yemeni rial = 100 fils
Size: 528,000 km^2 (25 Wales)

Complete history: The history of Yemen is so complex that nobody over the age of thirty has any chance of understanding it before they die.

Sad fact: Aden is a curious old place. The ancient part, known as Crater, is built entirely within an extinct (one hopes) volcano. For a hundred years the city was part of British India, until 1937 when someone in Whitehall looked at a map and realised where it was. Finally, it became the site of the first known Al Qaeda operation anywhere in the world when, on 29 December 1992, the Gold Mohor Hotel was bombed in an attempt to kill US troops staying there. **[10]**

Top spot: Sana'a, a city founded by Shem, one of the sons of Noah. It's quite a distance from Mount Ararat (in eastern Turkey), where the ark is said to have lodged when the waters finally abated, but perhaps Shem really needed to stretch his legs after forty days and forty nights of being cooped up with the tigers, giraffe, anteaters and what have you. **[11]**

Made in Yemen: The mini-skyscraper. Yemenis can build a six-storey house out of mud before you can say: 'Hi, Shem. Long walk was it?' **[13]**

Gift to the world: Mokha, on the Red Sea coast, gave its name to the first coffee beans Europeans ever clapped eyes on. It was also the busiest port in the ancient world for a time as a consequence. **[11]**

National anthem that could take some time: 'Repeat, O World, my song/Echo it over and over again.' **[7]**

SCORE: 52 WORLD RANKING: 125

colour key: 1 = turquoise 2 = blue 3 = green 4 = red 5 = yellow 6 = orange 7 = pink 8 = purple 9 = brown

AFRICA Zambia

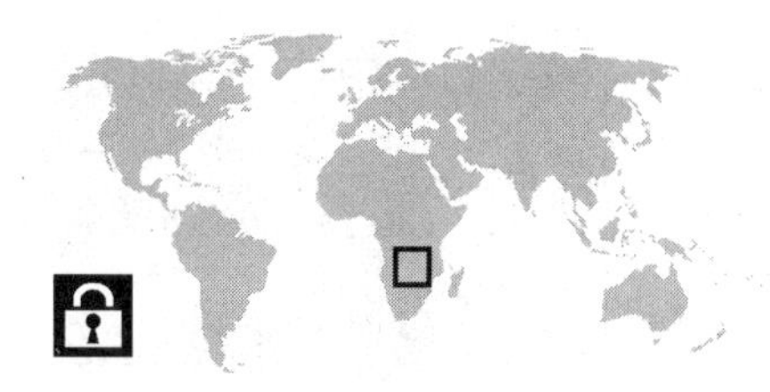

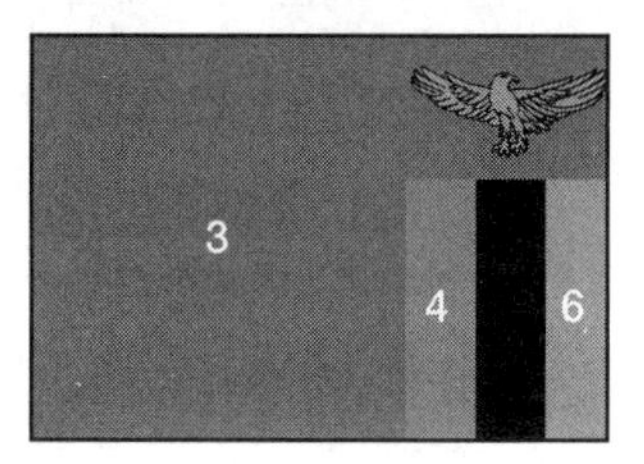

Name on driving licence: Republic of Zambia
Capital: Lusaka
Population: 11 million
Dosh: Zambian kwacha = 100 ngwee
Size: 752,600 km² (36 Wales)

Complete history: Khoisan types hunted and gathered their way through century after century before being ousted by Bantu-speakers. That was more or less it, save for the odd Portuguese explorer, until David Livingstone crossed the Zambezi and ushered in a new and exciting era of British meddling.

Top spot: Victoria Falls, the spray from which can famously be seen for 30 km, is the largest sheet of falling water on Earth, although that's about as useful as saying Wagner's *Ring Cycle* can be heard right round the block and has more notes in it than any other opera. Forget the scale for a moment and just enjoy the scene. **[15]**

Customs to treasure: The eating of *nshima*, a cornmeal hash that bulks up almost every meal. For those non-Zambians who find it a tad on the bland side, there's always *chigwangwa*, the burnt crust from the bottom of the pot in which the *nshima* is made. **[10]**

Pub fact: David Livingstone, when he wasn't being presumed (Burundi) or having his dog eaten (Malawi), was dying in Zambia. **[10]**

Made in Zambia: The Tonga basket, made by the women of Tonga in southern Zambia and not, as might be expected, in Tonga. In *that* Tonga they also make baskets but not of the ilala palm, and that's how you can tell the difference. **[11]**

Opening lines of national anthem: 'Stand and sing of Zambia, proud and free/Land of work and joy in unity.' It's happy happy happy all the way in Zambia. Also, the verses are nearly limericks. **[14]**

SCORE: 60 **WORLD RANKING:** 76

Zimbabwe

AFRICA

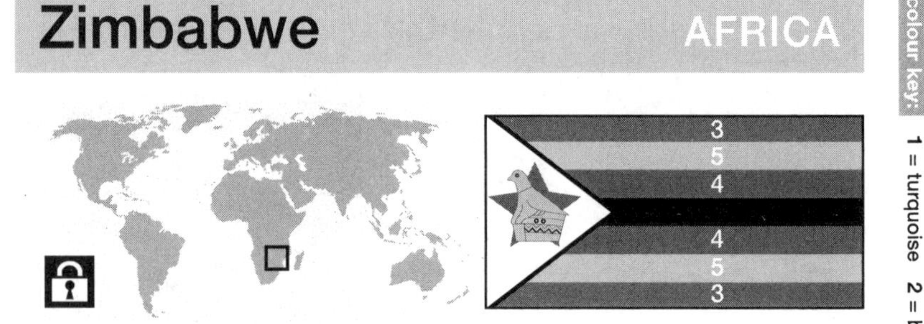

Name on driving licence: Republic of Zimbabwe
Capital: Harare
Population: 13 million
Dosh: Zimbabwean dollar = 100 cents
Size: 390,600 km^2 (19 Wales)

Complete history: The Shona people didn't much care for Cecil Rhodes, but then I doubt you or I would either, even if we haven't had our land stolen and named after him. In 1965, white minority prime minister Ian Smith announced a Unilateral Declaration of Independence from Britain, little knowing that most of the country was planning on living independently of him too.

Crowning achievement: Zimbabwe possesses the world's highest inflation rate, oscillating between 1,000 and 1,200 per cent. This means that something costing four Zimbabwean dollars on Monday will fetch about five by the weekend. **[3]**

Popular misconception: 'Robert Mugabe is just no fun anymore.' Oh yes he is. Just put on a Yorkshire accent and say his name backwards. Hours of childish mirth. **[6]**

Top spot: Just in case anyone should ever care to visit Zimbabwe ever again save on a human-rights fact-finding mission, one side of Victoria Falls falls within its territory. Rumours that the falls have been quietly inching their way into Zambia at night have been branded by the Zimbabwean government as the words of snake-tongued baby-chokers. **[8]**

Pub fact: 'Zimbabwe' is derived from the Shona *dzimba dzemabwe,* meaning 'houses of stone'. This is because the ancient city of Great Zimbabwe was built of stone. A literal lot, the Shona. **[10]**

National anthem: 'We praise our heroes' sacrifice/And vow to keep our land from foes.' There's no doubting that during the fight for independence, Mugabe did make certain sacrifices, the problem being that he has now become the foe. This is why all national anthems should be banned. **[5]**

SCORE: 32 **WORLD RANKING:** 187

colour key: 1 = turquoise 2 = blue 3 = green 4 = red 5 = yellow 6 = orange 7 = pink 8 = purple 9 = brown

NON-COUNTRIES

(WHETHER ASSOCIATED, DEPENDENT, UNINCORPORATED, AFFILIATED AND/OR COMMONWEALTH TERRITORIES, BUT JUST THE INHABITED ONES, FOR LIFE IS BRIEF)

American Samoa (USA) OCEANIA

Capital: Pago Pago **Population:** 69,000

Dosh: US dollar = 100 cents **Size:** 195 km² (0.01 Wales)

Pub fact: American Samoans revel in the sport of greased pig chasing. If you don't already know the rules, it's better not to ask.

Anguilla (UK) AMERICAS

Capital: The Valley **Population:** 12,500

Dosh: East Caribbean dollar = 100 cents **Size:** 95 km² (0.005 Wales)

Pub fact: On 12 June 2006, Anguillans celebrated the Queen's official birthday with a parade at Ronald Webster's Park. Five days too early.

Aruba (Netherlands) AMERICAS

Capital: Oranjestad **Population:** 71,000

Dosh: Aruban guilder/florin = 100 cents **Size:** 195 km² (0.009 Wales)

Pub fact: Arubans had their own gold rush in the 19th century but have little to show for it now but the world's second-largest desalination plant.

Ascension Island (UK) AFRICA

Capital: Georgetown **Population:** 1,200

Dosh: Saint Helenian pound = 100 pence **Size:** 88 km² (0.004 Wales)

Pub fact: Ascension is visited every other month by Britain's last working Royal Mail ship.

Azores (Portugal) AFRICA

Capital: Ponta Delgada **Population:** 243,000

Dosh: Euro = 100 cents **Size:** 2,330 km² (0.1 Wales)

Pub fact: Azores was used to stage a conference just before the second Gulf War so that George W. Bush could ignore the prime ministers of Britain, Spain and Portugal in person.

Bermuda (UK) AMERICAS

Capital: Hamilton

Population: 65,000

Dosh: Bermudian dollar = 100 cents

Size: 53 km^2 (0.003 Wales)

Pub fact: John Rolfe was shipwrecked and widowed on Bermuda before going on to marry Pocahontas and found America's tobacco industry, though not on the same day.

British Indian Ocean Territory (UK) ASIA

Capital: Diego Garcia

Population: 950

Dosh: Pound sterling = 100 pence

Size: 60 km^2 (0.003 Wales)

Pub fact: Between 1965 and 1973, Britain shamelessly removed all the inhabitants of Diego Garcia (a people called the Ilois), having secretly leased the island to the American military in return for a discount on a Polaris submarine. Despite a court ruling in the UK, the US refuses to allow the Ilois to return.

British Virgin Islands (UK) AMERICAS

Capital: Road Town

Population: 22,000

Dosh: US dollar = 100 cents

Size: 153 km^2 (0.007 Wales)

Pub fact: Columbus named the islands 'Las Virginas' because they reminded him of St Ursula and her fellow 11,000 virgins who were massacred by Huns. Apparently, at the time, no one thought this an odd comparison.

Canary Islands (Spain) AFRICA

Capitals: Las Palmas and Santa Cruz

Population: 1.7 million

Dosh: Euro = 100 cents

Size: 7,450 km^2 (0.4 Wales)

Pub fact: The Canaries may well have acquired their name not because of any exotic birdlife but because ancient Romans encountered large dogs on the islands and so named them Insularia Canaria (Isle of Dogs).

Cayman Islands (UK) AMERICAS

Capital: George Town
Population: 38,000
Dosh: Caymanian dollar
Size: 260 km² (0.01 Wales)

Pub fact: The average Cayman Islander enjoys one of the highest incomes in the world largely thanks to their offshore financial operations which, of course, are not at all murky. They also claim to have invented recreational scuba diving.

Christmas Island (Australia) ASIA

Capital: The Settlement
Population: 1,600
Dosh: Australian dollar = 100 cents
Size: 135 km² (0.007 Wales)

Pub fact: When the captain of Norwegian vessel MV *Tampa* rescued 438 asylum seekers from a shipwreck and tried to disembark them on nearby Christmas Island, the ship was boarded by Australian SAS troops who stopped him from doing so which, even in the times in which we live, seems a bit of an overreaction.

Cocos Islands (Australia) ASIA

Capital: n/a
Population: 650
Dosh: Australian dollar = 100 cents
Size: 14 km² (0.0007 Wales)

Pub fact: The Cocos Islands Mutiny resulted in the execution of three Ceylon Defence Force troops, the only Commonwealth soldiers to be executed for mutiny by the Allies during World War II.

Cook Islands (New Zealand) OCEANIA

Capital: Avarua
Population: 21,000
Dosh: New Zealand dollar = 100 cents
Size: 235 km² (0.01 Wales)

Pub fact: 95.7 per cent of the Cook Islands is forested, a greater proportion than any country in the world.

Faeroe Islands (Denmark) EUROPE

Capital: Tórshavn
Population: 47,000
Dosh: Danish krone = øre
Size: 1,400 km^2 (0.07 Wales)

Pub fact: Officially, the only fact anyone is allowed to know about the Faeroe Islands is that their goalkeeper Jens Martin Knudsen wore a bobble hat during their first-ever competitive international, a 1-0 win over Austria in 1990.

Falkland Islands (UK) AMERICAS

Capital: Stanley
Population: 3,000
Dosh: Falkland Islands pound
Size: 12,170 km^2 (0.59 Wales)

Pub fact: The Argentinean invasion of the Falkland Islands caused consternation all over Britain until it was realised that the islands weren't off the coast of Scotland. Then it caused consternation all over again when the ensuing conflict won a certain woman an election.

French Guiana (France) AMERICAS

Capital: Cayenne
Population: 191,000
Dosh: Euro = 100 cents
Size: 91,000 km^2 (4.3 Wales)

Pub fact: In 1885, the French government passed a law that anyone given three sentences for theft would be transported to French Guiana. This was tantamount to a death sentence. Alfred Dreyfus ('J'accuse!') and the murderer Henri Charrière were also sent there, the latter detailing his escape in the book *Papillon*.

French Polynesia (France) OCEANIA

Capital: Papeete
Population: 266,000
Dosh: CFP franc = 100 centimes
Size: 4,170 km^2 (0.2 Wales)

Pub fact: Paul Gauguin, who painted and is buried in French Polynesia, was once a worker on the Panama Canal.

Gibraltar (UK) EUROPE

Capital: Gibraltar Town
Population: 28,000
Dosh: Gibraltar pound = 100 pence
Size: 6.5 km² (0.0003 Wales)

Pub fact: The British tenure over Gibraltar is deeply galling to the Spanish who insist on its return and have refused on many occasions to accept Canvey Island in exchange.

Greenland (Denmark) AMERICAS

Capital: Nuuk
Population: 56,000
Dosh: Danish krone = 100 øre
Size: 2,175,500 km² (105 Wales)

Pub fact: Greenland is the largest island on Earth (since Australia is officially categorised as a continental landmass) and possesses the world's largest fjord (Scoresby Sund). The mosquitoes are something awful though.

Guadeloupe (France) AMERICAS

Capital: Basse-Terre
Population: 440,000
Dosh: Euro = 100 cents
Size: 1,780 km² (0.08 Wales)

Pub fact: Guadeloupe changed hands many times between the French and British before the islands became Swedish, somewhat by accident, in 1813. The Swedes, unsure quite what to do with them, handed them to the French a year later.

Guam (USA) OCEANIA

Capital: Hagatna
Population: 166,000
Dosh: US dollar = 100 cents
Size: 550 km² (0.03 Wales)

Pub fact: Japanese corporal Shoichi Yokoi lived in a tiny cave in the Guam jungle for 28 years before finally being discovered by two hunters in 1972. He revealed that he had become aware in 1952 that World War II was over but had been too afraid to come out of hiding.

Guernsey (UK) EUROPE

Capital: St Peter Port

Population: 65,000

Dosh: Pound sterling = 100 pence

Size: 65 km² (0.003 Wales)

Pub fact: A full 2 per cent of the population of Guernsey speak Dgèrnésiais, an old Norman language. Victor Hugo, however, stuck with French for *Les Misérables,* which he wrote to while away his exile on the island.

Hong Kong (China) ASIA

Capital: Victoria

Population: 6.9 million

Dosh: Hong Kong dollar = 100 cents

Size: 1,105 km² (0.05 Wales)

Pub fact: Hong Kong is saddled with the only Disneyland in Communist-held territory (although most Americans probably consider Euro Disney to be in this category too) and the only one to have been built along feng shui principles.

Isle of Man (UK) EUROPE

Capital: Douglas

Population: 75,000

Dosh: Pound sterling = 100 pence

Size: 572 km² (0.03 Wales)

Pub fact: In the Tynwald (established in 979), the Isle of Man lays claim to the oldest continuously sitting parliament in the world. Famed for its cutting-edge liberal ways, the Tynwald outlawed birching in the 1970s and legalised homosexuality in 1992.

Jersey (UK) EUROPE

Capital: St Helier

Population: 90,000

Dosh: Pound sterling = 100 pence

Size: 116 km² (0.006 Wales)

Pub fact: The Channel Islands are famously the only bit of British soil the Germans got around to invading during World War II. The Nazis were apparently under the impression that a nation shorn of its Jersey cows would soon crumble, and they were not far wrong.

Johnston Atoll (USA) OCEANIA

Capital: n/a
Population: 340
Dosh: US dollar = 100 cents
Size: 2.8 km^2 (0.0001 Wales)

Pub fact: The atoll was listed but then withdrawn from auction by the USA in 2005. The offer would have afforded a rare opportunity to purchase an atoll which has served as both a nuclear weapons test site and a base for the disposal of chemical agents. Form an orderly queue.

Madeira (Portugal) AFRICA

Capital: Funchal
Population: 253,000
Dosh: Euro = 100 cents
Size: 794 km^2 (0.04 Wales)

Pub fact: Madeira, known to the ancient Romans as the Purple Islands, got mislaid after the empire broke up and had to be discovered all over again, and quite by accident, by the Portuguese. Since then, the islanders have been busy inventing their own cake and wine, and developing an all-year-round 'toboggan' ride in a sort of larger wicker basket.

Martinique (France) AMERICAS

Capital: Fort-de-France
Population: 426,000
Dosh: Euro = 100 cents
Size: 1,100 km^2 (0.05 Wales)

Pub fact: The eruption of Mount Pelée in 1902 wiped out almost the entire population (c.40,000) of the former capital St Pierre. Just one person survived – Ludger Sylbaris, a prisoner under sentence of death, who was being held in an underground bombproof magazine used by the city's jail. He went on to find fame with a travelling circus as the 'Man who Lived Through Doomsday'.

Mayotte (France) ASIA

Capital: Mamoudzou
Population: 173,000
Dosh: Euro = 100 cents
Size: 375 km^2 (0.02 Wales)

Pub fact: The United Nations Security Council has drafted a number of resolutions attempting to recognise the sovereignty of neighbouring Comoros over Mayotte, but each time the French use their veto because they 'really really need Mayotte, wherever it is'.

Midway Islands (USA) ASIA (just)

Capital: n/a **Population:** 40

Dosh: US dollar = 100 cents **Size:** 5.2 km^2 (0.0002 Wales)

Pub fact: The Midway Islands – now little more than two airfields – were the scene of the naval battle in 1942 that ended Japanese domination of the Pacific Ocean. Their Hawaiian name, Pihemanu, means 'loud din of birds'.

Montserrat (UK) AMERICAS

Capital: Plymouth **Population:** 9,000

Dosh: East Caribbean dollar = 100 cents **Size:** 102 km^2 (0.005 Wales)

Pub fact: The Beatles' producer George Martin set up his famous AIR studios on Montserrat, unaware that in 1989 Hurricane Hugo would force it to close. Worse still, in 1995 the Soufrière Hills Volcano buried the capital, Plymouth, under 10 metres of mud and rendered the southern half of the island completely uninhabitable.

Netherlands Antilles (Netherlands) AMERICAS

Capital: Willemstad **Population:** 216,000

Dosh: Netherlands Antillean guilder = 100 cents **Size:** 960 km^2 (0.04 Wales)

Pub fact: One of the islands, Curaçao, is the home of the blue (usually) liqueur of the same name. It is made with the dried peel of Laharas, bitter oranges grown on the island (though not native to it). Disappointingly, most Curaçao is now made synthetically in other parts of the world.

New Caledonia (France) AUSTRALASIA

Capital: Nouméa **Population:** 214,000

Dosh: CFP franc = 100 centimes **Size:** 19,100 km^2 (0.9 Wales)

Pub fact: New Caledonians will get the chance to vote on independence from France in 2014, which is an excellent incentive to carry on living.

Niue (New Zealand) OCEANIA

Capital: Alofi **Population:** 2,200

Dosh: New Zealand dollar = 100 cents **Size:** 265 km^2 (0.13 Wales)

Pub fact: Impress your enemies and bore your friends with your ability to pronounce Niue (go for something sounding like 'new way') before going on to inform them that although the island's premier is called Young Vivian he was actually born in 1935.

Norfolk Island (Australia) AUSTRALASIA

Capital: Kingston **Population:** 1,880

Dosh: Australian dollar = 100 cents **Size:** 34 km^2 (0.002)

Pub fact: Captain Cook sighted the island on his second voyage to the South Pacific and named it after the Duchess of Norfolk, unaware that she had popped her clogs since his departure from England.

Northern Mariana Islands (USA) OCEANIA

Capital: Saipan **Population:** 78,000

Dosh: US dollar = 100 cents **Size:** 457 km^2 (0.02 Wales)

Pub fact: Saipan, the major island of the Northern Mariana Islands, has spawned a radical website (saipansucks.com) that rails against the nepotism, corruption and sleaze apparently rife within the political community. Still, it's not all bad: Saipan also features in the *Guinness Book of Records* for having the least variation in temperatures in the world.

Pitcairn Islands (UK) OCEANIA

Capital: Adamstown **Population:** 50

Dosh: New Zealand dollar = 100 cents **Size:** 47 km^2 (0.002 Wales)

Pub fact: Pitcairn Island was named in 1767 after fifteen-year-old midshipman Robert Pitcairn who was the first on board the British sloop HMS *Swallow* to sight it. Pitcairn soon found fame, if not fortune, as the island populated by the mutineers from the HMS *Bounty* (the burnt-out remains of which can still be seen in Bounty Bay).

Puerto Rico (USA) AMERICAS

Capital: San Juan **Population:** 4 million

Dosh: US dollar = 100 cents **Size:** 9100 km^2 (0.4 Wales)

Pub fact: Puerto Rico is at perpetual risk of becoming the 51st US state. Plebiscites on a possible move to statehood occurred twice in the 1990s and incorporation into wise old Uncle Sam (or the Great Devil of the North, depending on your outlook) was avoided by a mere few percentage points last time round.

Réunion (France) AFRICA

Capital: Saint-Denis **Population:** 766,000

Dosh: Euro = 100 cents **Size:** 2,500 km^2 (0.1 Wales)

Pub fact: In 2005/06, around a third of the population of Réunion contracted the mosquito-borne disease chikungunya which, though seldom fatal, is apparently extremely painful, and has no known cure. This was something of a plunge from the excitement gripping the island in 2002 when its easterliness allowed regional council president Paul Vergès to become the first person anywhere in the world to make a purchase with a euro (he bought a bag of lychees at a market stall).

St Helena (UK) AFRICA

Capital: Jamestown **Population:** 7,400

Dosh: Saint Helenian pound = 100 pence **Size:** 122 km^2 (0.006 Wales)

Pub fact: Hard though it tries, St Helena will only ever be famous for being the island on which the British held Napoleon Bonaparte from 1815 until his death in 1821 (probably from stomach cancer, though it's been suggested he could have been poisoned). His place of incarceration, Longwood House, was given by the British to the French government in 1858, which was jolly sporting.

St Pierre & Miquelon (France) AMERICAS

Capital: Saint-Pierre

Population: 7,000

Dosh: Euro = 100 cents

Size: 242 km^2 (0.01 Wales)

Pub fact: The British and French fought tooth and nail over this group of islands off the east coast of Canada until the former decided they couldn't be bothered any more since they owned the rest of the country anyway. The island of Langlade has been uninhabited since the death of its last resident, Charles Lafitte, in July 2006.

Svalbard (Norway) EUROPE

Capital: Longyearbyen

Population: 2,900

Dosh: Norwegian krone = 100 øre

Size: 62,900 km^2 (3 Wales)

Pub fact: Aside from the usual Svalbard fieldmouse, Svalbard reindeer and Arctic fox, Svalbard is awash with polar bears. This means that everyone going outside Longyearbyen has to carry a rifle. However, since polar bears are a protected species, they can't actually be shot. Expect some fun on the day they finally work this out.

Tokelau (New Zealand) OCEANIA

Capital: n/a

Population: 1,500

Dosh: New Zealand dollar = 100 cents

Size: 10 km^2 (0.0005 Wales)

Pub fact: None of the three tiny atolls that make up Tokelau has a port or harbour, so ships delivering supplies have to position themselves on the leeward side just off the reef, hope for fine weather and wait for whaleboats to be rowed out to them. As with many other atolls, a small rise in sea levels (about 2 metres) would see them disappear.

Tristan da Cunha (UK/St Helena) AFRICA

Capital: Edinburgh
Population: 271
Dosh: Pound sterling = 100 pence
Size: 98 km² (0.005 Wales)

Pub fact: A volcanic eruption in 1961 forced the entire population of the world's most isolated settlement (nearest neighbour, St Helena, 2,173 km away) into exile in the UK. They were housed at a former RAF base near Calshot, Hampshire, where they lived in a road called Tristan Close. Most of them returned two years later. The inhabitants boast just seven surnames, which raises a questioning eyebrow, although on the plus side the island has never had a divorce in its entire history.

Turks & Caicos Islands (UK) AMERICAS

Capital: Cockburn Town
Population: 19,000
Dosh: US dollar = 100 cents
Size: 430 km² (0.02 Wales)

Pub fact: Although the Caicos Islands form 96 per cent of the territory's landmass, they only get second billing because the Turks were settled first (by Bermudan salt-makers, of course). In 2004, Canada's Nova Scotia voted in favour of making Turks and Caicos part of their province should they ever fancy it. More curiously still, for a hundred years the T&C flag was emblazoned with igloos because the designer in London misinterpreted a sketch for the flag that featured heaps of salt.

United States Virgin Islands (USA) AMERICAS

Capital: Charlotte Amalie
Population: 125,000
Dosh: US dollar = 100 cents
Size: 355 km² (0.02 Wales)

Pub fact: The US Virgin Islands is the only American territory in which cars have to be driven on the proper (i.e. left) side of the road. This is a good thing. After all, Austria and Czechoslovakia only moved to right-side driving on the orders of Hitler, while Italy switched on a whim of Mussolini's. The US Virgin Islands' adoption of the left makes up in some small part for the ill fortune of no longer being ruled by the Knights of Malta.

Wake Island (USA) OCEANIA

Capital: n/a **Population:** 220

Dosh: US dollar = 100 cents **Size:** 6.5 km² (0.0003 Wales)

Pub fact: Annexed by the US in 1899, the first settlement on Wake Island was 'PAAville', a tiny village built by Pan American Airways in 1935 to serve the needs of its passengers on stopovers from America to China. It was still there when the Japanese started bombing it in the name of the glorious emperor. The Japanese later did for the atoll's only indigenous bird, the Wake Island rail.

Wallis & Futuna (France) OCEANIA

Capital: Matā'Utu **Population:** 16,000

Dosh: CFP franc = 100 centimes **Size:** 275 km² (0.01 Wales)

Pub fact: Wallis and Futuna is the only territory in the world to be named (at least in part) after a Cornishman, one Samuel Wallis, who breezed past in 1767. The third island in the group, Alofi, has just two inhabitants. The entire population is said to have been eaten by cannibals from Futuna in a raid in the 19th century.

Western Sahara (Disputed) AFRICA

Capital (unofficial): El Aaiun **Population:** 267,000

Dosh: Moroccan dirham = 100 centimes **Size:** 266,800 km² (13 Wales)

Pub fact: In 1975, Spain handed over two-thirds of Western Sahara to Morocco and the rest to Mauritania. The Mauritanians had the good grace to renounce their bit, but the Moroccans now claim they own that as well. This has not gone down well with the home-grown Polisario Front who have been brawling with them ever since.

NEW WORLD ORDERED

THE NATIONS WITH THEIR ACT TOGETHER

AND

THE ONES WHO HAVE LOST THE PLOT

Top Five European Nations

1 Netherlands (85)
2 Belgium (79)
3 Slovenia (76)
4 Czech Republic (75)
5 Latvia (73)

Bottom Five European Nations

1 Monaco (29)
2 Ukraine (40)
3 Albania (42)
4 Portugal (45)
5 Austria (46)

Top Five Asian Nations

1 East Timor (71)
2= Japan (70)
2= Kyrgyzstan (70)
4 Bhutan (69)
5 Afghanistan (67)

Bottom Five Asian Nations

1= Turkmenistan (28)
1= Maldives (28)
3 Burma (30)
4 Uzbekistan (33)
5 Brunei (36)

Top Five African Nations

1 Mali (72)
2= Congo (71)
2= Morocco (71)
4 Cape Verde Islands (67)
5 Botswana (66)

Top Five American Nations

1 Guatemala (81)
2 Brazil (80)
3 Chile (78)
4= El Salvador (71)
4= Guyana (71)

Top Five Island Nations

1 Kiribati (76)
2 Micronesia (74)
3 Northern Ireland (73)
4= East Timor (71)
4= Vanuatu (71)

Bottom Five Island Nations

1 Maldives (28)
2 Antigua and Barbuda (32)
3 Brunei (36)
4 Dominican Republic (42)
5 Solomon Islands (44)

Top Five Communist Nations

1 China (65)
2= Cuba (53)
2= North Korea (53)
4 Laos (51)
5 Vietnam (47)

Top Five Former Soviet Bloc Nations

1 Slovenia (76)
2 Czech Republic (75)
3 Latvia (73)
4= Kyrgyzstan (70)
4= Slovakia (70)

Bottom Five Landlocked Nations

1 Paraguay (25)
2 Swaziland (27)
3 Uzbekistan (33)
4 Central African Republic (38)
5 Austria (46)

Bottom Five Former Axis Powers

1 Romania (47)
2 Italy (59)
3 Germany (61)
4 Hungary (63)
5 Croatia (64)

Top Five 'Somewhere and Somewhere Else' Nations

1 Trinidad and Tobago (64)
2= St Kitts and Nevis (51)
2= São Tomé and Príncipe (51)
4 Bosnia and Herzegovina (50)
5 St Vincent and the Grenadines (45)

Top Five Countries Never To Have Won An Olympic Medal

1 Guatemala (81)
2 Kiribati (76)
3 Micronesia (74)
4 Jordan (73)
5 Mali (72)

ACKNOWLEDGEMENTS

The author would like to thank the following for their help in the writing of this book. They have given sacrificially of themselves so that knowledge and wisdom may triumph.

Clive Wills, Carl Palmer, Julia Castle, Ruta Lealaimatafao, Rebecca Adlington, Katy 'I have cut my hair' Nicholson, Carey Bowtell, Simon 'Fingers' Proctor, Mägda MacRobson, Ma and Pa Robson, Rob 'Bertie' Sturgess, Caroline Overy, Lorna Houston, Dylan 'The Animal Boy' Houston, Michael Houston, Eugene 'The Axe' Schoulgin, Mycroft Green and Elisabeth 'Heck' Whitebread.

Special thanks are due to Canada's David Kendall, the creator and curator of the most astonishingly thorough national anthem website in existence (nationalanthems.info).

This book is dedicated to Debden Jubilee.

OTHER BOOKS BY DIXE WILLS

- *The Z to Z of Great Britain*
- *Places to Hide in England, Scotland and Wales*